Possible Health Effects of Exposure to

RESIDENTIAL ELECTRIC AND MAGNETIC FIELDS

Committee on the Possible Effects of
Electromagnetic Fields on Biologic Systems

Board on Radiation Effects Research

Commission on Life Sciences

National Research Council

NATIONAL ACADEMY PRESS
Washington, D.C. 1997

National Academy Press • **2101 Constitution Avenue, N.W.** • **Washington, DC 20418**

The project that is the subject of this report was approved by the Governing Board of the National Research Council, whose members are drawn from the councils of the National Academy of Sciences, the National Academy of Engineering, and the Institute of Medicine. The members of the committee responsible for the report were chosen for their special competences and with regard for appropriate balance.

This report has been reviewed by a group other than the authors according to procedures approved by a Report Review Committee consisting of members of the National Academy of Sciences, the National Academy of Engineering, and the Institute of Medicine.

This report was prepared under Grant No. DE-FG01-92CE34100 between the National Academy of Sciences and the U.S. Department of Energy.

Possible Health Effects of Exposure to Residential Electric and Magnetic Fields is available from the National Academy Press, 2101 Constitution Ave., NW, Box 285, Washington, DC 20055 (1-800-624-6242; http://www.nap.edu)

Library of Congress Cataloging-in-Publication Data
National Research Council (U.S.). Committee on the Possible Effects of Electromagnetic Fields on Biologic Systems.
    Possible health effects of exposure to residential electric and magnetic fields.
      p. cm.
    Includes bibliographical references and index.
    ISBN 0-309-05447-8
    1. Electromagnetic fields—Health aspects. 2. Electromagnetism—Physiological effect.   I. Title.
    RA569.3.N378   1997
    612'.01442—dc21
                                      96-51230

COMMITTEE ON THE POSSIBLE EFFECTS OF ELECTROMAGNETIC FIELDS ON BIOLOGIC SYSTEMS

v

Preface

Can routine exposures to electric and magnetic fields found in homes and communities cause cancer, reproductive abnormalities, or neurobiologic disease? That question has been asked by a large number of people who live in today's highly industrialized world. It was asked of Congressman Joseph McDade by constituents of his Pennsylvania legislative district and prompted his proposal for a National Academy of Sciences study of the possible health consequences of exposure to low-strength low-frequency electric and magnetic fields. Subsequently, the Congress, in Public Law 102-104, designated the U.S. Department of Energy (DOE) as the lead agency for electric and magnetic field research and specified that DOE enter into an agreement with the National Research Council (NRC) to conduct an evaluation of the possible health effects of electric and magnetic fields. The interest of DOE and the Congress is a response to the public's concern about possible adverse health effects resulting from exposure to power-frequency electric and magnetic fields in the residential setting. The NRC Board on Radiation Effects Research responded to DOE's request by convening an expert committee to review and evaluate the literature on the possible health effects of exposure to electric and magnetic fields. This report is the result of nearly 3 years of committee study and numerous hours of committee deliberations.

Scientists often can be precise in pinpointing the cause and the remedy for a well-defined disease. Smallpox has become an extinct disease because its cause (a virus) was identified, and scientists developed an effective vaccine because of their knowledge of the human immune response system. The situation for power-frequency electric and magnetic fields and their effects on biologic systems is quite different. There is no widely accepted understanding of how extremely low-frequency electric and magnetic fields, such as those associated with the distribution and use of electric power, could cause a disease or whether it causes a disease. Considerable research has been conducted in this area, and numerous research data can be found on the subject, but given the lack of a specific disease end point to track or a well-accepted theory of how the fields might affect biologic systems, the data are discordant; they have been gathered using different exposure conditions and have resulted in conflicting observations of different effects or no effects.

Electricity has been a staple in U.S. homes for only the past 100 years. Questions about human health effects from exposure to electromagnetic fields (EMF) began only about 50 years ago, when persons in the armed forces were exposed to EMF from radar and other high-frequency radiative devices. Only in the past 15 years have claims arisen about the possible health effects of the extremely-low-frequency electric and magnetic fields encountered in residential environments. Meanwhile, research funding for studies of the possible health effects of electric and magnetic fields has fluctuated over the years—at various times research programs have been pursued actively by the DOE, the U.S. Environmental Protection Agency (EPA), and the Electric Power Research Institute; at other times, such programs have been given less priority.

Data are seldom sufficient to provide a definitive answer to the possible health effects of a physical or chemical agent in the environment. In such cases, professional judgment plays a large role in forming conclusions. It is especially important that the scientists selected for the evaluations be open to the evidence about the issues to be studied, wherever it might lead.

The committee that prepared this report comprised persons with expertise in cancer, reproductive and developmental effects, and neurobiologic effects. Some of them were experienced in epidemiology, risk, and exposure assessment, others in laboratory studies using animal and cellular systems. Members represented a wide range of disciplines; academic credentials included physics, engineering, chemistry, and biology, complemented by applied fields, such as risk perception. Some of the members had spent large parts of their scientific careers studying the effects of electric and magnetic fields; others, including myself, were newcomers to the field and to its data. We learned early in our work that our initial points of view ranged from belief that residential electric and magnetic fields had been shown to cause disease to skepticism that such effects had been proved or even clearly demonstrated. Brief biographic sketches of the members of this committee can be found at the end of this report.

Because some committee members were unfamiliar with the research on biologic effects of power-frequency electric and magnetic fields, informational workshops were held at several meetings. Scientists from the United States and abroad briefed the committee about their work and the state of knowledge concerning effects of these fields. Much of the remainder of our committee meetings was spent in assessing and evaluating the data and synthesizing our conclusions based on the data. We were assisted in our discussions by experts invited to join us for some of our meetings. In particular, we called on several biostatisticians to give us their views on the strength of individual scientific papers and the significance of the body of evidence as a whole. The use of biostatisticians outside the committee became especially important when the biostatistician originally appointed to the committee was forced to resign because of the press of duties at his home institution.

I wish to thank the members of the committee for their thoughtful and open-minded approach as they wrestled with the complexities of our charge and their hard work as we prepared the report. A significant burden of writing fell on those members who were experts in the field and were designated by the committee to prepare sections in their areas of expertise. We thank the staff of the Board on Radiation Effects Research, Larry Toburen, John Zimbrick, Doris Taylor, Lara Adamo, and Alvin Lazen, associate executive director of the Commission on Life Sciences. We also acknowledge the support of DOE and its program officer for this project, Imre Gyuk. We express our appreciation to the anonymous reviewers of this report whose comments and insights have sharpened the expression of the committee's views.

Charles F. Stevens, Chairman
Committee on the Possible Effects of
Electromagnetic Fields on Biologic Systems

OTHER REPORTS OF THE COMMISSION ON LIFE SCIENCES

Carcinogens and Anticarcinogens in the Human Diet: A Comparison of Naturally Occurring and Synthetic Substances (1996)

Toxicological and Performance Asepcts of Oxygenated Motor Vehicle Fuels (1996)

Upstream: Salmon and Society in the Pacific Northwest (1996)

A Review of EPA's Environmental Monitoring and Assessment Program: Overal Evaluation (1995)

Current Methodologies and Future Directions of the Fernald Dosimetry Reconstruction Project and Their Appropriateness and Soundness (1995)

EMF Research Activities Completed Under the Energy Policy Act of 1992: Interim Report (1995)

Evaluation of Centers for Disease Control/Marshall Islands Epidemiology Protocol: Letter Report (1995)

Nitrate and Nitrite in Drinking Water (1995)

Radiation Dose Reconstruction for Epidemiologic Uess (1995)

Science and the Endangered Species Act (1995)

Epidemiologic Research Program at the Department of Energy: Looking to the Future (1994)

The Hanford Environmental Dose Reconstruction Proejct: A Review of Four Documents (1994)

Health Effects of Exposure to Radon: Time for Reassessment? (1994)

Health Effects of Permethrin-Impregnated Army Battle-Dress Uniforms (1994)

Interpretation of Results of a Pilot Project by the Hanford Thyroid Disease Study and the Assessment of Feasibility of a Proposed Second-Phase Epidemiology Study (1994)

Radiological Assessments for Resettlement of Rongelap in the Republic of the Marshall Islands (1994)

Science and Judgment in Risk Assessment (1994)

Assessment of the Possible Health Effects of Ground Wave Emergency Network (1993)

Guidelines for Developing Community Emergency Exposure Levels for Hazardous Substances (1993)

Health Effects of Ingested Fluoride (1993)

Issues in Risk Assessment (1993)

Pesticides in Diets of Infants and Children (1993)

Research to Protect, Restore, and Manage the Environment (1993)

Setting Priorities for Land Conservation (1993)

Dose Reconstruction for the Fernald Nuclear Facility (1993)

Acknowledgments

The committee was assisted by many experts in the field during the course of its study. It is impossible to name all the individuals who contributed their time and talents in aiding the committee to complete its task.

In some instances the committee was able to build on previous reviews of possible health effects of electric and magnetic fields. We are especially indebted to William Mills and Diane Flack, managers of the Committee on Interagency Radiation Research and Policy Coordination's (CIRRPC) study of the health effects of electromagnetic fields, for their help in identifying important reports and lecturers to aid the committee in reaching an understanding of the potential health effects of power-frequency electric and magnetic fields. Informative discussions were also conducted with Robert McGaughy and James Walker of the Environmental Protection Agency and with Imre Gyuk and Robert Brewer of the Department of Energy.

At the beginning of the study the committee held a workshop to gain insight from a number of individuals with experience in the study of biologic effects of electric and magnetic fields. Among those who contributed to that workshop were Anders Ahlbom, Karolinska Institute, whose presentation was on "The Current Status of Epidemiological Studies of the Possible Health Risks of Residential Exposure to Electromagnetic Fields;" Edward P. Washburn, DOE, on "The Application of Techniques of Meta-analysis to the Investigation of the Potential Health Effects of Human Exposure to Electromagnetic Fields;" Keith Florig, Resources for the Future, on "Risk Perception and Public Policy;" Joseph V. Brady, Johns Hopkins University, on "Evidence for Neurobehavioral Effects

Resulting from Exposure to Electromagnetic Fields;'' Robert L. Brent, A. I. duPont Institute and the Jefferson Medical College, on "Evidence for Reproductive and Teratological Effects of Exposure to Low-Frequency Electromagnetic Fields;'' Gary S. Stein, University of Massachusetts, on "Evidence of Effects of Exposure to Electromagnetic Fields on Cellular Development, Growth, and Regulation;'' James Weaver, Massachusetts Institute of Technology, on "The Interaction of Electromagnetic Fields with Biological Cells;'' Ken McLeod, State University of New York, Stonybrook, on "Mechanisms for the Interaction of Electromagnetic Fields with Biological Tissue: Bone Healing and Other Effects;'' and Robert Tardiff, EA Engineering Sciences and Technology, Inc., on "Critical Factors in Risk Assessment.'' The committee is indebted to all these individuals for their participation in the workshop.

A major effort of the committee was the detailed analysis of the epidemiologic evidence of possible electric- and magnetic-field-induced diseases. Of special concern to the committee was the reliability of the statistical approaches used in the studies. The committee wishes to acknowledge Jay Lubin, National Cancer Institute, and John Tukey, Princeton University, for their contributions to the committee in their review of the committee's evaluation of the statistical methods used to analyze the epidemiologic data and to thank Dr. Lubin for providing his insights regarding this important question at one of its committee meetings. The committee also wishes to thank William Feero, Electric Research and Management Inc., for attending a committee meeting to share his insights on the issues in exposure assessment.

Contents

FIGURES

TABLES

The National Academy of Sciences is a private, nonprofit, self-perpetuating society of distinguished scholars engaged in scientific and engineering research, dedicated to the furtherance of science and technology and to their use for the general welfare. Upon the authority of the charter granted to it by the Congress in 1863, the Academy has a mandate that requires it to advise the federal government on scientific and technical matters. Dr. Bruce M. Alberts is president of the National Academy of Sciences.

The National Academy of Engineering was established in 1964, under the charter of the National Academy of Sciences, as a parallel organization of outstanding engineers. It is autonomous in its administration and in the selection of its members, sharing with the National Academy of Sciences the responsibility for advising the federal government. The National Academy of Engineering also sponsors engineering programs aimed at meeting national needs, encourages education and research, and recognizes the superior achievements of engineers. Dr. William A. Wulf is interim president of the National Academy of Engineering.

The Institute of Medicine was established in 1970 by the National Academy of Sciences to secure the services of eminent members of appropriate professions in the examination of policy matters pertaining to the health of the public. The Institute acts under the responsibility given to the National Academy of Sciences by its congressional charter to be an advisor to the federal government and, upon its own initiative, to identify issues of medical care, research, and education. Dr. Kenneth I. Shine is president of the Institute.

The National Research Council was organized by the National Academy of Sciences in 1916 to associate the broad community of science and technology with the Academy's purposes of furthering knowledge and advising the federal government. Functioning in accordance with general policies determined by the Academy, the Council has become the principal operating agency of both the National Academy of Sciences and the National Academy of Engineering in providing services to the government, the public, and the scientific and engineering communities. The Council is administered jointly by both Academies and the Institute of Medicine. Dr. Bruce M. Alberts and Dr. William A. Wulf are chairman and interim vice chairman, respectively, of the National Research Council.

*Possible Health Effects
of Exposure to*
RESIDENTIAL
ELECTRIC
AND
MAGNETIC
FIELDS

Executive Summary

CHARGE TO THE COMMITTEE

Public concern regarding possible health risks from residential exposures to low-strength, low-frequency electric and magnetic fields produced by power lines and the use of electric appliances has generated considerable debate among scientists and public officials. In 1991, Congress asked that the National Academy of Sciences (NAS) review the research literature on the effects from exposure to these fields and determine whether the scientific basis was sufficient to assess health risks from such exposures. In response to the legislation directing the U.S. Department of Energy to enter into an agreement with the NAS, the National Research Council convened the Committee on the Possible Effects of Electromagnetic Fields on Biologic Systems. The committee was asked "to review and evaluate the existing scientific information on the possible effects of exposure to electric and magnetic fields on the incidence of cancer, on reproduction and developmental abnormalities, and on neurobiologic response as reflected in learning and behavior." The committee was asked to focus on exposure modalities found in residential settings. In addition, the committee was asked to identify future research needs and to carry out a risk assessment insofar as the research data justified this procedure. Risk assessment is a well-established procedure used to identify health hazards and to recommend limits on exposure to dangerous agents.

CONCLUSIONS OF THE COMMITTEE

Based on a comprehensive evaluation of published studies relating to the effects of power-frequency electric and magnetic fields on cells, tissues, and

1

organisms (including humans), the conclusion of the committee is that the current body of evidence does not show that exposure to these fields presents a human-health hazard. Specifically, no conclusive and consistent evidence shows that exposures to residential electric and magnetic fields produce cancer, adverse neurobehavioral effects, or reproductive and developmental effects.

The committee reviewed residential exposure levels to electric and magnetic fields, evaluated the available epidemiologic studies, and examined laboratory investigations that used cells, isolated tissues, and animals. At exposure levels well above those normally encountered in residences, electric and magnetic fields can produce biologic effects (promotion of bone healing is an example), but these effects do not provide a consistent picture of a relationship between the biologic effects of these fields and health hazards. An association between residential wiring configurations (called wire codes, defined below) and childhood leukemia persists in multiple studies, although the causative factor responsible for that statistical association has not been identified. No evidence links contemporary measurements of magnetic-field levels to childhood leukemia.

STUDY FINDINGS

Epidemiology

Epidemiologic studies are aimed at establishing whether an association can be documented between exposure to a putative disease-causing agent and disease occurrence in humans. The driving force for continuing the study of the biologic effects of electric and magnetic fields has been the persistent epidemiologic reports of an association between a hypothetical estimate of electric- and magnetic-field exposure called the wire-code classification and the incidence of childhood leukemia. These studies found the highest wire-code category is associated with a rate of childhood leukemia (a rare disease) that is about 1.5 times the expected rate.

A particular methodologic detail in these studies must be appreciated to understand the results. Measuring residential fields for a large number of homes over historical periods of interest is logistically difficult, time consuming, and expensive, so epidemiologists have classified homes according to the wire code (unrelated to building codes) to estimate past exposures. The wire-code classification concerns only outdoor factors related to the distribution of electric power to residences, such as the distance of a home from a power line and the size of the wires close to the home. This method was originally designed to categorize homes according to the magnitude of the magnetic field expected to be inside the home. Magnetic fields from external wiring, however, often constitute only a fraction of the field inside the home. Various investigators have used from two (high and low) to five categories of wire-code classifications. The following conclusions were reached on the basis of an examination of the epidemiologic findings:

• Living in homes classified as being in the high wire-code category is associated with about a 1.5-fold excess of childhood leukemia, a rare disease.

• Magnetic fields measured in the home after diagnosis of disease in a resident have not been found to be associated with an excess incidence of childhood leukemia or other cancers.

The link between wire-code rating and childhood leukemia is statistically significant (unlikely to have arisen from chance) and is robust in the sense that eliminating any single study from the group does not alter the conclusion that the association exists. How is acceptance of the link between wire-code rating and leukemia consistent with the overall conclusion that residential electric and magnetic fields have not been shown to be hazardous? One reason is that wire-code ratings correlate with many factors—such as age of home, housing density, and neighborhood traffic density—but the wire-code ratings exhibit a rather weak association with measured residential magnetic fields. More important, no association between the incidence of childhood leukemia and magnetic-field exposure has been found in epidemiologic studies that estimated exposure by measuring present-day average magnetic fields.

• Studies have not identified the factors that explain the association between wire codes and childhood leukemia.

Because few risk factors for childhood leukemia are known, formulating hypotheses for a link between wire codes and disease is very difficult. Although various factors are known to correlate with wire-code ratings, none stands out as a likely causative factor. It would be desirable for future research to identify the source of the association between wire codes and childhood leukemia, even if the source has nothing to do with magnetic fields.

• In the aggregate, epidemiologic evidence does not support possible associations of magnetic fields with adult cancers, pregnancy outcome, neurobehavioral disorders, and childhood cancers other than leukemia.

The preceding discussion has focused on the possible link between magnetic-field exposure and childhood leukemia because the epidemiologic evidence is strongest in this instance; nevertheless, many epidemiologists regard such a small increment in incidence as inherently unreliable. Although some studies have presented evidence of an association between magnetic-field exposure and various other types of cancer, neurobehavioral disorders, and adverse effects on reproductive function, the results have been inconsistent and contradictory and do not constitute reliable evidence of an association.

Exposure Assessment

The purpose of exposure assessment is to determine the magnitudes of electric and magnetic fields to which members of the population are exposed.

The electromagnetic environment typically consists of two components, an electric field and a magnetic field. In general, for time-varying fields, these two fields are coupled, but in the limit of unchanging fields, they become independent. For frequencies encountered in electric-power transmission and distribution, these two fields can be considered independent to an excellent approximation. For extremely-low-frequency fields, including those from power lines and home appliances and wiring, the electric component is easily attenuated by metal elements in residential construction and even by trees, animals, and people. The magnetic field, which is not easily attenuated, is generally assumed to be the source of any possible health hazard. When animal bodies are placed in a time-varying magnetic field (as opposed to remaining stationary in the earth's static magnetic field), currents are induced to flow through tissues. These currents add to those that are generated internally by the function of nerve and muscle, most notably currents detected in the clinically useful electroencephalogram and the electrocardiogram. The currents produced by nerve and muscle action within the body have no known physiologic function themselves but rather are merely a consequence of the fact that excitable tissue (such as nerve and muscle) generate electric currents during their normal operation.

General conclusions from the review of the literature involving studies of exposure assessment and the physical interactions of electric and magnetic fields with biologic systems are the following:

• Exposure of humans and animals to external 60-hertz (Hz) electric and magnetic fields induces currents internally.

The density of these currents is nonuniform throughout the body. The spatial patterns of the currents induced by the magnetic fields are different from those induced by the electric fields. Electric fields generally are measured in volts per meter and magnetic fields in microtesla (μT) or milligauss (mG) (1 μT = 10 mG).

• Ambient levels of 60-Hz (or 50-Hz in Europe and elsewhere) magnetic fields in residences and most workplaces are typically 0.01-0.3 μT (0.1-3 mG).

Higher levels are encountered directly under high-voltage transmission lines and in some occupational settings. Some appliances produce magnetic fields of up to 100 μT (1 G) or more in their vicinity. For comparison, the static magnetic field of the earth is about 50 μT (500 mG). Magnetic fields of the magnitude found in residences induce currents within the human body that are generally much smaller than the currents induced naturally from the function of nerves and muscles. However, the highest field strengths to which a resident might be exposed (those associated with appliances) can produce electric fields within a small region of the body that are comparable to or even larger than the naturally occurring fields, although the magnitude of the largest locally induced fields in the body is not accurately known.

• Human exposure to a 60-Hz magnetic field at 0.1 μT (1 mG) results in the maximum current density of about 1 microampere per square meter (μA/m^2).

The endogenous current densities on the surface of the body (higher densities occur internally) associated with electric activity of nerve cells are of the order of 1 mA/m^2. The frequencies associated with those endogenous currents within the brain range from less than 1 Hz to about 40 Hz, the strongest components being about 10 Hz. Therefore, the typical externally induced currents are 1,000 times less than the naturally occurring currents.

• Neither experimental nor theoretic data on locally induced current densities within tissues and cells are available that take into consideration the local variations in the electric properties of the medium.

Because the mechanisms through which electric and magnetic fields might produce adverse health effects are obscure, the characteristics of the electric or magnetic fields that need to be measured for testing the linkage of these fields to disease are unclear. In most studies, the root-mean-square (rms) strength of the field, an average field-strength parameter, has been measured on the assumption that this measurement should relate to whatever field characteristics might be most relevant. As noted earlier, wire-code categories have been used in many epidemiologic studies as a surrogate measurement of the actual exposure.

• Exposure levels of electric fields and other characteristics of magnetic fields (harmonics,[1] transients,[2] spatial, and temporal changes) have received relatively little attention.

Very little information is available on the ambient exposure levels to environmental electric fields other than the rms measurements of field strength. Those might vary from 5 to 10 volts per meter (V/m) in a residential setting to as high as 10 kilovolts per meter (kV/m) directly under power transmission lines. Likewise magnetic-field exposures are generally characterized only in terms of their rms field strengths with little or no information on such characteristics as the frequency and magnitude of transients and harmonics. Residential exposures to power-frequency electric and magnetic fields are generally on the order of a few milligauss.

• Indirect estimates of human exposure to magnetic fields (e.g., wiring configuration codes, distance to power lines, and calculated historical fields) have been used in epidemiology.

These estimates of magnetic fields correlate poorly with spot measurements of residential 60-Hz magnetic fields, and their reliability in representing other characteristics of the magnetic field has not been established. Because of the many factors that affect exposure levels, great care must be taken in establishing electric- and magnetic-field exposures.

• Unless exposure systems and experimental protocols meet several essential requirements, artifactual results are likely to be obtained in laboratory animal and cell experiments.

---

[1] Signals of nf_0, where n is an integer and f_0 is the fundamental frequency. For example, the higher harmonics of a 60-Hz signal will be 120 Hz, 180 Hz, 240 Hz, and so forth.

[2] Short-duration signals containing a range of frequencies and appearing at irregular time intervals.

Many of the published studies either have used inferior exposure systems and protocols or have not provided sufficient information for their evaluation.

In Vitro Studies on Exposure to Electric and Magnetic Fields

The purpose of studies of in vitro systems is to detect effects of electric or magnetic fields on individual cells or isolated tissues that might be related to health hazards. The conclusions reached after evaluation of published in vitro studies of biologic responses to electric- and magnetic-field exposures are the following:

• Magnetic-field exposures at 50-60 Hz delivered at field strengths similar to those measured for typical residential exposure (0.1-10 mG) do not produce any significant in vitro effects that have been replicated in independent studies.

When effects of an agent are not evident at low exposure levels, as has been the case for exposure to magnetic fields, a standard procedure is to examine the consequences of using higher exposures. A mechanism that relates clearly to a potential health hazard might be discovered in this way.

• Reproducible changes have been observed in the expression of specific features in the cellular signal-transduction pathways for magnetic-field exposures on the order of 100 μT and higher.

Signal-transduction systems are used by all cells to sense and respond to features of their environments; for example, signal-transduction systems can be activated by the presence of various chemicals, hormones, and growth factors. Changes in signal transduction are very common in many experimental manipulations and are not indicative per se of an adverse effect. Notable in the experiments using high magnetic-field strengths is the lack of other effects, such as damage to the cell's genetic material. With even higher field strengths than those, a variety of effects are seen in cells.

• At field strengths greater than 50 μT (0.5 G), credible positive results are reported for induced changes in intracellular calcium concentrations and for more general changes in gene expression and in components of signal transduction.

No reproducible genotoxicity is observed, however, at any field strength. Again, effects of the sort seen are typical of many experimental manipulations and do not indicate per se a hazard. Effects are observed in very high field-strength exposures (e.g., in the therapeutic use of electromagnetic fields in bone healing).

The overall conclusion, based on the evaluation of these studies, is that exposures to electric and magnetic fields at 50-60 Hz induce changes in cultured cells only at field strengths that exceed typical residential field strengths by factors of 1,000 to 100,000.

In Vivo Studies on Exposure to Electric and Magnetic Fields

Studies of in vivo systems aim to determine the biologic effects of power-frequency electric and magnetic fields on whole animals. Studies of individual cells, described above, are extremely powerful for elucidating biochemical mechanisms but are less well suited for discovering complicated effects that could be related to human health. For such extrapolation, animal experiments are more likely to reveal a subtle effect that might be relevant to human health. The obvious experiment is to expose animals, say mice, to high levels of electric or magnetic fields to observe whether they develop cancer or some other disease. The experiments of this sort that have been done have demonstrated no adverse health outcomes. Such experiments by themselves are inadequate, however, to discount the possibility of adverse effects from electric and magnetic fields, because the animals might not exhibit the same response and sensitivities as humans to the details of the exposure. For that reason, a number of animal experiments have been carried out to examine a large variety of possible effects of exposure. On the basis of an evaluation of the published studies in this area, the committee concludes the following:

• There is no convincing evidence that exposure to 60-Hz electric and magnetic fields causes cancer in animals.

A small number of laboratory studies have been conducted to determine if any relationship exists between power-frequency electric- and magnetic-field exposure and cancer. In the few studies reported to date, consistent reproducible effects of exposure on the development of various types of cancer have not been evident. One area with some laboratory evidence of a health-related effect is that animals treated with carcinogens show a positive relationship between intense magnetic-field exposure and the incidence of breast cancer.

• There is no evidence of any adverse effects on reproduction or development in animals, particularly mammals, from exposure to power-frequency 50- or 60-Hz electric and magnetic fields.

• There is convincing evidence of behavioral responses to electric and magnetic fields that are considerably larger than those encountered in the residential environment; however, adverse neurobehavioral effects of even strong fields have not been demonstrated.

Laboratory evidence clearly shows that animals can detect and respond behaviorally to external electric fields on the order of 5 kV/m rms or larger. Evidence for animal behavioral response to time-varying magnetic fields, up to 3 μT, is much more tenuous. In either case, general adverse behavioral effects have not been demonstrated.

• Neuroendocrine changes associated with magnetic-field exposure have been reported; however, alterations in neuroendocrine function by magnetic-field exposures have not been shown to cause adverse health effects.

The majority of investigations of magnetic-field effects on pineal-gland function suggests that magnetic fields might inhibit nighttime pineal and blood melatonin concentrations; in those studies, the effective field strengths varied from 10 μT (0.1 G) to 5.2 mT (52 G). The experimental data do not compellingly support an effect of sinusoidal electric field on melatonin production. Other than the observed changes in pineal function, an effect of electric and magnetic fields on other neuroendocrine or endocrine functions has not been clearly shown in the relatively small number of experimental studies reported.

Despite the observed reduction in pineal and blood melatonin concentrations in some animals as a consequence of magnetic-field exposure, studies of humans provide no conclusive evidence to date that human melatonin concentrations respond similarly. In animals with observed melatonin changes, adverse health effects have not been shown to be associated with electric- or magnetic-field-related depression in melatonin.

• There is convincing evidence that low-frequency pulsed magnetic fields greater than 5 G are associated with bone-healing responses in animals.

Although replicable effects have been clearly demonstrated in the bone-healing response of animals exposed locally to magnetic fields, the committee did not evaluate the efficacy of this treatment in clinical situations.

1

Introduction

Electric and magnetic phenomena have been recognized since ancient times, but the means to measure, generate, control, and use the phenomena to develop practical devices became adequately understood only in the past 200 years. In little more than a century since the invention of the light bulb, society has become dependent on electricity and the myriad devices that are driven by it. It is relied on in nearly every aspect of everyday life. Electrically driven devices ease the workload in factories, farms, offices, and homes. Electricity is used to control the indoor climate, to clean clothes, to store and prepare food, and to perform many other tasks in the home and workplace. Electric devices are used in such diverse applications as medical imaging, cardiac pacemakers, cancer therapy, and communication. So widespread is the use of electricity that it is impossible to avoid exposure to the electric and magnetic fields produced in the transmission and distribution of electric power or to those fields generated by devices used in homes and workplaces.

Although the hazard of shocks and burns from coming into contact with energized electric conductors has been known since the first application of electric current, only during the past 15 years or so has public concern been raised about the more subtle effects of exposure to the fields generated by electric devices. To help determine whether a potential health risk from exposure to low-strength, low-frequency electric and magnetic fields might exist, the U.S. Department of Energy (DOE) asked the National Academy of Sciences to conduct a review. In response to the request, the Committee on Possible Effects of Electromagnetic

Fields on Biologic Systems was convened by the Board on Radiation Effects Research of the National Research Council's Commission on Life Sciences. The committee was to perform the appropriate review and report to the board on its findings.

SCOPE OF THE STUDY

The committee was asked to focus its attention on electric and magnetic fields typical of household frequencies and on the possible adverse health effects of cancer, reproductive and developmental abnormalities, and neurobiologic dysfunction, such as learning and behavioral disabilities. Those effects are the health-related end points most often suggested to be associated with exposure to power-frequency electric and magnetic fields. The committee also was asked to examine the scientific evidence for the effects of the electric and magnetic fields of household frequencies on biologic systems to determine if sufficient scientific data of adequate quality exist to perform a health risk assessment.

The DOE charge to the committee included the following:

• Review and evaluate the existing scientific information on the potential effects of exposure to electric and magnetic fields on cancer incidence, reproduction and development, and learning and behavior.

• Critically examine epidemiologic and laboratory data relating to those topics and assess potential health effects.

• Focus on electric- and magnetic-field frequencies and exposure modalities found in residential settings.

• Produce a report that contains a review of pertinent information on the effects of electric and magnetic fields, identification of research areas in which data are needed to better understand any potential health hazard, and recommendations for research in those areas and strategies for implementing research that would enhance understanding. If data of appropriate quality are available, include a health risk assessment of power-frequency electric- and magnetic-field exposures.

AREAS OF CONCERN

Concern about the possible health effects of exposure to electromagnetic fields first arose when military personnel were exposed to fields of relatively high strength from high-frequency radar systems and video screens during World War II. Since then, claims have risen of adverse health effects purportedly associated with high-frequency sources, such as radar units used by police, antenna systems used by the military, cellular phones used for communication, and microwave ovens and other appliances used in homes. Recently, attention has also focused on the potential for adverse health effects of low-frequency sources, such as transmission and distribution lines and electric appliances, including shavers, hair dryers, water beds, and electric blankets. It must be emphasized, however, that the effects of exposure to different sources of electric and magnetic fields can be quite different, depending on their frequency and strength. Possible effects of the fields generated by high-voltage transmission lines or electric blankets, operating at 60 hertz (Hz),[1] might be quite different from those generated by high-frequency (megahertz or gigahertz)[2] devices.

Of primary interest to this committee is the concern about sources of low-frequency electric and magnetic fields associated with the generation, distribution, and use of electric power, including transmission lines, substations, distributions lines, and numerous electric devices ranging from personal computers to electric clocks.

Questions of the possible adverse health effects of exposure to electric and magnetic fields from 60-Hz power lines were first raised by Wertheimer and Leeper (1979). They reported epidemiologic data suggesting an association between the configuration of power lines near homes and the incidence of leukemia and other types of childhood cancer. Similar studies have been published in succeeding years in the United States and in numerous other countries. The results of these studies have increased the scrutiny of the possible association between raised levels of electric and magnetic fields in residences, as one site of exposure, and the incidence of cancer—the adverse health effect of most concern. Much of the early laboratory research on biologic effects of very low-frequency electric and magnetic fields focused on the study of electric fields, but results of epidemiologic and other studies have caused a gradual shift of interest toward magnetic fields as a possible cause of disease.

DEFINITIONS AND DESCRIPTIONS OF TERMS

Electric and Magnetic Fields

The term "electromagnetic field," which is commonly used in the literature, applies to alternating fields. The electric and magnetic components of the fields generated by moving charged particles are formally linked and mathematically described by a set of coupled differential equations called Maxwell's equations. Electromagnetic fields are characterized by their wavelength, λ (expressed in meters), and their frequency, f (expressed in hertz); the frequency and wavelength

---

[1]Electric and magnetic fields are characterized by their frequency and strength. Power-line fields are sinusoidal, meaning they alternately change from positive to negative voltage in smooth variation, with a frequency of 60 cycles per second. That frequency is described as 60 Hz. Hertz is an internationally accepted unit of frequency, 1 Hz referring to 1 cycle per second.

[2]One megahertz (MHz) is 10^6 Hz; 1 gigahertz (GHz) is 10^9 Hz. A frequency of 10^9 Hz can also be represented as 1,000,000,000 Hz.

are related by the velocity of light, c, as $\lambda = c/f$. The full range of frequencies (or wavelengths) of natural and anthropogenic electromagnetic fields is described as the electromagnetic "spectrum." The electromagnetic spectrum, described in detail in textbooks on radiation, ranges from extremely low frequencies (ELF),[3] which are associated with common household electric current (50-60 Hz),[4] to radio waves (10^6-10^{10} Hz), microwaves (10^{10}-10^{12} Hz), infrared radiation (10^{12}-10^{14} Hz), visible light (10^{14} Hz), ultraviolet radiation (10^{15} Hz), and very high frequencies and very short wavelengths of X-rays and gamma rays ($>10^{17}$ Hz). In this list, which represents a hierarchy of increasing electromagnetic (or photon) energy, only the radiation with frequencies greater than about 10^{15} Hz is capable of ionizing atoms and molecules (i.e., producing charged particles from the atoms and molecules with which it interacts). Ionizing radiation (e.g., X-rays and gamma rays) is a well-understood source of damage to biologic systems through the reactions of the products of ionization with critical cellular components. ELF radiation, on the other hand, is nonionizing; it does not have sufficient quantal (photon) energy to produce ionization in the manner of high-frequency radiation, and its mode of interaction, if any, with molecules and biologic systems at low field strengths is speculative. Most equipment used for the generation, transmission, and distribution of electric power in the United States generates ELF (60-Hz) electric and magnetic fields. The components of the electric utility system that generate such fields include power plants (generating stations), which produce the electricity; high-voltage transmissions lines, which carry the electricity to major population centers; substations and their transformers, which reduce the voltage to levels suitable for distribution within a population center; distribution lines (distribution primaries), which commonly carry power along residential streets; distribution transformers, which reduce the voltage to amounts suitable for use in homes; and distribution secondaries (service drops), which carry electricity to individual residences. Transmission and distribution lines are commonly called "power lines," but the term can also include service drops.

Electric power that is used to operate devices in the home and workplace is also associated with the production of electric and magnetic fields. As electric charges move to produce a current, magnetic fields are created. An electric appliance connected to a source of electricity might have an electric field present even when it is turned off. When turned on and operating, a magnetic field is also present.

The exposures of interest in this report are limited to ELF electric fields (expressed in volts per meter) and magnetic fields (expressed in tesla (T) or gauss (G), where $1 \text{ T} = 10^4 \text{ G}$) associated with household use of electricity. Because

---

[3]The extremely-low-frequency designation is generally reserved for frequencies that range from 3 Hz to 3 kHz.

[4]Electric power in the United States is produced at 60 Hz, whereas power in Europe and other countries is generally produced at 50 Hz.

the range of magnetic fields encountered is usually quite small, the fields are generally described in units of microtesla (1 μT = 0.000001 T) or milligauss (1 mG = 0.001 G). For example, the earth's geomagnetic field is a static field of about 50 μT (0.5 G), and a current of 50 amperes (A) in a straight wire produces a magnetic flux density (magnetic field) of 100 μT at a distance of 10 centimeters (cm). Although household alternating current in the United States has a frequency of 60 Hz, other relatively low-frequency electric and magnetic fields can be induced when the current is used to operate appliances, such as electric razors, hair dryers, video-display terminals, and dimmer switches.

Electric fields from direct exposure to high-voltage power lines and electric appliances induce current on or just within the surface of an exposed person's body. Because the electric fields are perturbed by the tissue conductivity, the fields inside the body are very weak. On the other hand, magnetic fields pass through the body and can induce electric currents throughout the body. Magnetic fields can pass through most common building materials, including thin sheets of metal. However, magnetic materials, such as iron and some metallic alloys, can serve as convenient paths for the conduction of magnetic fields and can be used as magnetic shields in some cases. People can be shielded quite easily from exposure to electric fields, because most materials possess sufficient conductivity to attenuate the fields.

Although electric and magnetic fields are quite different in character, time-varying fields are generally described together as electromagnetic fields. As noted above, time-varying electric and magnetic fields are formally linked and described mathematically by Maxwell's equations. Through coupling, a time-varying magnetic field induces an electric field and vice versa. However, in the limit of unchanging (static) fields, the electric and magnetic fields are independent. At the low frequencies associated with electric-power use, the coupling is extremely weak, and electric fields and magnetic fields can be considered independent to an excellent approximation. In this report, the term electromagnetic field (EMF) is used when the electric and magnetic fields are substantially linked, usually only for high-frequency fields.

Biologic Effects

Very-low-frequency electric and magnetic fields are known or suspected to interact with biologic systems in a number of ways. Some biologic effects at high field strengths, such as nerve stimulation and tissue heating, are well understood and have been used to set standards for occupational and public exposure to fields. Other reported effects, particularly at low field strengths, are not as well understood; those include effects on cell metabolism and growth, gene expression, hormones, learning and behavior, and promotion of tumors. The reality of all those effects is the subject of scientific debate and an issue for discussion in this report.

The term biologic effect is intended to be a neutral term; it implies no judgment about whether an effect is good or bad. Some biologic effects of electric and magnetic fields have already been found to be beneficial; for example, the ability of fields to stimulate tissue and bone growth has been known and used for a number of years to speed the healing of fractures and burns. Other effects might be harmful.

Health Effects

The committee has focused its attention on three kinds of adverse health effects that have become the chief concerns of the public and of health officials. These are cancer, primarily childhood leukemia; reproductive and developmental effects, primarily abnormalities and premature pregnancy termination; and neurobiologic effects, primarily learning disabilities and behavioral modifications. Each of those effects has been reported in epidemiologic studies to be associated with exposure to some indirect estimates of the strength of power-frequency electric and magnetic fields. Childhood leukemia has attracted the most attention because of studies conducted in Denver, Los Angeles, and the Nordic countries that reported an increased risk of the disease in association with various indicators of exposure to electric or magnetic fields.

Exposure Assessment in Epidemiologic Studies

Determining the amounts and types of environmental agents to which an individual is exposed is often difficult. For example, persons exposed to environmental (secondhand) tobacco smoke might also be exposed to byproducts of tobacco as past or present smokers. That example is similar to the case of exposure to power-frequency electric and magnetic fields where there are multiple opportunities for exposure and almost no way to reconstruct the history of the exposure sources associated with any eventual adverse effect.

Multiple sources of possible adverse health effects are also difficult to separate in epidemiologic studies. For example, studies intending to determine if lead in the residential environment is a hazard can be altered by the fact that study subjects might be exposed to lead not only in their homes but also from outside air or in their workplaces. An individual who is exposed to one agent thought to be associated with an adverse effect might also be exposed to other agents that could contribute to the risk of the disease. Lung cancer might be assumed to be caused by an individual's smoking, even though the individual was exposed instead or in addition to another potential causative agent, such as radon or asbestos. Determining exposure to power-frequency electric and magnetic fields can be confounded similarly, and thus it is difficult to associate accurately the purported exposure with health effects in individuals.

Several dissimilar methods have been used to assess exposure to electric and magnetic fields and adverse health effects in epidemiologic studies. In only a small number of studies, actual electric-field or magnetic-field measurements were made of an individual's exposure. But even then, the measurements were made of present-day field strengths rather than the strengths to which individuals were exposed when they developed health problems or, for cancer, over the years when the cancer might have been induced. One study used records of power-line loads to calculate the average prediagnosis magnetic field. Several studies used the distance from the power line to judge whether a residence should be considered in the high-exposure category. In a number of the key studies used in this report, the association between power-frequency electric and magnetic fields and adverse health effects was made on the basis of exposed persons living in houses with power transmission or distribution lines in particular locations or configurations. The authors of those studies used the terms "wiring configuration" or "wire code" to describe a particular configuration of wires nearby a residence; the configurations were defined in such a way that they were expected to correlate approximately with electric- and magnetic-field measurements.

Risk Assessment

The formal techniques used for risk assessment were developed primarily for assessment of risk from chemicals in the environment. Risk assessment is generally divided into four distinct steps: hazard identification, dose-response assessment, exposure assessment, and risk characterization (NRC 1983). The last step implies the quantification of the proportion of persons in a population who might be adversely affected. For example, one in 1 million persons might be affected by exposure to a given agent at a certain concentration. Quantitative risk assessments have been the topic of discussion of numerous authors. One recent publication developed for the lay reader by Morgan (1995) describes quantitative assessments that place an upper limit on the lifetime risk of cancer due to exposure to electric and magnetic fields and compares that risk with other common risks.

Risk assessment can be complete (i.e., encompassing all four steps described above and leading to a quantitative risk estimate), or it can be partial (i.e., using only some of the steps). The completeness of risk assessment is governed by the available data as well as the purpose of the risk assessment.

At each step of the risk-assessment process, the assessor makes judgments about the "weight" of particular pieces of the evidence. An epidemiologic study might be deemed well designed and authoritative because the population studied is large, and thus amenable to meaningful statistical analysis, and because the study is free of confounding by other exposures. Such an epidemiologic study will be given more weight than another study that fails in some of these aspects. Laboratory tests using cell systems or whole animals will be given more weight if they have been replicated and validated by other investigators in other laboratories.

Experiments that appear to show an effect but have no biologic or physical explanation might be given less weight. In weighing the evidence, some assessors might weigh evidence differently than others in reaching their conclusions.

At the end of the risk-assessment process, the body of evidence is weighed together to reach an overall assessment of a possible hazard. If the results from several areas of research (e.g., epidemiologic studies, tests in cell systems, or whole-animal studies) are consistent and have been replicated and if a biologically plausible mechanism of action for the effect is evident, the evidence for the effect is given great weight. In contrast, a body of evidence that includes inconsistent and conflicting results, no replication of results, and effects that are often at the threshold of detection might be given little weight in reaching a conclusion.

In this report, the committee attempts to explain carefully why individual studies are given more or less weight and how the body of evidence is weighed in reaching its conclusions.

SOURCES OF EXPOSURE

Electric Power Lines

Electric transmission lines are generally built on "rights-of-way" and vary in voltage, height, and configuration of suspension from the towers. Electric power lines commonly observed in typical neighborhoods are not *transmission* lines as described above, but are lower-voltage *distribution* lines. The magnetic fields immediately under distribution lines are generally of the order of 0.5 μT (5 mG), although in some densely populated areas, fields as high as 5.0 μT (50 mG) have been measured. The fields decrease rapidly as the distances from the power lines increase.

Electric substations are installations where the voltages used with transmission lines are stepped down to lower voltages used with distribution lines. Electric and magnetic fields produced by substation equipment are generally not appreciable beyond the substation boundaries, but the fields can be somewhat stronger near them than in other parts of the neighborhood, because the power lines converge at the substation and might be closer to the ground as they go in and out of the substation.

Electric Appliances

Electric appliances in residences and workplaces are all potential sources of exposure to electric and magnetic fields. The strengths of magnetic fields vary widely: the magnetic fields from household appliances might be as low as a few tenths of a microtesla or as high as 150 μT (1,500 mG). The magnetic fields from all sources tend to decrease rapidly as the distance from the source increases. It has been noted that the earth's natural static magnetic field of about 50 μT (500 mG) is about 100 times stronger than residential magnetic fields normally associated with the alternating current (ac) of power lines and electric appliances; exposure to the earth's field is constant, and exposure to constructed alternating fields is intermittent. However, that comparison might be of little relevance, because it is not known which aspect of the magnetic fields could be of significance to health.

Transportation Systems

Some transportation systems, including subways and intercity trains, operate on ac current and generate ac electric and magnetic fields. Measurements in the Baltimore-Washington commuter trains indicate exposures to magnetic fields at 25 Hz with peak strengths as large as 50 μT (500 mG) in the passenger areas at seat height. The fields vary greatly with the position in the car as well as with the particular type and model of the car; detailed measurements indicate an average field of approximately 12.5 μT (125 mG).

WHY KNOWLEDGE ABOUT ELECTRIC AND MAGNETIC FIELDS IS IMPORTANT

Because electricity is used so extensively and sources of electric and magnetic fields are everywhere, every person in modern society is unavoidably exposed to them. Thus, understanding any biologic effects that might be associated with exposure to electric and magnetic fields is fundamentally important.

It is easy to see why so much attention has been given to the possibility that power-frequency electric and magnetic fields are associated with adverse effects. People who study how individuals respond to risk have learned that certain types of risks elicit stronger responses than others (Slovic 1987). One of the health effects that has been associated with exposure to electric and magnetic fields is an especially dreaded one, namely, cancer. Children, a group of particular concern, are a reported target for leukemia and possible reproductive and behavioral effects. The sources of the reported electric- and magnetic-field risks are largely imposed on people and not under their control. Furthermore, the fields that are the source of the reported risks are invisible and mysterious to many. All these factors cause many people to respond with concern and anxiety to potential risks associated with exposure to electric and magnetic fields (MacGregor et al. 1994).

When such an indispensable resource as electricity is reported to be associated with adverse health effects, it is not difficult to understand why concerns have arisen. It is also apparent that the potential health effects are only one part of the concern. If extreme steps are taken to reduce exposure to power-frequency electric and magnetic fields, large sums of money will need to be expended (e.g., to bury transmission and distribution lines, to redesign residential wiring and electric appliances, or to retrofit existing ones). Every citizen would contribute to the implementation of these measures through higher utility bills and greater personal expenditures on appliances. If the concerns are misplaced, but measures are taken to satisfy the public, enormous costs would be incurred unnecessarily. If the concerns are real, those who have called attention to them will have made an important contribution to public health. The charge to this committee is to assess the scientific information that will aid the government agencies, public utilities, and the general public to conduct more fully informed discussions regarding the potential health risks of exposure to power-frequency electric and magnetic fields and ultimately to help shape the appropriate policy.

RELATED REPORTS

There have been numerous studies of the possible health effects of nonionizing radiation and specifically of the very-low-frequency electric and magnetic fields on which the committee has focused. Perhaps the most quoted of the national and international reviews of the possible adverse health effects from exposure to power-frequency electric and magnetic fields is that published by the Oak Ridge Associated Universities in 1992 (ORAU 1992). The ORAU report concluded that there was "no convincing evidence in the published literature to support the contention that exposures to extremely-low-frequency electric and magnetic fields (ELF-EMF) generated by sources such as household appliances, video display terminals, and local power lines are demonstrable health hazards." They noted that the results of their review did not justify an expansion of the national research effort to investigate the health effects of exposure to electric and magnetic fields and that "in the broad scope of research needs in basic science and health research, any health concerns over exposures to ELF-EMF should not receive a high priority."

In 1989, the Office of Technology Assessment published a 103-page background paper prepared for them by Carnegie Mellon University's Department of Engineering and Public Policy and entitled "Biological Effects of Power Frequency Electric and Magnetic Fields" (Nair et al. 1989). The report called for more research on the potential effects of power-frequency electric and magnetic fields on the central nervous system and on the possibility of cancer promotion. Nair et al. (1989) recommended a policy of "prudent avoidance," which was defined as avoiding exposure by formulating strategies that were prudent from the standpoint of cost and the best understanding of risks.

Great Britain's National Radiological Protection Board (NRPB) published reports on the biologic effects of nonionizing EMF and radiation. These reports contain summaries of experimental investigations. The board's assessment of power-frequency electric and magnetic fields and the risk of cancer was published in 1992 (NRPB 1992). In summary, no firm evidence of a carcinogenic hazard was found from exposure of paternal gonads, the fetus, children, or adults to ELF electric and magnetic fields. A follow-up to that report reaffirmed its earlier conclusions (NRPB 1994).

A panel on EMF and health was established by the Victorian government of Australia in 1991 to review the range of approaches that are taken in relation to power-line fields and, when appropriate, to recommend appropriate courses of action. As a part of its activities, the panel reviewed the literature on health effects of exposure to low-frequency electric and magnetic fields. In its report (Peach et al. 1992), the panel concluded that the uncertainties in the data were so great as to preclude the possibility of establishing an association of risk with exposure. Although it noted that such fields have not been proven scientifically to be harmful, the panel recommended adoption of a policy of prudent avoidance.

ORGANIZATION OF THIS REPORT

In Chapters 2, 3, 4, and 5 of this report, the committee summarizes the research on the biologic effects of exposure to power-frequency electric and magnetic fields. Emphasis is placed on the data that pertain to the focus of the charge, studies of electric and magnetic fields at strengths and frequencies typical of residential settings. In Chapter 2, exposure is discussed: how it is measured, how it is estimated in epidemiologic and laboratory studies, and what types of exposures occur in various environments. This chapter also describes the way ELF magnetic fields interact with biologic systems. In Chapter 3, the committee summarizes what has been learned about the biologic effects of exposure to power-frequency electric and magnetic fields from cellular and molecular studies in the laboratory. Chapter 4 focuses on studies involving whole animals. Chapter 5 describes the epidemiologic data relating to the possible effects of power-frequency electric and magnetic fields on the development of cancer, reproductive and developmental abnormalities, and learning and behavioral disabilities. In Chapter 6, the evidence presented in the preceding chapters is evaluated and synthesized, and the committee's findings and conclusions are presented concerning the possible health risks of exposure to power-frequency electric and magnetic fields. The committee's recommendations for areas of possible research appear in Chapter 7. In chapters in which numerous tables of data were included in the evaluation, the tables were placed in Appendix A to improve the readability of the text. Appendix B contains a detailed description of the wire codes used in the various epidemiologic studies.

In the course of its work, the committee examined a vast amount of scientific literature. The criterion for including a paper in the deliberations was that it be published in a recognized peer-reviewed journal (unless otherwise noted). Technical reports and abstracts of papers delivered at scientific meetings were read as background information, but were not relied upon in forming judgments. The U.S. Environmental Protection Agency and Electric Power Research Institute publications used in Chapter 2 and Appendix B are notable exceptions to this rule; these technical reports are a major source of exposure data and were relied upon after careful review by committee members. As in any committee deliberation, committee members individually and collectively ascribed different weights to the importance of the papers evaluated and the quality of the research therein. The committee members often had specialized criteria based on their individual disciplines for evaluating the literature; the criteria are described in the report.

2

Exposure and Physical Interactions

SUMMARY AND CONCLUSIONS

This chapter provides an overview of various aspects of exposure to electric and magnetic fields that are considered relevant to the understanding of biologic interactions evaluated in other chapters. No attempt was made to be comprehensive; rather the goal was to provide a brief introduction to the tools necessary for exposure analysis. References to published comprehensive reviews and reports are given.

The following general conclusions are derived from the review of the literature presented in this chapter:

• Ambient levels of 60-Hz (or 50-Hz in Europe and elsewhere) magnetic fields in residences and most workplaces are typically in the range of 0.01-0.3 μT (0.1-3 mG). Higher levels are encountered directly under high-voltage transmission lines and in some occupational settings. Some appliances produce magnetic fields of up to 100 μT (1 G) or more in their vicinity.

• Exposure levels of electric fields and other characteristics of magnetic fields (harmonics,[1] transients,[2] spatial, and temporal changes) have received relatively little attention in the studies of possible biologic and health effects.

---

[1] Signals of frequency $n \times f_0$, where n is an integer, and f_0 is the fundamental frequency. For example, the higher harmonics of a 60-Hz fundamental frequency are 120 Hz, 180 Hz, 240 Hz, and so forth.

[2] Short-duration signals containing a range of frequencies and appearing at irregular intervals.

• Indirect estimates of human exposure to magnetic fields have been commonly used in epidemiology (e.g., calculations, "wire codes" (see Glossary and Appendix B for definitions of wire codes), distance, and contemporary measurements).

• Wire codes, the most commonly used estimates of possible exposure to electric and magnetic fields, are not strong predictors of magnetic-field strengths in homes. Within a given geographic region, however, wire codes do tend to distinguish relatively well between the higher and lower field strengths in homes.

• Exposure of humans and animals to 60-Hz electric and magnetic fields induces currents internally. The density of these currents is nonuniform throughout the body. Also, the spatial patterns of the currents induced by the magnetic fields are different from those induced by the electric fields.

• The endogenous current densities on the surface of the body (higher densities occur internally) associated with electric activity of nerve cells (as measured by electroencephalograph) are of the order of 1 milliampere per square meter (mA/m^2) and are chiefly of even lower frequency than 60 Hz, peaking at 5-15 Hz. Human exposure to a 60-Hz field of 100 μT (1 G) is needed to produce an equivalent current density in the body. The induced current densities caused by typical residential fields (about 1 mG) are therefore about 1 $\mu A/m^2$, or 1,000 times less than endogenous current densities.

• Microscopic heterogeneity has not been accounted for in evaluations (experimental or theoretic) of local current densities within tissues and in and around cells and cell assemblies.

• Several features are required for laboratory field-exposure systems to help eliminate potential experimental artifacts. Relatively few experimental studies have reported using systems that satisfy all these requirements. One requirement, obtaining and analyzing the data blind to the status of the exposure conditions, is extremely important. A large fraction of the published reports either do not provide sufficient information or do not satisfy many of the requirements for appropriate exposure-system design and operation.

DEFINITION OF TERMS

Electric and magnetic fields are produced by electric charges and their motion. A static electric field is produced by electric charges whose magnitude and position do not change in time. A static magnetic field can be produced either by a permanent magnet or by a steady flow of electric current (moving electric charges). The magnetic field produced by the latter means is frequently called a direct-current (dc) magnetic field. Alternating-current (ac) magnetic fields are produced by electric currents alternating in time. Electric and magnetic fields are vector quantities and thus are characterized by their magnitude and direction at every point in space and time. The behavior of electric and magnetic fields and their interrelationship are comprehensively described by Maxwell's equations (see Peck 1953; Kraus 1992; Iskander 1993; or other general texts on electromagnetic theory). One of the primary features of electric- and magnetic-field behavior is that a time-varying electric field produces a magnetic field and vice versa; therefore, reference is often made to the electromagnetic field. This field behavior, and simultaneous existence of both field components, occurs at all frequencies. However, for slowly varying fields (low frequencies), either the electric field or the magnetic field can predominate (i.e., much stronger in terms of the energy associated with it). Frequencies associated with power lines and their common harmonics are low enough for electric fields and magnetic fields generated by them to be considered separately (i.e., uncoupled). The physical reason for this simplification is that the electric field induced by the magnetic field (or vice versa) is proportional to the time rate of change. Quantitatively, one can consider the fields separately, if the magnetic field produced by the original magnetic field via the induction of the electric field is only a very small fraction of the original field. Furthermore, the common sources of the fields of low frequencies are generally separated from the exposed person, experimental animal, or cells by distances much smaller than the wavelength of the exposure field. (The electric and magnetic fields are not related through the plane wave intrinsic impedance, because no such waves are formed.) At frequencies above a few kilohertz, more careful consideration needs to be given to the coupling of the electric and magnetic fields.

An electric field is described by its strength (designated $\overline{E}$) (a bar over the field symbol indicates a vector) and its displacement vector ($\overline{D}$), also called the electric flux density. The two vectors are interrelated by the electric properties of the medium:

$$\overline{D} = \epsilon\overline{E}, \tag{2-1}$$

where ϵ is the medium permittivity; for free space $\epsilon = \epsilon_0$. For biologic materials, the permittivity is a complex number consisting of a dielectric constant and a loss factor (related to the conductivity). The electric field is measured in volts per meter (V/m), and the electric flux density in coulombs per square meter (C/m^2).[3]

A magnetic field is described by its strength ($\overline{H}$) and the magnetic flux density ($\overline{B}$). The two vectors are related by magnetic properties of the medium:

$$\overline{B} = \mu\overline{H}, \tag{2-2}$$

where μ is the medium permeability; for free space $\mu = \mu_0$. For most biologic materials (except magnetite found in small quantities in some tissues), $\mu \cong \mu_0$. The most commonly used magnetic-field descriptor is its flux density $\overline{B}$ presented either in units of tesla (T), an internationally approved (SI) unit, or an older and

---

[3]The coulomb is a unit of charge such that an ampere of current is 1 coulomb per second; the charge carried by an electron is then 1.6×10^{-19} coulomb.

more common unit of gauss (G), (1 G = 10^{-4} T; also 1 T = 1 Wb/m^2, where Wb = weber). The magnetic-field strength $\overline{H}$ has units of amperes per meter (A/m).

One of the characteristics of an ac electric or magnetic field is its waveform (i.e., the change in amplitude and phase with time). Sinusoidal (harmonic) fields of 50 or 60 Hz are the most commonly encountered ac fields in the environment, and they are often used in biologic experiments. They can also contain small distortions resulting in harmonics (multiples of the fundamental frequency, e.g., 120 Hz, 180 Hz, etc., for the 60-Hz fundamental frequency). Another common waveform used in the laboratory is a "rectangular" pulse or a time series of pulses either bipolar or unipolar. A whole range of frequencies are associated with a pulse or a series of pulses. The exact frequency spectrum depends on the pulse characteristics, such as its duration, repetition rate (for multiple pulses), the rise time (of the leading edge), and the fall time (of the trailing edge). In technical terms, these frequencies are determined by Fourier analysis. A brief discussion of the spectra of simple waveforms is given in a report by ORAU (1992). Waveforms associated with some phenomena, such as lightning and on and off switching of electric devices, are frequently referred to as transients and are very complex and unique for a given event. Their frequency spectra are broad and extend into the megahertz range.

A parameter characterizing the field and related to its frequency (for harmonic fields) is the wavelength. The wavelength in free space is related to frequency in free space as

$$\lambda_0 = \frac{c}{f}, \qquad (2\text{-}3)$$

where c is the velocity of light ($c = 3 \times 10^8$ m/sec). In media, such as biologic tissues, the wavelength is shorter than that in free space and equal to

$$\lambda = \lambda_0/\sqrt{\epsilon/\epsilon_0}, \qquad (2\text{-}4)$$

where ϵ is the permittivity of the tissue in question; it should be noted that various tissues have different permittivities.

The range of frequencies in which the power-line fundamental frequencies of 50 or 60 Hz fall is referred to as extremely low frequencies (ELF). ELF are generally considered to extend from 3 Hz to 3 kHz.

The electric field at power-line frequencies produced by specific voltages on high-voltage transmission lines can be accurately evaluated by analytic or numeric methods. Similarly, for distribution lines and any other known configuration of wires and other shape conductors, it is possible to evaluate the strength and direction of the electric field in any point of the surrounding space. Simple cases, such as a plate, a single straight wire (in free space), a wire above ground, two wires (infinitely long), three phase wires, and similar configurations, can be solved analytically. Several examples are given in the ORAU (1992) report. It is important, however, to realize that the electric field is significantly perturbed by any conducting or dielectric object that is placed in it. Thin objects placed perpendicularly to the direction of the field introduce only a minimal field perturbation. That feature of electric fields has a significant bearing on correct measurements of the fields. People and animals greatly affect the field (Kaune and Gillis 1981). Therefore, the measured field in the presence of a person is significantly different from the exposure field without the person present.

Similarly to the electric field, the magnetic field can be accurately evaluated (analytically or numerically) for various configurations of current-carrying conductors. Examples of simple calculations are given in the ORAU (1992) report. For any arbitrary but known configuration of conductors, the magnetic field can be computed numerically. In cases of motors and other devices of complex geometry, particularly those containing magnetic materials, theoretic evaluation of the exposure field is impractical. Unlike the electric field, the ELF magnetic field is not affected by the presence of humans and animals. Therefore, the measured field represents the actual exposure field.

METHODS OF EXPOSURE ASSESSMENT

General Problems

Electric and magnetic fields at 60 (or 50) Hz can be calculated or measured in practically any environment. Even their more complex characteristics (e.g., harmonics and temporal and spatial changes) can be determined. Similarly, transients can be measured, albeit only with sophisticated instrumentation. Determination of human exposure and, in particular, determination of human exposure as it relates to epidemiologic studies is much more difficult. An average adult or child encounters a variety of environments of electric and magnetic fields in a day, not to mention in a month or a year.

The original interest in possible health effects of power-line fields was precipitated by an epidemiologic report (Wertheimer and Leeper 1979), which suggested that the strength of 60-Hz magnetic fields, as classified or estimated by a wire code, correlates with increased rates of childhood cancers, including leukemia. In subsequent studies, wire codes, or other presumed indicators of the average root-mean-square (rms) strength of the 60-Hz magnetic field, have been used.

Various characteristics of electric and magnetic fields, other than their rms magnitude at 60 Hz, might be responsible for their interaction with biologic systems (e.g., harmonics, transients, and temporal and spatial changes). Knowledge of which characteristic (if any) of the exposure fields is important in the interaction would permit reliable exposure assessment in epidemiologic studies. Lack of knowledge of the relevant field characteristic makes comprehensive human exposure assessment nearly intractable. Nevertheless, a majority of studies have been conducted with the tacit assumption that the 60-Hz magnetic field (rms averaged and cumulative over time) is directly related to the exposure of interest.

Exposure can be assessed by direct measurements or by indirect modeling and estimation of the electric and magnetic fields present in the spaces occupied by humans or experimental animals. In most cases, such evaluations have been made at 60 (or 50) Hz only.

Measurement Methods and Instrumentation

Without any clear guidance about what aspect of a field is biologically relevant, most commonly available field-measurement devices today have been designed to determine the average rms field strength (magnetic flux density or electric-field strength) over a specific time. The minimum averaging time is usually about 1 sec, and some instruments can average for hours. More elaborate equipment has the capability to measure the detailed time variation or frequency spectrum of a field, but analyzing or choosing a simple metric from the enormous amount of information collected is difficult.

As a compromise, some of the more popular field-measurement devices today are able to record many samples of the magnetic field over a long period; for example, they can be set to record a sample every 10 sec for 24 hr. The resulting amount of data is manageable and permits the calculation of a limited range of summary metrics (such as average rms field, peak field, median field, difference between successive measurements, and time above a specific threshold).

Most of the monitoring devices for electric and magnetic fields use filtering to limit the range of frequencies measured. Such a device can measure a narrow band of frequencies from 50 to 60 Hz or cover a broad band of frequencies from 20 to 2,000 Hz. Regardless of the frequency range measured, the instruments report a single number reflecting the sum of all fields in that frequency range.

The most common method used to determine the electric field is to measure the voltage between two conductors. In one popular instrument, the two conductors are the upper and lower halves of the device case. Because the presence of the instrument user can alter the electric field, the sensing probe of the measurement device must be held away from the body by means of a long nonconducting rod. To make a reading, the user rotates the probe until its axis is parallel to the direction of the electric field (the maximum reading).

The most common method used to determine the magnetic field is to measure the voltage induced in a coil of wire by the alternating field. To make a reading, the coil must be rotated until its axis is parallel to the direction of the magnetic field. Some devices use a coil that is separate from the instrument electronics package; others incorporate the coil in the instrument case so that the entire device must be rotated. More expensive devices for the measurement of magnetic fields use three orthogonal coils in the instrument case. Instead of having to rotate a single coil, the devices sense the three mutually perpendicular components of the field by these coils and calculate the vector sum of the fields. Procedures for measuring electric and magnetic fields in the environment are described in detail by ANSI/IEEE (1987).

Field Calculations

For well-defined sources, magnetic flux densities can be calculated accurately, and measurements support the accuracy of such calculations. Electric fields can also be calculated, but because the fields are perturbed by conducting objects, calculations are often of limited value unless the perturbations by such objects can be modeled. Electric- and magnetic-field calculations, when properly performed, are more accurate than measurements; in fact, field-measuring devices are frequently calibrated against the calculated field of a simple geometric arrangement of conductors.

For most environments (in the home or workplace), conductor geometries are complex or unknown, so measurements must be used. For distribution lines, even though the geometry is relatively simple, the currents are not the same in each wire (not balanced) and are generally not known accurately enough to rely on calculations. For transmission lines, however, the amount of power transmitted is generally recorded, and the line currents are usually balanced sufficiently to be estimated accurately; therefore, the field can be calculated accurately, assuming no other sources or shielding materials are near.

TYPICAL EXPOSURES

Electric Fields

Exposure to electric fields in the home and workplace environments can be attributed primarily to electric equipment used in those environments. Because of the ease with which very-low-frequency 60-Hz electric fields are shielded or perturbed, electric fields in the home and workplace have not been characterized satisfactorily. Attempts have been made to measure personal exposure to electric fields (e.g., by the Electrical Power Research Institute (EPRI 1990)), but the measurements are heavily dependent on where the exposure meter is worn, the orientation of the exposure meter, and the presence of any conductors near the exposure meter. With that caveat, EPRI found that the mean personal exposure to 60-Hz electric fields in the home or office typically ranges from 5 to 10 V/m.

Power-line electric fields have been well characterized; depending on line voltage, ground-level electric fields under a line might be as high as 10 kV/m, which is sufficient to cause fluorescent tubes to glow and to induce noticeable shock currents in a person who touches a vehicle parked under a high-voltage line. Mean personal exposure to electric fields for substation, distribution-line, and transmission-line workers ranges from 50 to 5,000 V/m (EPRI 1990).

Residential and Environmental Magnetic Fields

Residence

The three most common sources of residential 60-Hz magnetic fields are electric appliances, the grounding system of the residences (most often, water pipes), and nearby power lines (most commonly, low-voltage distribution lines). Normally, unless there are wiring anomalies, internal residential wiring is not a significant source of personnel exposure. Although high-voltage transmission lines produce relatively high magnetic fields directly under them, they contribute relatively little to the residential and environmental levels at distances greater than 100 m, as illustrated in Table 2-1.

Background Fields

The background magnetic fields in the center of the rooms (away from most appliances) of the home are most likely caused by power lines, grounding systems, or some combination of the two. In an EPRI (1993a) study in which extensive magnetic-field measurements were taken in the center of the rooms of 992 homes, only 5% of the homes had average magnetic fields exceeding 0.29 μT (2.9 mG).

The distribution of fields observed for selected rooms in the homes and the all-room averages are shown in Table 2-2. Assuming people's activity patterns are uniformly distributed within the residence living space, the all-room average field shown in Table 2-2 should provide a slightly better characterization of exposure than the all-room median field (not shown in the table) (EPRI 1993a). Household energy consumption and the presence of electric heating did not explain variations in the measured magnetic fields between homes (EPRI 1993a).

TABLE 2-1   Magnetic Fields as a Function of Distance from Power Lines

Transmission Lines, kV	Maximum Magnetic Field on Right-of-Way, μT (mG)	Representative Magnetic Fields at Different Distances from Lines, μT (mG)			
		15.24 m (50 ft)	30.48 m (100 ft)	60.96 m (200 ft)	91.4 m (300 ft)
115	3.0 (30)	0.7 (7)	0.2 (2)	0.04 (0.4)	0.02 (0.2)
230	5.8 (58)	2.0 (20)	0.7 (7)	0.18 (1.8)	0.08 (0.8)
500	8.7 (87)	2.9 (29)	1.3 (13)	0.32 (3.2)	0.14 (1.4)

SOURCE: EPA 1992.

TABLE 2-2   Average and Percentile Values for the Magnetic Field in Homes According to Room

Room Center (n = 992 homes)	Magnetic Flux Density, μT			
	5%	50%	95%	Mean
All rooms (average)	0.01	0.06	0.29	0.09
Kitchens	—	0.07	0.35	—
Bedrooms	—	0.05	0.29	—
Highest room	—	0.11	0.56	—

—, not provided.
SOURCE: EPRI 1993a.

In the EPRI study, the power-line fields were found to be the dominant source of average and median fields. For most low-voltage power lines, the load current on the three wires is not always balanced. To analyze the effect of the imbalance, the load current can be mathematically divided into the balanced and unbalanced parts; the balanced part of the current produces a field that decreases approximately as the inverse square of the distance from the power line; the unbalanced part (the zero-sequence current) causes a field that decreases with the inverse of the distance. Therefore, at greater distances, the field associated with the zero-sequence current dominates. When the median field in a house is greater than 0.16 μT (1.6 mG), the home in question is usually near a power line and the main field source is usually the balanced part of the power-line load current.

Appliances

The strongest magnetic fields in homes are generally caused by appliances. However, the fields usually decrease rapidly with distance. When the main source of a magnetic field in an appliance is a coil of wire, the field decreases approximately as the inverse cube of the distance. Some of the magnetic-field values measured near household and other appliances are shown in Tables 2-3 and 2-4. The values show the range of all the measurements made (e.g., 95% of the color television sets measured emitted magnetic fields less than 0.33 μT at a distance of 56 cm). The values are based on measurements of the rms fields averaged over time from about 1 sec or more for the spot measurements to 24 hr for long-term and personal exposure measurements. Different appliances of the same type can produce different magnetic fields because of differences in their design. Important differences are the amount of current they use, the size and shape of conducting parts, the number of turns of wire in coils, and whether shielding or field-canceling technology was used.

TABLE 2-3   Magnetic-Field Strengths of Common Household Appliances

Sources	Magnetic Field at 6 in (0.15 m), μT	Magnetic Field at 1 ft (0.3 m), μT
Bathroom sources		
Hair dryers	0.1-70.0	Bkg[a] to 7
Electric shavers	0.4-60.0	Bkg to 10
Kitchen sources		
Blenders	3-10	0.5-2
Can openers	50-150	4-30
Coffee makers	0.4-1	Bkg to 0.1
Dishwashers	1-10	0.6-3
Food processors	2-13	0.5-2
Garbage disposals	6-10	0.8-2
Microwave ovens	10-30	0.1-20
Mixers	3.0-60	0.5-10
Electric ovens	0.4-2	0.1-0.5
Electric ranges	2.0-20	Bkg to 3
Refrigerators	Bkg to 4	Bkg to 2
Toasters	0.5-2	Bkg to 0.7
Laundry and utility-room sources		
Electric clothes dryers	0.2-1	Bkg to 0.3
Washing machines	0.4-10	0.1-3
Irons	0.6-2	0.1-0.3
Portable heaters	0.5-15	0.1-4
Vacuum cleaners	10-70	2-20
Office sources		
Air cleaners	11-25	2-5
Copy machines	0.4-20	0.2-4
Fax machines	0.4-0.9	Bkg to 0.2
Fluorescent lights	2-10	Bkg to 3
Electric pencil sharpeners	2-30	0.8-9
Video-display terminals	0.7-2	0.2-0.6
Workshop sources		
Battery chargers	0.3-5	0.2-0.4
Drills	10-20	2-4
Power saws	5-100	0.9-30

[a]The magnetic field of the device producing the lowest level could not be distinguished from background (Bkg) levels.

SOURCE: EPA 1992.

In addition to the appliances listed in Tables 2-3 and 2-4, magnetic fields from electric blankets might contribute a large part to the magnetic-field exposures in the home. When measured about 5 cm from the surface of the blanket, approximately the distance of internal organs, the magnetic fields average about 2.2 μT (22 mG) for conventional electric blankets and about 0.1 μT (1 mG) for positive-temperature-coefficient blankets.

TABLE 2-4 Percentile Values for the Magnetic Fields of Typical Appliances

Appliances (number measured)	Magnetic Flux Density, μT, at 22 1/3 in. (56 cm)		
	5%	50%	95%
Color TVs (343)	0.09	0.18	0.33
Microwave ovens (485)	0.49	1.0	1.7
Analog clocks (118)	0.05	0.19	0.36
Can openers (13)	0.18	2.48	3.18

SOURCE: EPRI 1993a.

Personal Monitoring

Because the subject moves around the house during personal-exposure measurements, personal exposures include contributions from electric appliances as well as from power lines and grounding systems. The contribution of appliances to total exposure can be seen in Table 2-5. The values shown include personal exposures and long-term stationary measurements made using the same instrumentation. The personal-exposure measurements are consistently higher than the measurements made at fixed positions in the rooms. The fixed-position measurements were made in a frequently used room, other than the bedroom, and away from any identifiable local magnetic-field source (EPRI 1993b).

Several studies have examined the personal exposures of children (e.g., Kaune et. al. 1994). The personal exposures were approximately log-normally distributed with both residential and nonresidential geometric means of 0.1 μT (1 mG). The correlation between log-transformed residential and total personal-exposure levels was 0.97. This close correlation might be due to the fact that children spend most (two-thirds to three-fourths) of their time at home.

Environmental Fields

In the EPRI residential study (EPRI 1993b), short-duration measurements were made in the center of the rooms and outdoors around the perimeter of the

TABLE 2-5 Average and Percentile Values for Personal Exposure and Spot and Long-term Measurements of the Magnetic Field in Homes

Measurements (n = 380 homes)	Magnetic Flux Density, μT			
	5%	50%	95%	Mean
Personal exposure	0.038	0.133	0.519	0.198
Long term	0.027	0.102	0.479	0.169
Inside spot	0.019	0.101	0.399	0.141
Outside spot	0.020	0.097	0.385	0.144

house. The two sets of measurements were similar; both included contributions from the grounding system and nearby power lines. At greater distances from a home, the magnetic field is controlled primarily by components of the power-delivery system, including distribution-system primary and secondary wires and transmission lines.

Transformers and substations are usually not important sources of magnetic fields in a community (beyond the substation boundary). However, because substations represent a point of convergence for power lines, residences near substations have a greater chance of being near power lines and, therefore, have a greater chance of exposure to higher magnetic-field levels from the power lines. Also, some power lines might be closer to the ground near the substation, thus causing the ground-level fields to be greater. In addition, there might be higher ground-level return currents in the earth near substations and therefore higher field levels in the area.

Power Lines (Transmission and Distribution)

For homes near transmission rights-of-way, transmission lines can be important sources of magnetic fields. Typical values for the magnetic fields from transmission lines were illustrated in Table 2-1 (EPA 1992). At peak usage, the magnetic fields could be double the average figures shown in the table. Note that the fields fall off roughly as the inverse square of the distance from the line, and the magnetic field is greater for the higher-voltage lines. There are two reasons the magnetic field is greater for the higher-voltage lines: higher-voltage lines generally have thicker wires to carry more current (magnetic field is directly proportional to current); and higher-voltage lines have a greater separation between wires to avoid arcing. The magnetic fields produced by the three conductors of a three-phase transmission line tend to cancel one another. Greater wire spacings result in less cancellation; closer spacings result in more cancellation. If the three phases could occupy the same space, the fields would exactly cancel each other and a balanced (in current and phase angle) three-phase transmission line would produce zero magnetic field. That situation is approximated when the wires are insulated and placed together in an underground pipe.

The electric power lines commonly observed in neighborhoods are usually not transmission lines as described above, but are lower-voltage distribution lines. The magnetic fields near distribution lines are generally around 0.5 μT (5 mG), but in some densely populated areas, field levels as high as 5.0 μT (50 mG) have been measured.

There are two major types of underground power lines: direct burial (the individual wires are buried separately) and pipe-type cables (all the wires are placed in a single metal pipe). Direct-burial underground power lines can produce ground-level magnetic fields as large as equivalent capacity overhead lines (but over a more limited area). Although the underground wires might be closer together than overhead wires (tending to decrease the field), they are buried at a depth of only about 5 ft and, therefore, are much closer to the surface of the ground than overhead lines (a fact that tends to increase the field). In underground pipe-type transmission lines, the close spacing of the wires in the pipe and the metal pipe itself decrease the magnetic field, so that the resulting ground-level field is typically less than 0.1 μT (1 mG).

Occupational Magnetic Fields

Workplace

The magnitude of the magnetic field for the office environment is similar to that for the home; however, differences might exist in the extent and pattern of exposure to electric devices. Typical magnetic fields from workplace equipment are illustrated in Table 2-6 (EPA 1992). Note that the fields for a particular type of device (e.g., copy machines) vary greatly from one machine to another because of the amount of current used and other design features. The data are missing for the lowest-field fluorescent light because its field was less than the background field. Compact fluorescent lights, which are being used extensively in the workplace as a more efficient replacement for incandescent lights, produce negligible magnetic fields at 2 ft because of their compact size and low current use.

Personal Monitoring

In a study of occupational exposures (EPRI 1990), personal-exposure measurements were performed in the home and workplace. As Table 2-7 shows, the home and office are similar magnetic-field environments; however, office exposures are somewhat higher, possibly reflecting more frequent (or continuous) and proximal use of electric equipment. Measurements of the home environment provided in Table 2-7 are somewhat different from those in Table 2-5, because

Table 2-6   Typical Magnetic-Field Levels Measured Near Workplace Devices

Devices	Magnetic Flux Density, μT, at 2 ft (0.61 m)		
	Lowest	Median	Highest
Copy machines	0.1	0.7	1.3
Personal computers with color video-display terminals	0.1	0.2	0.3
Power saws	0.1	0.5	4.0
Fluorescent lights	—	0.2	0.8

SOURCE: EPA 1992.

Table 2-7   Typical Magnetic-Field Personal Exposures in the Home and Workplace

Occupied Environment	Magnetic-Field Distribution, μT			
	5%	50%	95%	Mean
Substation	0.045	0.7	5.957	3.443
Generation	0.022	0.124	2.661	0.835
Office	0.017	0.075	0.596	0.207
Home	0.017	0.061	0.432	0.147

SOURCE: EPRI 1990.

the protocols of the two studies were different. Those differences make it difficult to make detailed comparisons between the studies.

In work environments involving high currents and large distances between the current-carrying conductors, such as substations and generating stations, magnetic fields are at high levels over large areas. In the occupational project by EPRI (1990), which included such utility-specific environments as substations, generating stations, and work areas involving transmission and distribution lines, the highest exposures were of workers in substations.

In another study that focused on the home exposure of adults (Kavet et al. 1992), the average exposure while away from home was relatively uniform at about 0.2 μT (2 mG) for the study population. That amount is consistent with the data for office workers shown in Table 2-7. However, the subjects spent a large fraction of their time at home, so their total exposures were similar to their at-home exposure.

Transportation

Some transportation systems, including subways and intercity trains, operate on alternating current and generate ac electric and magnetic fields. As shown in Tables 2-8 and 2-9, magnetic fields in transportation environments are extremely variable. The temporal and spatial variations are great, and for some trains, large frequency variations occur as the speed of the trains changes.

Transients

Transients are present in all environments where switched equipment is used. A recent study of transients (EPRI 1994a) found that transients within the home can be random, originating from manual and automatic switching devices (such as those associated with lights and appliances), or repetitive, originating from electronic controllers (such as dimmer switches that produce transients twice per

TABLE 2-8   Typical Magnetic Fields from Commuter Trains[a]

| | Magnetic Flux Density, μT | | |
Train	Minimum	Average	Maximum
Boston subway (electric)	0.022	0.343	1.671
MAGLEV (electric)	0.992	3.055	7.692
Washington, D.C., subway (electric)	0.6	6.023	14.592
Amtrak (nonelectric)	0.092	0.642	1.26

[a]Measurements were made at 3.5 ft (110 cm) above the floor and include the frequency range of 5 to 2,560 Hz.

SOURCE: EPA 1993.

TABLE 2-9   Typical Personal Exposures Estimated for Transportation Workers

| | Magnetic Flux Density, μT | | | |
Type of Travel	5%	50%	95%	Mean
Work related	0.017	0.079	0.716	0.309
Non-work related	0.017	0.073	0.473	0.156

SOURCE: EPRI 1990.

power-frequency cycle). The rate of occurrence of transients is greater in industrial areas than in residential areas.

The frequencies associated with switching transients range from less than 60 Hz to about 500 MHz. Switching transients begin with high-frequency components radiated directly from the vicinity of the switch, followed by low-frequency components resulting from currents in residential and distribution wiring. Repetitive transients typically have a single dominant frequency component.

The measured amplitude of the magnetic flux density ranged from 0.001 μT (0.01 mG) to 10 μT (100 mG), depending on source, frequency, and position. Transients can propagate in residential wiring through secondary distribution lines to neighboring residences and along ground pathways (conductive plumbing pipes). Switching operations in primary distribution lines can generate transients within the residences (EPRI 1994a).

Personal exposure to transient fields is extremely complex because of their strong spatial dependence. Exposure to high-frequency fields is typically due to switching operations within the residence (e.g., light dimmer switches), but exposure to low-frequency fields can also be due to external sources in the neighborhood. The frequency and strength of the fields are such that transients can induce currents that are larger than thermal noise; thus, magnetic fields associated with transients to which individuals might be frequently exposed could be considered in future research.

EXPOSURES IN EPIDEMIOLOGIC STUDIES

Residential

The primary hypothesis tested in most residential epidemiologic studies of electric- or magnetic-field effects is that the presence of power lines near the home is related to occurrence of disease. With few exceptions, the studies have focused on indirect estimates of the magnetic fields rather than electric fields. At low frequencies, such as 60 Hz, electric fields are substantially shielded by the shell of a house and by surrounding trees, so that residential exposure to electric fields is difficult to describe and nearly impossible to model with any accuracy. In the study by London et al. (1991), nearby power lines appeared to have no influence on indoor electric fields.

On the other hand, magnetic fields are largely unaffected by intervening structures and therefore might reflect more directly the operation of nearby power lines. Exposure assessments in the residential studies (e.g., wire codes and distance to power lines) have been viewed as estimates or indirect measurements of some aspect of magnetic-field exposure experienced by the subjects before their disease was diagnosed. Because of limitations in instrumentation or available data, the magnetic-field characteristic examined is usually a short- or long-term rms average field. Each of the exposure assessment methods has its strengths and weaknesses. The following sections briefly describe the major types of exposure assessment used.

Wire Codes

This indirect measure was used in the first epidemiologic study of presumed exposure to magnetic fields and cancer risk (Wertheimer and Leeper 1979). This means of exposure quantification was subsequently used in later studies with several refinements and modifications. Categories of wire codes and how they relate to the high-voltage transmission and distribution lines are illustrated in Figure 2-1. Detailed information on wire codes can be found in Appendix B, and more details on their use in epidemiology are given in Chapter 5.

The use of the wire code illustrated in Figure 2-1 and its modifications has a qualitative physical rationale but also significant quantitative limitations. Its use in previous epidemiologic studies was based on the assumption that the wire code reflects the average exposure to 60-Hz magnetic fields. The rationale of this assumption is that larger power lines with thicker wires, which serve more residences and other consumers of electricity, carry more current and therefore provide a measure of exposure in the past and over a prolonged period. The

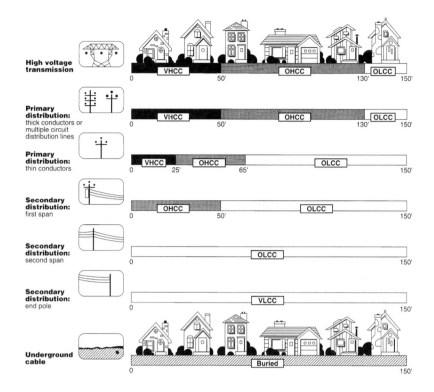

FIGURE 2-1 A simplified schematic of the basic features of the differences in the wire codes as defined to support epidemiologic studies. VHCC, OHCC, OLCC, and VLCC stand for very high, ordinary high, ordinary low, and very low current configurations. Figure provided courtesy of Robert S. Banks Associates, Inc. (originally prepared for EPRI and revised for this report).

merit of this rationale is that it considers (albeit in a qualitative manner) several of the factors used to calculate power-line magnetic fields; however, the reliability of the wire codes as a quantitative measure of exposure to 60-Hz magnetic fields is very limited. The following is a summary of some of the findings in a review of the characteristics of the wire codes as used in the epidemiologic studies:

• Although the rank ordering of fields in homes is predicted reasonably well by wire codes, the wire code accounts for only 15-20% of the variance in magnetic-field measurements.

• A large overlap exists between the estimated ranges of the magnetic fields for various categories (e.g., very high current configuration and ordinary high current configuration). There are large differences between the magnetic fields for different studies in different geographic locations for the same categories of the wire code. For instance,

—In a comparison of Los Angeles, California, and Denver, Colorado, the differences between the cities within one category are greater than the differences between categories.

—Total exposure to 60-Hz magnetic fields in single-family residences is affected by other factors (appliance fields and grounding system fields). The field from a power line adjacent to the residence is a dominant factor for high wire codes but not for low wire codes.

Figure 2-2 compares mean residential magnetic-field measurements by wire-code category for different studies (EPRI 1994b). The data were obtained by several different procedures. Indoor spot measurements were usually taken with hand-held meters for a brief time at a limited number of locations in the residence (usually at the center of rooms, where the fields from appliances and house wiring are not very significant). Stationary measurements were made with instruments left in the residence for an extended time (usually 24 hr or more); appliances and house wiring might contribute somewhat to the measured field. For personal-exposure measurements, the meter is always near the subject (either worn or placed on a bedside table); the contribution of appliances to the measured field is usually significant.

Six studies are represented in the graphs—three are epidemiologic studies, and three are exposure assessment studies. The author of the EPRI (1994b) report cautions that Figure 2-2 includes multiple data points of the same type of measurement from several studies to show some of the variation possible in arithmetic mean measurements under different conditions (e.g., spot measure-

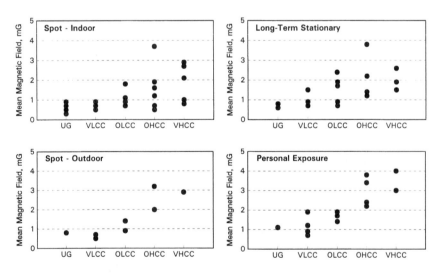

FIGURE 2-2   A comparison of mean residential magnetic-field measurements by wire-code category for six studies. Source: EPRI 1994b.

ments under low- and high-power usage conditions or personal exposures for children and adults in the same homes). Several studies using different methods are shown in Figure 2-2; therefore, some of the differences between measurements are undoubtedly due to the differences in study populations or study methods.

The possibility that wire codes better represent another characteristic of magnetic fields rather than their magnitude at 60 Hz is relatively unexplored so far. For instance, they might reflect the numbers and magnitude of transients. If laboratory studies had clearly indicated that 60-Hz magnetic fields were carcinogenic at such low levels as those considered in the epidemiologic studies, then there might be less concern about the limitations of the wire codes as measures of exposure.

Calculations

Magnetic fields from power lines can be calculated accurately if the relevant variables are known. The most important variables for calculations are wire spacing and height, current in each wire, and distance to the residence. Utilities rarely have records of the current in each wire of a power line for past years (the prediagnosis period). More often, there is a record of power-line load from which current in each wire can be calculated, if it is assumed that the currents are balanced. Balanced current is usually a good assumption for high-voltage transmission lines but not for the more common low-voltage distribution lines. Because of these limitations, calculations are not often used. Historical records of power-line loads are sufficiently complete in Sweden, however, for Feychting and Ahlbom (1993) to use the calculated magnetic fields in their study.

Distance

Because the magnetic field decreases with distance from an operating power line, distance can be used as a crude predictor of the field. The utility usually has records showing whether a particular line was operating during the prediagnosis period. Because the actual field also varies with line load and the geometric configuration of the power line, the use of distance can result in significant misclassification, particularly if a study involves power lines of several different designs. Because of its simplicity, distance is frequently used as an exposure metric to assign subjects to two or more exposure groups.

Measurements

Two types of measurements are used most often: spot measurements (the rms average magnetic field over a period of seconds or minutes), and a long-term rms average (a measurement extending 24-48 hr). The long-term average is frequently called the time-weighted average. A personal-exposure measurement is a long-term average recorded by a measuring device worn by a subject. Other measurements, such as the peak field or field exceeded 10% of the time, are used occasionally.

Transmission-line measurements correlate well with calculated fields, but they only reflect line currents during the measurement period, which might or might not be typical for the power line. Measurements are the only practical way to determine the field from complex sources, such as distribution lines, appliances, and ground currents. A major criticism of contemporary magnetic-field measurements is that they might not reflect conditions accurately that prevailed years before—during the period the disease developed.

Occupational

Occupational studies typically rely on job title as an indicator of a subject's magnetic-field exposure or on magnetic fields measured at representative work locations (either personal exposure or spot measurements) and combined with each subject's work history. Both methods are likely to result in misclassifications because of the large overlapping range of fields measured for a given job title (e.g., see the measurements reported by Theriault et al. 1994).

EXPOSURES IN LABORATORY EXPERIMENTS

Animal and Cellular Studies

In Vivo Studies

Electric Fields   The most commonly used exposure apparatus consists of parallel plates between which an alternating voltage (50 or 60 Hz, or other frequencies) is applied to produce the electric field. Typically, one of the plates (bottom) is grounded. When proper dimensions of the plates are selected (large area in comparison to their separation), a uniform field can be produced within a reasonably large volume between the plates. Distribution of the electric-field strength within a volume of interest can be calculated. The field uniformity deteriorates close to the plate edges.

Electric fields of up to 100 kV/m have been used in experiments with animals. This field strength for a rodent corresponds to about 10 kV/m for a human, as described in the following section on in vitro studies.

The original reasonably uniform field in an animal-exposure system can be significantly perturbed by two factors. One factor, which is unavoidable but controllable, is due to the presence of test animals and their cages. A considerable amount of information is available on proper spacing of animals to ensure the same exposure field for all test animals (Kaune 1981a) and to limit the mutual shielding of test animals (Kaune 1981b; Creim et al. 1984). Animal cages, drinking bottles, food, and bedding cause additional perturbation to the electric field (Kaune 1981a,b). Particularly, any metallic or other highly conductive objects or substances (e.g., animal excrement) must be eliminated or kept to a minimum. One of the most critical problems resulting in artifactual results of some studies is induction of currents in the nozzle of the drinking-water container. If the induced currents are sufficiently large, animals experience electric microshocks while drinking. Corrective measures have been developed to deal with this problem (Free et al. 1981; Kaune 1981b). Perturbation of the exposure field resulting from nearby metallic objects is easy to prevent.

Improper design, construction, and use of the electric-field-exposure facility can result in unreliable measurements. Therefore, unless sufficient information is given about the exposure system or an independent expert evaluates the system, the results of the studies might not be given full weight in this assessment. Table 2-10 summarizes essential specifications for any animal or in vitro experiment, whether exposure is to electric, magnetic, or combined electric and magnetic fields. The critical and desirable specifications that should be reported for each animal experiment in electric fields are listed in Table 2-11.

Magnetic Fields   A magnetic field in an animal-exposure apparatus is produced by current flowing through an arrangement of coils. The apparatus can vary from a simple set of two Helmholtz coils (preferably square or rectangular because of the geometry of cages), to an arrangement of four coils (Merritt et al. 1983), to more complicated coil systems (Stuchly et al. 1991; Kirschvink 1992; Wilson et al. 1994; Caputa and Stuchly 1996). The main objectives in designing exposure apparatus for magnetic fields are (1) to ensure the maximal uniformity of the field within the largest volume encompassed by the coils, and (2) to minimize the stray fields outside the coils, so that sham-exposure apparatus can be placed in the same room. Square coils with four windings arranged according to the formulas of Merritt et al. (1983) are optimal in satisfying the field-uniformity requirement. Limiting the stray fields is a challenge, as shielding magnetic fields is much more complex than shielding electric fields. Nonmagnetic metal shields provide only a small reduction in the field strength. Only properly designed multilayer shielding enclosures made of high-permeability materials are effective. An alternative solution relies on partial field cancellation. Two systems of coils placed side by side or one above the other form a quadrupole system that results in a substantially faster decrease of the magnetic field as distance from the system is increased (Harvey 1987; Stuchly et al. 1991). An even greater decrease is obtained with a doubly compensating arrangement of coils. Four coils (each consisting of four windings) are arranged side by side and up and down; coils placed diagonally are in the same direction as the field, and the neighboring coils are in the opposite direction (Caputa and Stuchly 1996).

TABLE 2-10   Typical Exposure Specifications to be Defined, Evaluated, and Reported in Any Experiment[a]

Critical Specifications	Desirable Specifications
Frequency of the field	Transients (when turning on/off); the time or frequency characteristic
Waveform type: sinusoidal, pulsed	Waveform: exact time or frequency domain characterization
Magnitude of the field in the exposure and sham apparatus	Magnitude of the other field (magnetic for exposure to the electric field and vice versa)
	A record of the amplitudes of the exposure and sham-exposure fields throughout the experiment
Direction of the field in exposure apparatus	Direction of the background field (undesired) in the sham-exposure apparatus
Type of control (e.g., true sham, unexposed treated identically) and their location	Identical exposure of control except the field magnitude
Duration of exposure	Time of day of exposure (animal studies only)
Estimated uniformity of the field	Measured uniformity of the exposure and sham-exposure fields
Dimensions of exposure apparatus	Design sketches or photos of exposure apparatus
Location and type of cages	Size and material of cages; number of animals per cage
Number of animals or cell culture dishes per exposure apparatus	Rotation of cages (schedule and details)
Temperature	Illumination and noise level in exposure and sham apparatus (only for animal studies)

[a]Applies to in vivo and in vitro exposures to electric and magnetic fields.

Common potential artifacts associated with magnetic-field-exposure systems are heating, vibrations, and audible or high-frequency (nonaudible for humans) noise. These factors can be minimized but not entirely eliminated with careful design and construction, which can be costly. The most economic and reliable way of addressing and solving these problems is through essentially identical design and construction of the field- and sham-exposure systems except for the current direction in bifilarly wound coils (Kirschvink 1992; Caputa and Stuchly 1996). This solution provides for the same heating in both systems. Vibration and noise are usually not exactly same but are similar. To limit the vibration and noise, the coil windings should be restricted mechanically in their motion.

TABLE 2-11   Typical Exposure Specifications for Animal Exposures to
Electric Fields

Critical Specifications	Desirable Specifications
Arrangement for drinking (materials, geometry)	Measured currents during drinking
Spacing between animals	Type of feed and bedding
Measures to prevent microshocks	Distance and shielding (if any) between sham and exposed groups

Furthermore, the animal-cages support system must be physically separate from the coils and their support system. Vibration and noise are inherently limited but not eliminated in a plate type of exposure system (Miller et al. 1989). In general, the higher the desired magnetic flux density in the exposure system and the larger the required volume of the uniform field, the more severe the problems of heating and vibrations become.

Another important feature of a properly designed magnetic-field system is shielding against the electric field produced by the coils. Depending on the coil shape, the number of turns in the coil, and the diameter of the wire, a large voltage drop can occur between the ends of the coils. That produces exposure to the electric field in addition to the magnetic field. The electric-field strengths are different in the sham-exposure systems from those in the field-exposure systems. The electric field can be eliminated easily by electrostatic shielding of the coil windings. Shielding of the connecting wires is also recommended. Additional specifications for animal exposures to the magnetic field are listed in Table 2-12. The critical specifications listed, as well as those listed in Table 2-10, must

TABLE 2-12   Typical Exposure Specifications for Animal and In Vitro
Exposures to Magnetic Fields

Critical Specifications	Desirable Specifications
Magnitude of stray fields in the location of the sham-exposure apparatus	Means to limit stray fields
Means to prevent cage or culture-dish vibration	Measured vibrations of cages (culture dishes) and coils
Means to limit noise from the coils (only for animal experiments)	Audible (animal studies only) and high-frequency (only for coils operating above 20 kHz) noise levels
Electric-field magnitude in the field and sham systems	Direct current (dc) magnetic field

be clearly described for a study to be considered reliable from the point of view of appropriateness of the exposure systems. Typical magnetic flux densities used in animal studies range from 10 μT to 1 mT, and densities up to 5 or 10 mT have been used in a few studies. The rationale for selecting a high flux density in some studies is given in terms of interspecies scaling of the induced current values from human to rodent.

In Vitro Studies

Electric Fields　Theoretically, exposure of cell and tissue cultures can be accomplished by placing them in an electric field produced between parallel plates identical to those used for exposures of animals. In practice, this procedure is hardly ever followed, because the electric fields in the in vitro preparation produced this way are very weak, even for strong applied fields. For instance, an externally applied field of 10 kV/m at 60 Hz results in only up to a fraction of a volt per meter in the culture (Tobey et al. 1981; Lymangrover et al. 1983). Furthermore, the field strength is usually not uniform throughout the culture, unless the culture is thin and placed perpendicularly or parallel to the field. A viable practical solution involves a placement of proper electrodes in the cultures. A comprehensive review of various systems has recently been published (Misakian et al. 1993). The most essential characteristics, in addition to those applicable to other exposure systems (Table 2-10), are related to the properties of electrodes. Two types of problems with electrodes are critical. First, electrode polarization causes most of the applied potential difference between the electrodes to occur within an ion layer adjacent to the electrodes. Use of special electrodes (e.g., silver or silver chloride), platinum black, and a very low resistance current source can partly alleviate the problem. Second, the chemical reactions taking place at the electrodes can affect the culture. Agar or other media bridges can be used to eliminate the contamination problem (McLeod et al. 1987).

Shape and size of the electrodes define the electric-field uniformity and associated spatial variations of the current density. Either accurate modeling or measurements, or preferably both, should be performed. Additional potential problems associated with this type of exposure system are the medium heating and accompanying induced magnetic fields. Both of these factors can be evaluated (Misakian et al. 1993).

Additional critical and desirable specifications for experiments involving electric fields are listed in Table 2-13.

Magnetic Fields　Similar types of coils to those used for animal studies can be used for in vitro studies (Misakian et al. 1993). However, illumination is generally not important in cellular experiments, so the investigator can use solenoidal coils

TABLE 2-13   Typical Exposure Specifications for In Vitro Exposures to the Electric Field

Critical Specifications	Desirable Specifications
Electrodes (material shape) in contact with biologic medium	Exact configuration and materials of electrodes
Estimated current density (and the method of estimation)	Measured current density at various locations within the medium and method of measurement
Conductivity of the biologic preparation	CO_2
Cell density	Low-density cells

(which can interfere with illumination), rather than Helmholtz coils (which are more open and do not interfere with illumination).

In addition to the specifications listed in Tables 2-10 and 2-12, in vitro exposures need to be characterized with respect to chambers, such as a Petri dish, flask, or tubes, holding the preparation and their placement in the field (orientation with respect to the direction of the applied magnetic field). This information is required for evaluation of induced electric fields and currents. These field and currents can be calculated for simple chamber geometries provided that the medium conductivity is known and the cell density is low (Misakian et al. 1993). The evaluation of the induced currents and fields is more complex for high-density cells in a monolayer, a confluent monolayer, or a tissue preparation.

In in vitro studies, special care has to be devoted to ambient levels of 60 Hz and to other magnetic fields. It is not uncommon that magnetic flux densities in incubators exceed the desired ambient level of approximately 0.1 μT by 10-fold or more. Similarly, some other laboratory equipment with electric motors might expose biologic cells to high, but unaccounted for, magnetic flux densities. The potential problems with exposures that are unaccounted for or that are at incorrect levels, as well as the critical influences of temperature and CO_2 level on some cell preparations, can result in unreliable findings in laboratory experiments.

The differences between the actual sham-exposure conditions and the assumed "no-exposure" conditions, as well as differences in cell density, suspending medium, and treatment of the cell preparation, quite likely explain the difficulties in reproduction of the test results by other laboratories and in corroboration of various experiments.

In some of the studies, simultaneous exposures to alternating and static magnetic fields are used. That procedure is used to test the hypothesis of possible "resonant" effects. All requirements for controlled exposure to the alternating field have to be applied to the static field. Some requirements, for example, vibration prevention, do not apply to static field systems, which have no vibrations, with the possible exception of on and off switching. In experiments involving static magnetic fields, the earth's magnetic field needs to be measured and controlled locally. Depending on the hypothesis to be tested, the earth's magnetic field might need to be controlled for sham-exposed cells and for the field-exposed cells. In addition to the exposure specifications listed in Tables 2-10 and 2-12, Table 2-14 lists the specifications needed for in vitro exposures.

INDUCED FIELDS AND CURRENTS

Placement of a biologic system or a cell preparation in an ELF electromagnetic field induces internal electric currents and fields and surface charges at the interfaces of electrically dissimilar media. That behavior is described by Maxwell's equations. In the case of ELF fields, major simplifications to the solution of the equations can be made. The solutions are quasi-static. Because of the size of the objects and the electric properties of biologic tissue, consideration of penetration depth can be neglected. Furthermore, when electric permittivity of tissues is evaluated, it becomes apparent that, for frequencies up to a few kilohertz, the induced conduction current is much greater than the induced displacement current, because $\sigma/\omega\epsilon \gg 1$, where σ is the volume conductivity, ϵ is the media permittivity, and ω is given by 2π times the frequency of the radiation (consult Foster and Schwan (1986) for the dielectric properties of tissues and cells). Therefore, an ELF electromagnetic field produces currents and electric fields in the exposed biologic system and causes oscillating (at ELF) charges at interfaces (i.e., for the interface between the external biologic body and air and for internal interfaces, such as those between different tissues and the cell and cell medium). The magnitudes and spatial patterns of those currents and fields depend on the type of exposure field, its characteristics (frequency, magnitude,

TABLE 2-14   Typical Specifications for In Vitro Exposures to the Magnetic Field

Critical Specifications	Desirable Specifications
Dimensions of cell culture dishes	Dimensions (height) of the medium in the exposure dish
Orientation of the field with respect to the culture dish	CO_2
dc field level and orientation where applicable to the experimental design	Computed or measured currents (electric fields) in the cell preparation
Medium conductivity	Ambient 60-Hz field in all areas occupied by the cell preparation during the experiment

orientation, etc.), and the size, shape, and electric properties of the exposed system. There is an important difference between the physical interaction of the electric field with a biologic system and the interaction of the magnetic field with a biologic system.

Electric-Field Exposure

Fundamental analyses (e.g., Kaune and Gillis 1981; Polk 1986) indicate that biologic bodies produce considerable perturbation of the external electric field. The internal fields induced by exposure to 50- and 60-Hz electric fields are typically 10^{-6}-10^{-7} times lower than the external fields for a conductive body, such as a culture medium or an animal. The charge density at the tissue-air interface is substantial, and the external electric field is approximately perpendicular to the surface of a biologically conductive body. Local higher-than-average electric fields, but about 10^{-5} lower than the exposure field, can occur at sharp edges within biologic objects.

Induced electric fields and currents have been computed, as well as measured, for simple and more realistic animal models, including humans. The early analyses of grossly simplified models of humans and animals represented as spheres (Spiegel 1976) or spheroids (Shiau and Valentino 1981) provide only order-of-magnitude estimates. More reliable information is obtained from analysis of more realistic models as conducted by several investigators (Spiegel 1981; Chiban et al. 1984; Chen et al. 1986; Dimbylow 1987, 1988; Hart 1990). Results of several measurements of people and animals and their models are also available (Deno 1977; Kaune and Phillips 1980; Kaune 1981a,b; Kaune and Forsythe 1985; Hart 1992a,b; Gandhi and Chen 1992). Recent reviews of these topics are available (Tenforde and Kaune 1987; Bracken 1992; Misakian et al. 1993). Collectively, these investigations confirm quantitatively the general features of the physical interaction between biologic bodies and externally applied electric fields. Predictably, they also indicate that the induced internal fields and external field perturbation depend on whether and how the conductive body is grounded. For grounded humans and animals, the total induced current (short-circuit current) can be reliably evaluated using a simple formula (Deno 1977; Kaune and Phillips 1980). For this report, the most important aspect of these dosimetric investigations is the differences among various animal species in various parameters (e.g., body-surface electric field, average induced electric field or current density, or maximum-induced current density). To illustrate, scaling factors based on some parameters are shown in Table 2-15 (Kaune and Phillips 1980; Kaune 1981a; Kaune and Forsythe 1988; Bracken 1992).

It must be noted that the scaling factors provide only approximate guidance, if at all, for drawing conclusions from animal and in vitro studies. The values are approximate, and they are for homogeneous models. An analysis using more refined models and considering the different positions a person might assume in the

TABLE 2-15   Typical Scaling Factors to Produce Equivalent Induced Currents for Grounded Animals Compared with a Grounded Person 1.7 m in Height Standing in a Vertical Field of 1 kV/m (Homogeneous Models)

Metric	Human	Equivalent Exposure Field, kV/m	
		60-kg Pig	0.5-kg Rat
Maximum surface field	18.0 kV/m	2.7	4.9
Average surface field	2.7 kV/m	1.9	2.2
Current density			
Neck	0.5 mA/m^2	14.0	20.0
Torso	0.25 mA/m^2	7.3	12.0
Ankle	2.0 mA/m^2	1.8	1.4
Short-circuit current	16.0 μA	2.3	100

exposure field shows large differences in induced current densities (Dimbylow 1987). More important, it has not been determined what characteristics of the exposure field or the internal field are responsible for the biologic interaction. Nevertheless, the induced current densities and the corresponding induced electric fields $\bar{E} = \bar{J}/\sigma$ are used and are likely to be useful for comparing various species and in vitro preparations. In some experiments, they are also used as guidance on the magnitude of the exposure field. Use of such scaling and reference to the induced electric field and current density are not unreasonable. This is a well-established physical-interaction mechanism that might well aid in developing hypotheses and, eventually, in understanding the biophysical interactions involved.

Magnetic-Field Exposure

Induced electric fields and currents from exposure to 50-60 Hz and other ELF magnetic fields can be found by solving Maxwell's equations under the same simplifying conditions as those for the electric field (i.e., quasi-static case, large penetration depth). The main difference is that currents induced by ELF magnetic fields form closed loops. They are frequently referred to as "eddy currents" (Polk 1986). For simple geometries and uniform magnetic fields, current densities or electric-field strengths can be easily found from analytic expressions derived from Faraday's law. The induced voltage (electromotive force) around a closed path in a conductive medium is

$$V = \oint_l \bar{E} \cdot d\bar{l} = -\int_s \frac{\partial \bar{B}}{\partial t} \cdot \bar{n} ds, \qquad (2\text{-}5)$$

where $\bar{E}$ is the induced electric-field vector, $d\bar{l}$ is a vector incremental length along the closed contour l enclosing surface s, $\bar{n}$ is a unit vector perpendicular to the surface element ds, and $\overline{B}$ is the magnetic-flux-density vector. If the surface s is perpendicular to $\overline{B}$ and $\overline{B}$ is uniform, then the induced electric field for a circular path of radius r is

$$\overline{E} = \frac{\omega B r a_l}{2}, \qquad (2\text{-}6)$$

where $\omega = 2\pi f$ and the induced electric field is in the direction of $\overline{a}_l$, a unit vector along the closed circular path. The electric-current density $\overline{J}$ is then given by

$$\overline{J} = \sigma\overline{E} = \sigma\pi f B \overline{r} \overline{a}_l, \qquad (2\text{-}7)$$

with σ being the volume conductivity; σ is a scalar quantity for isotropic conducting media and a second rank tensor for anisotropic conducting media.

The secondary magnetic field induced by the current in the conductive medium (given by Eqs. 2-2 to 2-7) is neglected. The error due to this simplification is less than a small fraction of a percent as long as the following condition is satisfied (Polk 1986):

$$f\mu_0\sigma L_2 = \;<<\; 1, \qquad (2\text{-}8)$$

where L is the largest dimension of the biologic body.

Simplified analyses have been used to estimate the order of magnitude of induced currents and fields in experimental animals and humans. They have also been used to evaluate the induced currents and fields in various cell preparations used in laboratory studies. Although useful in many cases, such simplified analyses might be misleading under some conditions.

Homogeneous spheroids and ellipsoids of sizes and shapes representing humans and rodents have been analyzed (Spiegel 1977; Hart 1992a,b). Numerical analyses have also been applied to a heterogeneous representation of a human body in a uniform field (Gandhi and Chen 1992; Xi et al. 1994), and calculations with high spatial resolution have been conducted for the head (Xi and Stuchly 1994). Limited measurements have been performed on rats that generally confirm the results of rodent models (Miller 1991). Tissue heterogeneity, however, significantly alters the analysis (Polk 1990; Polk and Song 1990). Representative data for a heterogeneous human model with calculations conducted on a grid of 1.3 cm^3 cells and similar calculations for homogeneous rodents are given in Table 2-16 to provide a reference for scaling and a comparison with the currents and fields induced by exposure in the electric field. In all cases, the magnetic-field orientation is selected to give maximum values of induced current densities. These conditions mean that the magnetic field is directed front to back (and vice versa), which translates into a horizontal magnetic field for a person and a vertical magnetic field for a rodent in its usual position. The interspecies scaling values are different from those frequently used and are based on the estimated maximum currents paths for various species. For instance, comparing maximum currents,

TABLE 2-16   Typical Induced Currents and Fields for a 1-μT, 60-Hz Uniform Magnetic Field

Subject	Current Density, μA/m^2		Electric Field, μV/m	
	Average	Maximum	Average	Maximum
Human, 1.7 m, 70 kg	1.3-1.9[a]	8 (20)[b]	14-17.7	161 (296)[b]
Rat, 0.3 kg	0.3	1.31	4.4	17.7
Mouse, 0.02 kg	0.12	0.4	1.7	5.7

[a]The average current density depends on the electric properties used for the muscle tissue.
[b]Values in parentheses obtained from the analysis with an improved resolution of 0.65 cm instead of 1.3 cm.

SOURCES: Xi et al. 1994; Xi and Stuchly 1994.

a ratio of 1:9 is obtained for humans versus rats from the modeling results and 1:6 from the weight-to-volume ratios (maximum current path).

Induced current densities have also been computed for a lineman working near power lines (Stuchly and Zhao 1996). Predictably, both average and maximum values are much greater in this case than in environmental exposures. A comparison is given in Table 2-17. A range of maximum current densities induced locally by hand-held appliances is also given in the same table (Cheng et. al. 1995).

It is interesting to compare the induced currents for human exposure to 60-Hz electric versus magnetic fields. Referring to Tables 2-15 and 2-17, approximately the same maximum current densities (of 2 μA/m^2) are obtained for an exposure to a 4-V/m electric field and a 0.1-μT (1-mG) magnetic field. (These results are for equal current densities in the head for the magnetic-field exposure and in the neck for the electric-field exposure.) Another comparison can be made by considering the average and maximum induced electric fields. For the electric-field exposure, the reduction factor is about 10^{-7} for the average field and 10^{-5} for the maximum field as compared with the external electric field. Therefore,

TABLE 2-17   Current Densities Induced in a Person by a 60-Hz Magnetic Field Under Various Exposure Conditions

Exposure	Induced Current Density	
	Average	Maximum
0.2 μT, uniform	0.56 μA/m^2	4.2 μA/m^2
500 kV, 1,000 A, 0.5 m away	0.4 mA/m^2	2.8 mA/m^2
138 kV, 500 A, 0.5 m away	0.24 mA/m^2	1.7 mA/m^2
25 kV, 200 A, 0.5 m away	83 μA/m^2	0.6 mA/m^2
Hair dryers	—	0.1-8 mA/m^2
Electric shavers	—	1.5-11 mA/m^2

for the average electric field, exposures to 14-18-V/m electric fields and 0.1-μT (1-mG) magnetic fields are equivalent, and for the maximum induced electric fields, 3-V/m electric fields and 0.1-μT (1-mG) magnetic fields are equivalent. In comparing the maximum values of either the induced current density or the electric fields, very close environmental levels of electric (4 V/m and 3 V/m) and magnetic (0.1 μT) fields are obtained.

Endogenous current densities associated with action potentials of excitable tissues are of the order of 1 mA/m^2 or an electric field of approximately 1 mV/m. To obtain similar induced current densities from exposure to external 60-Hz fields, human exposures of about 2-kV/m electric fields or 100-μT (1-G) magnetic fields would be required. Those fields are considerably larger than are commonly encountered in the residential environment.

The induced currents and fields have been evaluated so far for a grossly simplified structure of tissues by considering only its bulk electric properties. Inclusion of cellular structure, including anisotropies, presents a formidable task, so far unsolved (McLeod 1992; Polk 1992a,b).

Evaluation of induced current and electric fields is also important in quantifying and interpreting results of in vitro laboratory studies. It is especially important when determining whether the biologic effect observed is due to the magnetic field or the electric currents and fields induced in the test sample by the magnetic field. When results of a study in one laboratory are not corroborated by other data from other laboratories, evaluation of induced fields might also be useful in finding the differences in apparently identical experiments.

For low-density biologic cells placed in a conductive medium, the induced current density can be computed solely on the basis of geometry of the medium contained in the exposure dish and the magnetic-field characteristics (Misakian et al. 1993). Methods of calculation for several dish shapes, including an annular ring, have been published (McLeod et al. 1983; Misakian and Kaune 1990; Misakian 1991; Misakian et al. 1993; Wang et al. 1993). Some dish configurations and magnetic-field orientations facilitate obtaining the same current density in most of the medium volume occupied by cells. However, even at low densities, the presence of biologic cells affects the spatial pattern of the induced currents and fields because of the low conductivity of the cell membranes. The effect of cell density is much more pronounced when density is high and when cells form a confluent monolayer (Hart et al. 1993; Stuchly and Xi 1994).

3

Cellular and Molecular Effects

In this chapter, the published literature on the exposure of cultured cell systems to power-frequency electric and magnetic fields is discussed. The committee's conclusions, based on a review of the data, are presented first, followed by a discussion of the utility and limitations of in vitro studies and an examination of three categories of in vitro effects. Those effects are genotoxicity and carcinogenicity, changes in intracellular calcium concentrations, and changes in gene expression with emphasis on signal-transduction pathways.

SUMMARY AND CONCLUSIONS

From its review of the data on cellular and molecular effects from exposure to power-frequency electric and magnetic fields, the committee concludes the following:

• Magnetic-field exposures of 50-60 Hz delivered at field strengths similar to those measured for residential exposure (0.01 to 1.0 μT) do not produce any significant in vitro effects that have been replicated in independent studies.

Although a few studies have reported positive effects at those field strengths, most of the studies reported negative results. Those few studies that had positive results provided no evidence that superior methods or cell systems were used that would give them precedence over the greater number that had negative results. All the studies, as with any others using cultured cell systems, are subject to experimental artifacts that can affect the results. However, the number and quality of studies with negative results are impressive, and in the absence of compelling findings in animals or human beings, the committee finds no justification for any conclusion other than that magnetic fields with field strengths from 0.01 to 1.0 μT have no significant effects in cultured cell systems.

• Magnetic-field exposures of approximately 100 μT (1 G) have been observed in independently replicated studies to produce effects on ornithine decarboxylase (ODC) activity, one of the membrane-mediated signal-transduction pathways, and numerous peer-reviewed reports have been done on the effects of magnetic fields on other components of the signal-transduction pathways. However, a mechanism through which these magnetic fields produce such biologic effects is unknown.

Positive results in the field-strength range of 50 to 500 μT (0.5 to 5.0 G) have been observed and reproduced for only the signal-transduction effect on ODC activity. For other effects—genotoxicity, intracellular calcium concentrations, and general patterns of gene expression—no convincing and reproducible results have been observed.

• Magnetic-field strengths greater than 500 μT (5 G) have induced changes in intracellular calcium concentrations and general patterns of gene expression as well as in several components of signal transduction. However, no reproducible genotoxicity is observed at any field strength.

Again, effects of the sort seen here are typical of many experimental manipulations and do not, per se, indicate a health hazard. For the positive results that have been observed, as in the bone-healing studies, the effects cannot be extrapolated to lower field strengths; therefore, it is not known whether the effects observed at higher field strengths are induced by mechanisms distinctly different from those that might cause effects at residential and occupational field strengths.

The committee's overall conclusion based on analysis of in vitro experimentation is that magnetic-field exposures at 50-60 Hz have been shown to induce changes in cultured cells only at field strengths that exceed residential exposure levels by factors of 1,000 to 100,000.

UTILITY AND LIMITATIONS OF IN VITRO STUDIES

In vitro studies are useful for documenting responses of selected cell systems to chemical and physical agents. Interpreting probable responses of cells in culture in terms of potential or putative target-cell response in the body is problematic, however, and requires similar exposures and appropriate surrogates for target-cell populations in vivo. For example, a number of extracellular signals induce a common set of early-response genes. The early response is not, however, sufficient to determine the biologic outcome. The initial response of cells to the phorbol ester 12-O-tetradecanoylphorbol 13-acetate (TPA) is activation of protein kinase C (PKC) and induction of the early-response genes *fos*, *myc*, and *jun*. Phorbol esters were originally identified as promoters of papilloma development in mice. However, TPA has also been reported to inhibit the growth and suppress the tumorigenicity of human HT29 colon cancer cells that overexpress the β-1 isoform of PKC. Thus, phorbol ester can either stimulate or inhibit cell proliferation, depending on cell type and signal-transduction pathway (e.g., see Huang et al. 1995).

The choice of an appropriate surrogate is an absolutely crucial consideration to extrapolate from in vitro to in vivo response. If the goal in using cultured cells is to document cellular response to a general phenomenon (e.g., induction of DNA damage), then almost any cell system can be considered appropriate, provided the agent under consideration reaches the DNA in its active form. As the induced phenomenon becomes more specific, it will be observable in fewer cell systems. For instance, in vitro transformation systems used to measure induced transition from a nontumorigenic phenotype to a malignancy are specialized systems, because the genotype and phenotype of these cells can only represent limited phases in multiphase carcinogenic processes, which vary between different human somatic cells. When experiments are initiated to determine whether such agents as power-frequency electric or magnetic fields produce patterns of response in multiple-cell systems that are similar to those produced by carcinogens, neurotoxins, or developmental toxins, the results can be regarded with confidence when those systems exhibit responses similar to those produced by documented toxins and when the responses are consistent with the mechanism that is hypothesized to underlie the hazard.

GENOTOXICITY AND CARCINOGENIC POTENTIAL OF POWER-FREQUENCY ELECTRIC AND MAGNETIC FIELDS

Substantial numbers of experiments have been performed in which cultured mammalian cells or prokaryotic cells were placed in an electric or magnetic field, a putative exposure delivered, and the cell response examined for a variety of end points. Scientists conducting those experiments interpreted their results as both negative and positive mimicries of results from known carcinogens tested in the same biologic systems. The goal of this discussion is to compare the results and determine with relative confidence whether the data, taken as a whole, present a compelling argument for or against the probability that power-frequency electric and magnetic fields are a possible carcinogen. To accomplish that goal, data obtained from exposures to electric or magnetic fields must be compared with those from exposure to known carcinogens. Such a comparison must be done in the framework of a detailed knowledge of the carcinogenic process.

Using the analytic power of molecular biologic techniques, human cancer cells can be described in great detail at the molecular level. The essential biology of cancer cells is now most often described as an accumulation of multiple genetic changes that in combination lead to loss of proliferative control, loss of responsivity to differentiation signals, and loss of controlled interaction with other tissues. The in vivo processes leading to the cancer state are less well understood. Those processes must generally be inferred from changes observed in the characteristics of cells at different stages of carcinogenesis, from the molecular action of agents that induce transition from one stage of carcinogenesis to the next, and from characteristics of cells in humans who are genetically susceptible to cancer. In vitro systems have been used effectively to describe the molecular sequelae of different agents demonstrated to induce cancer in animals and humans. In special cultured cell systems, some aspects of carcinogenesis can be mimicked, and assays for potency of in vitro transformation can be measured quantitatively. However, it must be realized that in vitro assay systems are artificial representations of target cells in vivo, and confidence in extrapolating from in vitro results to determine the relative potency for inducing cancer in vivo is limited. Using a battery of tests produces a profile of results for particular agents, and that profile can be considered a descriptor of carcinogens or types of carcinogens. Descriptors have been most widely used and accepted as indicators of carcinogenic potency for agents that are directly genotoxic. Directly genotoxic agents are generally considered to be those that can interact directly with DNA, inducing chemically stable changes in that molecule. Those changes must be repaired or restituted in some way to prevent heritable changes in DNA sequence or function; some such changes are observed in the cancer cell. Thus, biologic systems that manifest changes in DNA sequence, such as mutations or chromosomal aberrations, are useful in identifying potentially genotoxic agents, most being potential carcinogens if acting on susceptible target cells in vivo.

EXPERIMENTAL STUDIES OF IN VITRO EFFECTS

Data on the biologic effects observed in cultured mammalian cells or selected prokaryotic cells exposed to electric and magnetic fields are divided in this section into two categories: heritable changes and transient changes. Those two categories are relevant for the three categories of hazards under evaluation: carcinogenic, neurobehavioral, and reproductive effects. Of these three, the cancer phenotype is not only most clearly associated with heritable changes in the genotype of the target cells but is also clonal in origin. Neurobehavioral effects, except those induced during ontogeny, represent multicell alterations in the nervous system. Although heritable susceptibility to neurologic disease has been documented, environmental agents that induce such disease are invariably active through alteration in neurometabolism or through direct toxic effects on cell populations of the central or peripheral nervous systems. Developmental toxins need to act only for a short period during gestation, and therefore developmental effects are the most susceptible to transient changes from exposure to electric or magnetic fields. Cultured cells are limited as direct surrogate systems for the three effects of concern, but can act as a surrogate of general effects, such as mutation, growth retardation, or some changes in gene expression. If the biologic changes induced by electric and magnetic fields are highly selective and specific, cultured cells are more limited as descriptors of induced effects.

Heritable Changes in Cells Exposed In Vitro to Electric and Magnetic Fields

Cultured cells have been widely used to detect genotoxicity of different environmental agents. Generally, end points can be divided into two categories: those that measure the induction of heritable genetic changes directly, such as mutation and certain heritable chromosomal aberrations, and those that are accepted as indicative of heritable changes, such as induction of DNA damage, DNA repair, nonheritable chromosomal aberrations, and sister chromatid exchanges (SCE).

Genotoxic effects of electric and magnetic fields have been reviewed recently by McCann et al. (1993) and by Murphy et al. (1993). According to McCann et al., there is no convincing evidence that power-frequency fields induce direct genotoxic effects and that positive effects observed for high static electric fields might result from corona, arc, or spark. They express concern that some unconfirmed results suggest that magnetic fields can act to enhance the effects of other carcinogens. According to Murphy et al., the preponderance of the data suggest that no genotoxic effects are induced by power-frequency electric or magnetic fields, but the presence of some positive data and the lack of breadth of the studies suggest the need for further study.

Table A3-1, provided in Appendix A, contains a list of published studies in which measurements were made of genotoxic responses of cultured cell systems exposed to electric and magnetic fields at 50-60 Hz. A number of additional references are included in the table of studies in which cells were exposed to static fields, high-frequency electromagnetic fields, or fields applied intermittently as modulated or unmodulated pulses. Many of those studies were discussed and analyzed by McCann and co-workers (1993) in their review of genotoxic studies of these fields. The studies also illustrate the type of data that supplements the literature strictly limited to electric- and magnetic-field exposures at 50-60 Hz. Table A3-1 lists 29 articles that report effects of exposure to electric and magnetic fields at frequencies of 2-100 Hz, with most using 50-60 Hz. In 24 of the studies, cells were exposed to fields at 50-60 Hz that are sinusoidal or approximately such. Only two of the reports, by the same laboratory (d'Ambrosia et al. 1985, 1988-1989), report positive induction of chromosomal aberrations in bovine lymphocytes; the other 22 studies, including those that are similar to the d'Ambrosia studies, are negative. From those studies, the committee concludes that genotoxic effects were not reproducible for exposures to electric and magnetic fields at 50-60 Hz delivered in a form that approximates sinusoidal exposure. Significant genotoxic effects have been reported for exposure conditions in which magnetic fields of greater than 30 μT are delivered by intermittent exposure

(Nordenson et al. 1994) or in which fields of 1 mT are delivered by pulsed exposure (Khalil and Qassem 1991). Those exposure levels are orders of magnitude larger than those encountered in everyday life. Rosenthal and Obe (1989) reported that magnetic fields induced excess SCE in human lymphocytes exposed to 50-Hz sinusoidal magnetic fields and *N*-nitroso-*N*-methylurea (NMU) or trenimon.[1] However, the authors question whether their results reflect a true genotoxic effect. Other authors reported no increase in genotoxic effects when cells were exposed to magnetic fields at 1 mT and ionizing radiation at 5 grays (Gy) (Frazier et al. 1990) or to electric fields and ultraviolet light at 254 nanometer (nm) (Whitson et al. 1986). Rosenthal and Obe (1989) also produced negative results when they used diethylbenzene (DEB) rather than NMU or trenimon.

Balcer-Kubiczek and Harrison (1991) reported an interaction in producing transformed foci in C3H/10T½ cells between pulse-modulated microwaves of 2.45 GHz and TPA or with this combination following or preceding X-rays (0.5, 1.0, or 1.5 Gy). No effect was induced by electromagnetic fields alone. Although these data are difficult to interpret in terms of effects from exposure to fields at 50-60 Hz, they and data showing copromotion between pulse-modulated high-frequency electromagnetic fields and tumor promoters in some animal systems, as will be discussed subsequently, might warrant further study.

Although some examples of positive effects exist, the great majority of effects are negative, and for those published studies that describe positive effects, similar studies describe negative effects. There seems no imperative that would give precedence to any set of positive data. Further, a major characteristic of genotoxic agents is their induction of positive effects across a range of cell types, at least when the agent reaches DNA in its active form. Thus, the data must be interpreted as strong evidence that electric and magnetic fields are not genotoxic as defined experimentally. The positive result reported when power-frequency electric or magnetic fields were combined with certain genotoxic and nongenotoxic carcinogens is an extremely interesting observation, but one that is also extremely difficult to interpret in terms of its implications, if any, for potential carcinogenesis in human populations. Cultured cells stressed with nonphysiologic concentrations of agents might not be good models for human somatic cells in populations exposed to electric and magnetic fields.

The committee concludes from these studies that positive data either are (1) specific to particular laboratories (i.e., other laboratories using similar systems and similar exposure conditions have negative, rather than positive, results) or are (2) specific to particular biologic systems (i.e., positive results are peculiar to biologic factors in the systems used in particular laboratories). Those factors that determine the differences observed are extremely difficult to identify in retrospect.

The general conclusion from these studies is that power-frequency electric and magnetic fields are not directly a genotoxic agent; if they were, a wider

---

[1] An alkylating agent used in the treatment of ovarian cancer.

range of positive responses would have been observed. No consistent pattern is found across biologic systems or exposure conditions.

Transient Changes in Cells Exposed In Vitro to Electric and Magnetic Fields

The data showing that magnetic fields can induce transient changes in cell expression are significant. Those data fit into three categories: changes in the signal-transduction pathway including changes in concentrations of ODC, changes in gene expression, and changes in intracellular calcium levels.

Signal Transduction

Signal refers to molecular systems, both at the cell membrane and inside the cell, in which signals from the environment and from other cells are received, and which regulate intracellular processes, such as metabolic activities, gene expression, differentiation, and cell proliferation in response to the signals received. Signal-transduction processes present an interesting possible mechanism for electric and magnetic fields to influence cell function. In particular, membrane signal-transduction processes have been an area of intense focus. One reason for the interest is that the cell membrane presents a substantial barrier to electric fields, especially in the range of field strengths and frequencies present in the ambient environment. Attenuation of electric fields by the plasma membrane of mammalian cells has been estimated at 10^3-10^5 between the external plasma membrane surface and the interior of the cell (Polk 1992b). For all intents and purposes, no significant penetration of information-containing electric signal across the cell membrane can be postulated for the 60-Hz ambient fields encountered in ordinary household exposures (Polk 1992a,b). Because membrane-mediated signal transduction by hormones and other signaling agents involves the transmission of signals across the plasma membrane without requiring that the signal itself penetrate the membrane, low-frequency electric or magnetic fields have been postulated to act on intracellular processes by influencing only the initial extracellular steps of signal transduction (Adey 1992a). A number of studies have been interpreted by the investigators to indicate that weak electric or magnetic fields can produce changes in membrane signal-transduction pathways. Numerous reviews of low-frequency, low-energy electric- and magnetic-field interactions with biologic systems, including cells, animals, and humans, have been conducted (Adey 1992a,b; Cardossi et al. 1992; Cleary 1993; Liburdy 1992a; Luben 1991, 1993; Tenforde 1991, 1992; Walleczek 1992).

Although many types of signals can be found in biologic systems, the mechanisms for transmitting the information contained in those signals across the cell membrane are relatively few. In all known signal-transduction systems, a signal interacts with a cellular protein (a receptor or voltage-sensitive ion channel) and triggers conformational changes in the protein that result in other signals or modifications of cellular metabolism. Signaling agents with limited ability to cross the cell membrane (e.g., peptide hormones, neurotransmitters, and growth factors) interact with receptor proteins that span the cell membrane. These ligand-activated receptors have an extracellular domain that is exposed to the medium surrounding the cell, and signaling agents interact with this extracellular domain. Interaction of the signal with the extracellular portion of the receptor produces conformational changes which are propagated across the membrane to the intracellular portions of the receptor molecule. Interaction of the intracellular portion of the receptor with other intracellular (effector) molecules causes changes in the activities of cellular pathways. In addition to receptors for soluble molecules (ligands), such as neurotransmitters, transduction mechanisms also exist for non-chemical signals, such as mechanical deformation, temperature, and light.

There are three well-understood methods by which signals associated with a membrane protein conformational changes are propagated across the cell membrane: (1) opening and closing of ion channels and resultant current flow; (2) changes in an intrinsic enzymatic activity of the receptor; and (3) changes in affinities of the receptor for intracellular proteins, which might have enzyme activity or be enzyme regulators. Because membrane proteins can be difficult to study, additional mechanisms undoubtedly remain to be discovered. In contrast to the relatively few signal-transduction mechanisms known, the variety of biologic responses in the cells being regulated is almost infinite. In nearly all cases, the mechanism of signal transduction distal to the receptor involves intracellular pathways being influenced either by changes in ionic composition of the cytosol (e.g., changes in intracellular ion concentrations, such as calcium or sodium) or by changes in phosphorylation of intracellular proteins (e.g., changes in enzymes, enzyme regulators, or transcriptional regulatory factors). The cellular responses to signals can be either short-term, with little or no persistence of the effect after removal of the signal, or long-term, involving persistent changes in the function of cells, such as increased or decreased proliferation, changes in gene expression or differentiation, and in some cases, apoptosis (programmed cell death). Short-term changes are generally mediated by modification of cytosolic or membrane enzyme activities of the cell; long-term changes invariably involve alteration of nuclear functions, such as transcription, cell division, cell-cycle regulation, or cytosolic and membrane effects.

A distinction should be made between the direct (biophysical) interactions of electric and magnetic fields with atoms or molecules in cells and the indirect (more general biochemical) effects, which are produced as a result of the direct interactions. All the data reported in this section deal with biologic or biochemical measurements of indirect effects on cells or tissues, although in many cases the stated purpose of the experiments was to infer the nature of postulated direct mechanisms. Several potential direct interaction mechanisms are described elsewhere in this report and will not be analyzed in detail here.

In general terms, two possible scenarios have been addressed experimentally: (1) the interaction of electric fields (either ambient or induced in the medium adjacent to the cell by oscillating magnetic fields) with ions or charged molecules at the surface of the membrane; and (2) the interaction of magnetic fields with atoms, ions, or molecules in the membrane or within the cytosol or nucleus of the cell. Either of these possible interaction mechanisms is postulated to modulate a step (or steps) in some signal-transduction event, leading to further changes in the function of the cell. The focus of some experiments was on the observable effects of exposure to electric and magnetic fields alone (i.e., attempt to show the existence of receptors for such fields). The focus of other work was on possible interactions between electric or magnetic fields and the existing signal-transduction systems for other ambient signals, such as hormones or neurotrans-mitters (i.e., to look for electric- or magnetic-field effects on receptors for other agents).

A number of laboratories have examined the effects of electric and magnetic fields on bone and connective tissue cells, including studies of signal transduction as well as other regulatory and differentiation processes. Those studies are summarized in Chapter 4 of this report in the section on bone healing and will not be repeated here. It is important to emphasize, however, that most of the studies on bone and connective tissue have used field strengths much higher than those encountered in either residential exposures or most occupational exposures. The lowest magnetic-field strength that has been shown to have reproducible effects on connective-tissue cells is approximately 100 μT (1 G), and most of the studies relating to clinical effectiveness of magnetic fields have been at strengths of 500-2,000 μT (5-20 G) (Brighton and McCluskey 1986).

Other examples have been presented of interactions between magnetic fields and already-recognized signal-transduction pathways. For example, Walleczek and Liburdy (1990) observed that 60-Hz magnetic fields caused increased ^{45}Ca influx during concanavalin-A (Con-A)-induced signal transduction in lymphocytes. A 60-min exposure of rat thymic lymphocytes to a 22-mT (220-G) magnetic field (induced electric field = 1.0 mV/cm) at 37°C was performed in the presence or absence of Con-A. Nonactivated cells (no mitogen) were unresponsive to the magnetic field; ^{45}Ca influx was not altered. When Con-A was present, the magnetic field led to an increase in ^{45}Ca influx of 50-200%. In these studies, as in those of Luben's group on bone cells (Luben et al. 1982; Luben 1991, 1994), the effects of magnetic-field exposure were prevalent mainly at low concentrations of the signaling molecule, suggesting that power-frequency magnetic-field exposures could cause changes in the affinity of the receptor for the ligand or in the effectiveness of the transduction process at low field strengths, but could not produce a change in the maximal responsiveness of the cells to the signal. Liburdy et al. (1993a) also recently reported that cell-surface antibody binding to human lymphocytes could be altered by a 60-Hz 200-G magnetic field. T lymphocytes were reported to exhibit an approximate doubling of anti-CD3 antibody released from cells compared with that released from the sham-treated cells (Liburdy 1992b). Again, the changes induced by power-frequency magnetic fields were most significant at low concentrations of antibody—a result consistent with a change in receptor affinity but not in the total number of receptors expressed on the cells. Those results and the anti-CD3 antibody results mentioned above suggest that the binding of ligands to their receptors would be fruitful to investigate in other cell systems. It should also be pointed out that none of the studies described have been independently replicated, and this type of replication is a pressing research need.

Blank's group has reported effects of very-low-frequency electric fields on the Na, K-ATPase ion pump in membranes (Blank 1992; Blank and Soo 1992). Electric fields at 30-300 Hz were applied for 15 min to membrane preparations at current densities of 0.05-50 mA/cm^2 (0.001-1 mV/cm); the response was complex, with either increases or decreases in enzyme activity, depending on the level of Na and K ions in the medium. Field inhibition of ATPase activity occurred when the enzyme was in a medium containing optimal concentrations of activating cations, and field stimulation occurred when the enzyme activity was reduced by using ouabain or by lowering temperature. Blank estimated the threshold for effects at electric-field strengths of approximately 5 μV/cm (5 $\times$ 10^{-4} V/m) across the membrane, and this threshold was associated with a current density of 8 mA/cm^2. This threshold value, although low by comparison with ambient electric fields in air near power lines, is much higher than those believed to be induced by environmental exposures to electric fields. The results can be interpreted in terms of the electric field inducing changes in the binding of substrate ions (Na$^+$ and K$^+$) to the ion pump at high and low concentrations of the ligands, similarly to the studies of Liburdy and Luben described above. Blank's work has not been replicated by other investigators, but neither have failures to replicate been reported.

ODC activity is modulated by membrane-mediated signaling events, and its activation is associated with the activity of mitogens and tumor-promoting agents of various types during carcinogenesis. Byus et al. (1987) reported that three cell lines—human lymphoma cells (CEM), mouse myeloma cells (P3), and rat hepatoma (Reuber H35) cells—exhibited increases of 50-300% in ODC activity when exposed to sinusoidal 60-Hz electric fields at 10 mV/cm. Increases in ODC were detected as low as 0.1 mV/cm in Reuber H35 cells. For comparison, phorbol ester at doses associated with tumor promotion produced activation of ODC levels by more than 1000%. The investigators interpreted these results as indicative of an electric-field effect on the cell membrane, resulting in a signal-transduction effect on ODC activation by mechanisms not directly investigated in these or subsequent studies. These findings have been used as basis for a hypothesis that electric fields might act as a copromoter with tumor-promoting agents, producing more activation of ODC and more growth promotion of carcinogen-induced cells in the presence of low electric-field strengths than in the absence of electric fields. Several in vivo studies have been initiated to test this hypothesis directly; they are discussed in Chapter 4 of this report. Litovitz et al. (1991) also reported enhancement of ODC activity in mouse L929 cells by 60-Hz magnetic-field exposure for 8 hr at strengths of 1, 10, or 100 μT. Maximal enhancement of approximately 100% above controls was produced by 4 hr of exposure to a magnetic field at 10 μT. The studies with ODC have the distinction among signal-transduction investigations of having been independently replicated by at least two laboratories in addition to the original findings of Byus et al. (1987), although the conditions and signaling agents were slightly different in the various studies. Little effort was made in these studies to isolate the specific changes in membrane receptor mechanisms that were bringing about the observed changes in ODC activity.

The enzyme PKC is believed to be the receptor for tumor-promoting phorbol esters (Kikkawa et al. 1989), and several current research projects are investigating the possible effects of magnetic fields on PKC. Luben's group reported that exposure to 60-Hz magnetic fields at 100 μT (1 G) produced transient activation, followed by down-regulation, of PKC activity in mouse bone cells (presented as a non-peer-reviewed preliminary report by Luben 1994). Those data suggest that modification of the cellular response to tumor promoters might be an effect of the magnetic field at these relatively high field strengths. Uckun et al. (1995) also reported that PKC activity increased in human pre-B leukemia cells exposed to 60-Hz 100-μT magnetic fields; they showed that the activation of PKC was dependent on the activation of *lyn* kinase, a *src* family tyrosine protein kinase, which is known to be involved in the proliferation of leukemia cell clones. Like most other work on the cellular effects of low-frequency electric and magnetic fields, these studies have not been replicated by independent laboratories, and the positive findings are at field levels well above the constant background fields in households. However, these lines of evidence if confirmed could provide possible mechanistic links between magnetic-field exposures at 100 μT and changes in pathways known to regulate cell proliferation, differentiation, and tumorigenesis. For example, one possible mechanism to explain the apparent copromotion of tumor growth by magnetic fields and TPA in vivo (see Chapter 4) would be alteration by magnetic-field exposure of the sensitivity of PKC-dependent pathways to TPA treatment.

A potential correlation between cancer-cell growth and magnetic-field exposure was described by Liburdy et al. (1993b). In these studies, human estrogen-responsive breast cancer cells (MCF-7 cell line) were used. These cells grow rapidly in the presence of normal concentrations of female sex hormones, but their growth rate is decreased by a hormone produced by the pineal gland, melatonin. Other studies have reported that melatonin synthesis is altered by exposure of whole animals to extremely-low-frequency (ELF) EMF (Wilson et al. 1990a), and it has been proposed that disruptions of the normal daily cycles of melatonin synthesis are a risk factor for human breast cancer (Stevens 1987a,b).

Liburdy et al. (1993b) confirmed that melatonin at normal physiologic concentrations could decrease the growth rate of MCF-7 cells. However, application of a 1.2-μT (12-mG) sinusoidal magnetic field at 60 Hz prevented the oncostatic action of melatonin on the breast cancer cells. A lower field of 0.2 μT (2 mG) did not have any significant effect, suggesting that a threshold might exist between 0.2 and 2 μT. If these studies were replicated by other laboratories, they would be an exception to the observation that magnetic-field strengths near those encountered in households do not produce significant effects on cells in tissue culture.

In summary, the body of work on signal transduction suggests that power-frequency electric and magnetic fields, with magnetic fields at 100 μT (1 G) and above or electric fields at 10 kV/m and above, are likely to have some effect on a number of signal-transduction-related pathways in mammalian cells. These effects have been used clinically and might be of additional use in future treatment of disease. However, the evidence for such effects at field strengths resembling those in households (0.1 μT or 10 V/m) is essentially absent. There has been little demonstration of receptors for electric or magnetic fields in cells outside of the known pathways for signal transduction by signals that normally affect cells. Most of the studies, even those that appear to be carefully done and reliable, have not been independently replicated and thus cannot be considered conclusive. Little of the work can be interpreted in terms of possible direct (biophysical) mechanisms of interaction between cells and electric- and magnetic-field exposure. Research needs in this area include replication, more precise mechanistic studies, and studies designed to show the presence or absence of effects of exposure to environmental levels of magnetic fields.

Gene Expression and Protein Synthesis

Considerable attention has been focused on the possibility that low-strength electric and magnetic fields might produce changes in the transcription of genes, the processing or lifetime of mRNAs, or the synthesis of specific proteins by cells. Although most studies have shown that low-strength, low-frequency fields were unlikely to change the structure or function of DNA, the possibility that transcription-related events are influenced by exposure to electric or magnetic fields was raised by findings reported by Goodman and Shirley-Henderson (1991). Their results showed an increase in transcript activity for selected chromosome loci of salivary-gland cells of *Drosophila* and *Sciara* after brief exposures (<60 min) to sinusoidal fields (60-72 Hz, 0.5-1 mT, 0.3-5 $\times$ 10^{-4} V/m). Individual chromosomes in exposed cells were reported to possess loci with 10-1,000-fold increases in transcription over those in control cells. In general, the chromosome loci in exposed cells exhibiting field-induced changes in transcription were those loci that were active in the control cells; a change in amount of activation but not pattern of activation was the main effect postulated for exposure to electric and magnetic fields. These original studies were carried out using bone-healing appliances that produced complex pulsed magnetic fields at field strengths up to 2 mT, four orders of magnitude above normal environmental exposures. In later studies, the fields used were sinusoidal 60-Hz fields, but the field strengths required for consistent observations remained as high as 100 μT. The original observations in the chromosome studies of Goodman and Shirley-Henderson (1991) have not been replicated, but neither have failures to replicate these studies been reported.

In subsequent studies, increases in mRNA transcripts were induced in the human promyelocytic cell line HL-60 by sinusoidal 60-Hz magnetic fields at 0.57-570 μT, 0.011-11 × 10^{-4} V/m (Goodman et al., 1992) and by 60-Hz sinusoidal electric fields at 0.3 × 10^{-4} V/m (Blank et al. 1993). Dot blots of mRNA probed with radiolabeled transcripts for c-*myc*, $-actin, histone H2B, and several other markers were reported to have significant increases in transcription over those observed in control cells. The c-*myc* proto-oncogene is a member of the family of immediate, or early-response, genes, which respond to a wide variety of mitogenic or differentiation signals; changes in activity of c-*myc* are potentially related to the development of tumorigenic characteristics in cells. Thus, the results from these studies were considered to be of great potential interest.

Phillips et al. (1992) also reported that 1-G 60-Hz magnetic-field exposure of a T-lymphoblastoid cell line, CEM-CM3, increased c-*myc* mRNA. Nuclear run-off assays, rather than the less discriminatory dot-blot technique used by Goodman et al. (1992), were used in the Phillips study to assess alterations in specific gene transcription rates for *myc, jun, fos* and PKC. In general, the results of Phillips et al. (1992) revealed small (30-80%), transient changes in transcription rates; the direction of the changes (increases or decreases in the activity) was dependent on the time of exposure and cell density. It is difficult to relate these results to the consistent apparent increases in total mRNA abundance reported by Goodman et al. (1992). In subsequent studies, Goodman and co-workers (Lin et al. 1994) analyzed transcription of c-*myc* by northern blots as well as dot blots and reported not only that the findings based on dot blots were confirmed, but also that selected regions of the upstream promoter region of the c-*myc* gene were required for the increased transcription induced by electric and magnetic fields, a finding that the investigators interpret as indicating that the activity of some transcriptional regulatory factor, or factors, might be altered by such exposures. This hypothesis would be consistent with a possible signal-transduction effect caused by the magnetic-field exposure (Goodman and Shirley-Henderson 1991, Goodman et al. 1992, Phillips et al. 1992). Consistent with this hypothesis, Liburdy et al. (1993c) correlated alterations in calcium influx and in c-*myc* mRNA gene induction in experiments that measured both parameters in the same exposed cell population—i.e., thymocytes exposed to a 220-G 60-Hz magnetic field. In the presence of the magnetic field, a 1.5-fold increase in calcium ions was observed for Con-A-treated cells; in the same cell population, the presence of the magnetic field resulted in an approximate 2-fold increase in c-*myc* mRNA

Errata: Page 64, line 12, replace "$-actin" with "β-actin."

appliances that produced complex pulsed magnetic fields at field strengths up to 2 mT, four orders of magnitude above normal environmental exposures. In later studies, the fields used were sinusoidal 60-Hz fields, but the field strengths required for consistent observations remained as high as 100 μT. The original observations in the chromosome studies of Goodman and Shirley-Henderson (1991) have not been replicated, but neither have failures to replicate these studies been reported.

In subsequent studies, increases in mRNA transcripts were induced in the human promyelocytic cell line HL-60 by sinusoidal 60-Hz magnetic fields at 0.57-570 μT, 0.011-11 $\times$ 10^{-4} V/m (Goodman et al. 1992) and by 60-Hz sinusoidal electric fields at 0.3 $\times$ 10^{-4} V/m (Blank et al. 1993). Dot blots of mRNA probed with radiolabeled transcripts for c-*myc*, β-actin, histone H2B, and several other markers were reported to have significant increases in transcription over those observed in control cells. The c-*myc* proto-oncogene is a member of the family of immediate, or early-response, genes, which respond to a wide variety of mitogenic or differentiation signals; changes in activity of c-*myc* are potentially related to the development of tumorigenic characteristics in cells. Thus, the results from these studies were considered to be of great potential interest.

Phillips et al. (1992) also reported that 1-G 60-Hz magnetic-field exposure of a T-lymphoblastoid cell line, CEM-CM3, increased c-*myc* mRNA. Nuclear run-off assays, rather than the less discriminatory dot-blot technique used by Goodman et al. (1992), were used in the Phillips study to assess alterations in specific gene transcription rates for *myc, jun, fos* and PKC. In general, the results of Phillips et al. (1992) revealed small (30-80%), transient changes in transcription rates; the direction of the changes (increases or decreases in the activity) was dependent on the time of exposure and cell density. It is difficult to relate these results to the consistent apparent increases in total mRNA abundance reported by Goodman et al. (1992). In subsequent studies, Goodman and co-workers (Lin et al. 1994) analyzed transcription of c-*myc* by northern blots as well as dot blots and reported not only that the findings based on dot blots were confirmed, but also that selected regions of the upstream promoter region of the c-*myc* gene were required for the increased transcription induced by electric and magnetic fields, a finding that the investigators interpret as indicating that the activity of some transcriptional regulatory factor, or factors, might be altered by such exposures. This hypothesis would be consistent with a possible signal-transduction effect caused by the magnetic-field exposure (Goodman and Shirley-Henderson 1991, Goodman et al. 1992, Phillips et al. 1992). Consistent with this hypothesis, Liburdy et al. (1993c) correlated alterations in calcium influx and in c-*myc* mRNA gene induction in experiments that measured both parameters in the same exposed cell population—i.e., thymocytes exposed to a 220-G 60-Hz magnetic field. In the presence of the magnetic field, a 1.5-fold increase in calcium ions was observed for Con-A-treated cells; in the same cell population, the presence of the magnetic field resulted in an approximate 2-fold increase in c-*myc* mRNA

as measured by quantitative microdensitometry of northern blots. The field strengths used by Liburdy and co-workers were much higher than those used by Goodman and co-workers and were several orders of magnitude above those reported for residential exposures.

Studies by Czerska et al. (1992) also indicated that human lymphocytes (Daudi cell line) showed increased *myc* expression when exposed to field strengths similar to those used by Goodman and co-workers; in Czerska et al. (1992), the internal control β-actin was found to be unaltered (although in the Goodman et al. studies, β-actin appeared to be stimulated at the same field strengths as *myc* and other markers). None of the experiments reported in this section appears to have been carried out using blinded experimental protocols, and there has been considerable criticism of the lack of precision of some methods and the lack of consistent internal or external controls in many of the gene-expression experiments, particularly those in the Goodman et al. studies.

In response to those criticisms, at least two groups set out to perform rigorous replication studies of the findings reported by Goodman and co-workers. As reported in two current publications (Saffer and Thurston 1995; Lacy-Hulbert et al. 1995), as well as in several abstracts presented at meetings, these groups used the original equipment, exposures, and cellular end points (including experiments done in the laboratories of Goodman and Shirley-Henderson) and a more-sophisticated exposure apparatus, along with more-specific and sensitive detection techniques, to examine the claims of increased gene transcription induced by power-frequency magnetic fields. In Saffer and Thurston's (1995) study, the original exposure conditions and an extended range of exposures were used (5.7-100 μT); a number of refinements were also introduced to eliminate possible evaluator bias (double blinding) and nonuniformity of exposures (double-wrapped, double-blinded exposure coils) and to improve experimental techniques (loading techniques for RNA and internal and external standards) and detection methods (northern blots, differential display, and ribonuclease protection assays). In addition, a number of different strains of HL-60 cells (including the strain provided by Goodman and Shirley-Henderson used for the original findings) as well as Daudi cells were used. The net result of these studies was that no significant effect of magnetic-field exposure could be detected in any of the genes examined under any of the exposure conditions used. The positive control TPA was used to demonstrate that small effects on transcription could be observed under the conditions used, and a number of internal controls were used to verify the minimum levels of changes in expression that could be detected. Changes in the range of 10% could have been detected reliably, but none was found. This finding is in contrast to the increases of 30-280% reported by Goodman and co-workers in their studies. In addition to in-depth studies with c-*myc* and other genes used by Goodman and co-workers, Saffer and Thurston (1995) carried out differential-display polymerase-chain-reaction (PCR) analysis on HL-60 cells exposed to magnetic fields to determine whether any gene transcripts in the cell could be shown to be affected by the exposure. The results indicate that (1) neither the exposures reported by Goodman and co-workers nor extended exposures (both in time and field strength) could be verified as producing any transcriptional effect on any of the genes reported to be modulated; and (2) no gene transcript modulation could be produced in HL-60 cells under any of the conditions tested using differential-display PCR. The authors analyzed several possible differences in technique that could have led to erroneous findings in the original studies, including insufficient controls for loading of RNA, lack of internal and external controls, and poor discrimination ability of the assay techniques. The original experiments cannot be replicated precisely because of those imprecisions; therefore, no complete explanation can be given for the complete lack of magnetic-field effects in the more-recent, more-careful experiments in contrast to the original findings.

Lacy-Hulbert et al. (1995) incorporated many of the same refinements of exposure and measurement techniques in their study as in the Saffer and Thurston (1995) study. Also in the Lacy-Hulbert et al. study, a complete lack of demonstrable effects of magnetic-field exposure on transcription of c-*myc* and β-actin was found. Extensive attempts were made to force a positive result by improving the reliability of the exposure system, the assay system, and the measurement of mRNA transcripts on northern blots. The failure of these two groups to replicate the reported gene-expression effects despite elaborate precautions (including, at least in the early stages, the cooperation of Goodman and Shirley-Henderson) is difficult to analyze in any way other than to conclude that the original results are in error.

In summary, the ultimate resolution of the controversy involving the magnetic-field-induced gene expression in mammalian systems remains to be determined; however, for the purposes of this report, it can be concluded that no effects of electric- and magnetic-field exposure on gene expression have been convincingly replicated by independent laboratories. It should also be pointed out that even the originally reported results were done mainly under conditions of 2 to 4 orders of magnitude higher field strengths than those encountered in residential households. Evidence for electric- and magnetic-field effects on gene expression at residential field strengths is completely lacking.

Calcium Changes

Calcium is an important and ubiquitous inorganic ion that serves as a messenger in numerous biochemical events (Rasmussen and Barrett 1984). For example, it is involved in muscle contraction, bone formation, cell attachment, hormone release, synaptic transmission, maintaining membrane potentials, function of ion channels, and cellular regulation. It also serves as a second messenger in neural function in which the concentration of calcium inside the cell regulates a series of enzymatic events caused by kinases. Thus, any exogenous agent that affects the flow of calcium ions either into or out of the cell could potentially have a major impact on biologic function.

Given the importance of calcium in numerous biologic processes, it is understandable that hundreds of measurements have been made and dozens of hypotheses have been set forth involving the interaction between power-frequency electric and magnetic fields and calcium. Results appearing in peer-reviewed journals (studies reviewed by experts before acceptance) from 1990 to October 15, 1994, are summarized in Table A3-2. Earlier work has been reviewed in numerous reports (e.g., Brighton and McCluskey 1986; Nair et al. 1989; EPA 1990; Goodman and Shirley-Henderson 1991; Chernoff et al. 1992; ORAU 1992) and will not be discussed here, with the exception of some of the early research on calcium efflux from chick brains.

Calcium Efflux from Chick Brains   A large body of pre-1992 literature on the effects of electric and magnetic fields on calcium efflux from chick brains originally led to the concept of a complex set of "frequency windows," "power-density windows," "temperature windows," and a dependence on the local geomagnetic field. Although that concept is currently discounted because the effects were largely attributable to temperature or pH fluctuations that occurred during tissue analysis, the work is reviewed here to place into context the concept of windows and the concept that calcium is modulated by low-frequency electric and magnetic fields. The work consists primarily of a set of peer-reviewed papers by Blackman and colleagues (Blackman et al. 1979, 1980a,b, 1982, 1985a,b 1988a, 1989, 1991; Joines and Blackman 1980; Joines et al. 1981).

The experiment, performed on thousands of chick brains, is outlined in Figure 3-1. The experiment was conducted as follows: the forebrains of four chicks (*Gallus domesticus*) between 1 and 7 days of age were severed along the midline and labeled by immersing them in a test tube containing Ringer's solution to which radioactive $^{45}Ca^{2+}$ was added. The brains were incubated in these tubes for 30 min at 37°C. After incubation, the brains were thoroughly rinsed with nonradioactive solution and each half was placed in fresh solution; one half of the brain served as a control for the treated half. The control brains were incubated at 37°C for 20 min; the treated brains were either exposed to EMF of calibrated frequency and intensity, or they were sham exposed. After 20 min, 0.2 milliliter (mL) of the liquid from each control tube was added to scintillation fluid and its radioactivity counted. Typical counting times were 10 min, and typical counts were 1,800 counts per minute. This number for the control was called V_c. Samples of liquid from the tubes containing treated brains were counted in an identical fashion; this number was V_t. The results from a number of runs were pooled to produce ratios of $V_t:V_c$ for the sham-exposed and the exposed samples.

Typical data for 16-Hz exposures are shown in Figure 3-2 (Blackman et al. 1982). The statistically significant data, marked with a single asterisk (*) ($p < 0.05$) and double asterisk (**) ($p < 0.01$), show peaks at field strengths at

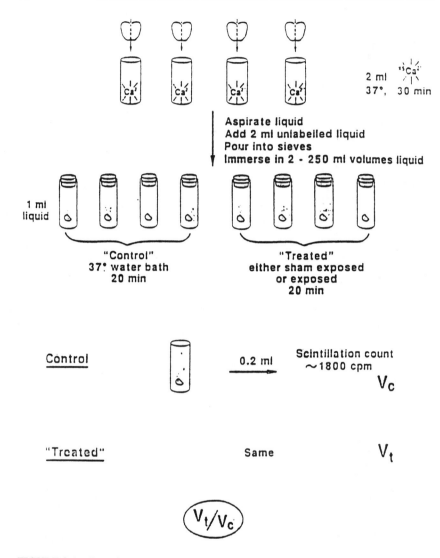

FIGURE 3-1   Experimental protocol for studies of the effects of EMF on calcium efflux in chick brain.

about 6 V/m and at about 45 V/m, respectively. Data, such as those shown in Figure 3-2, were used to compile the map of frequency and power-density windows shown in Figure 3-3 (Blackman et al. 1985a).

If accepted at face value, those and other data in the Blackman papers suggest that more is not necessarily worse when exposure to low-frequency EMF is concerned. The data exhibit sharp resonant-like responses in both frequency and

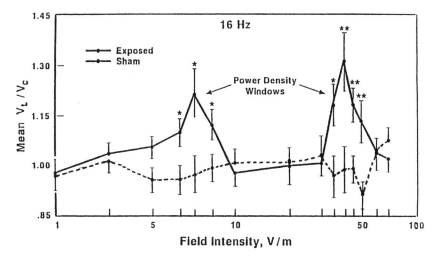

FIGURE 3-2   The ratio of Ca^{2+} ions in the culture media of chick-brain cultures for electric-field exposures and controls shown as a function of the electric-field strength for exposure at 16 Hz. SOURCE: Blackman et al. 1982. Reprinted with permission; copyright 1982, *Radiation Research*, Academic Press, New York.

field strength, and the responses are further complicated by the orientation between the oscillating ELF magnetic field and earth's dc magnetic field. It should be noted that although Blackman and co-workers observed an *increase* in calcium efflux, similar experiments performed earlier by Bawin and Adey (1976) showed a *decrease*. The disagreement between the results of Blackman et al. and those of Bawin and Adey was originally attributed to the smaller size of the ac magnetic-field component used by Bawin and Adey than that used by Blackman et al. More recently, the existence of temperature windows was proposed as one of the variables necessary to explain the differences in results between the two laboratories (Blackman et al. 1991). Either an increase or a decrease or a null result is possible depending on the temperature of the tissue samples before and during exposure.

The chick-brain preparation used in these experiments was not a particularly robust one. The solution was only weakly buffered and the large brain samples, once excised, were not maintained in a stable living state. Consequently, small variations in temperature (perhaps caused by variations in the temperature of the medium used to wash the tissue (Lee et al. 1987)) or insufficient removal of metabolic wastes by rinsing could produce large swings in pH, affecting the results. Blasiak et al. (1990) were unable to replicate specific Blackman experiments using the Blackman chick-brain preparation, and Albert et al. (1987) was unable to obtain the Blackman results using a more stable preparation with small

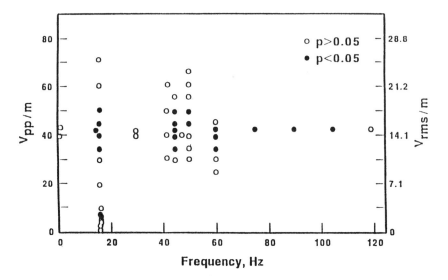

FIGURE 3-3   The ratio of Ca^{2+} ions in the culture media of chick-brain cultures for electric-field exposures and controls shown as a function of the electric-field strength and frequency. The electric field is given in volts per meter peak-to-peak (V_{pp}/m) (*left*) and volts per meter root-mean-square (V_{rms}/m) (*right*). SOURCE: Blackman et al. 1985a. Reprinted with permission; copyright 1985, *Bioelectromagnetics*, Wiley-Liss, a subsidiary of John Wiley & Sons, New York.

tissue slices, adequate media to support the high metabolic rate of nervous tissue cells, and culture under an atmosphere of 5% carbon dioxide to 95% oxygen.

Newer Methods to Visualize Cytosolic Calcium Changes in Individual Cells
During the past decade, significant advances occurred in the study of cell calcium metabolism. The advances are in two main areas: one is the development of intracellular calcium probes, coupled with ultra-sensitive imaging technology, and the second is the development of calcium-selective microelectrodes (Borle 1990; McLeod 1992). These methods have allowed the direct measurement of calcium and, in some cases, direct spatial localization within the cell of intracellular calcium concentrations (Loew 1992).

Comment on Ion Cyclotron Resonance Model Proposed to Explain How a Weak Magnetic Field Could Affect Calcium at the Membrane   The ion cyclotron resonance (ICR) mechanism has been the subject of study as a possible source of interaction of low-frequency EMF with biologic systems for more than 20 years (Liboff et al. 1990). Most efforts to explain biologic effects, however, have met with little success. As an example of these studies, Lednev (1991) used the ICR model to explain how a weak magnetic field could affect calcium ions.

Calcium ions (Ca^{2+}) either occur inside a hydration shell or are bound to a protein, producing a charged spatial oscillator. In the model, either of those states resonantly absorb energy from the appropriate alternating fields and are spiraled into or out of the cell through the protein channels in the membrane. Although a charged ion of the mass of calcium has a resonance at nearly 60 Hz in the earth's static magnetic field, such a resonance can only be observed if the state decays through an electromagnetic transition. Lednev's calculations neglected the effects of collision damping of the states (Adair 1992) and have been shown to violate the laws of physics (Halle 1988). Therefore, whether or not 60-Hz EMF cause changes in calcium concentrations, the ICR model, as currently set forth, is not a viable mechanism for biologic systems.

Even though the ICR mechanism is not viable, a number of experiments have been performed at charge-to-mass ratios (q:m) corresponding to the calculated ICR frequency for calcium (e.g., see Parkinson and Hanks 1989a; Walleczek and Budinger 1992; Yost and Liburdy 1992). One such set of resonant conditions occurs at a frequency of 16 Hz and a dc magnetic field (B) of 23.4 μT (234 mG) according to the relationship frequency (Hz = $1/2\tau(q/m)B_{dc}$).

The results are conflicting. Parkinson and Hanks (1989a) see no effect on changes in cytosolic calcium concentrations (sweeping through both resonant and nonresonant EMF conditions) for BALB/c3T3, L929, V-79, and ROS cells, but other investigators (Liburdy 1992b; Walleczek and Budinger 1992; Yost and Liburdy 1992) see changes when the low-frequency magnetic-field challenge is combined with a mitogen.

Cytosolic Calcium Oscillators It is well established that the intracellular calcium concentration can display an oscillatory behavior in response to an external stimulus (Fewtrell 1993; Meyer and Stryer 1991, and references cited therein). The period of these oscillations is typically between 1 sec and several minutes. Recently, a model based on nonlinear dynamics and on the theory of self-sustained (limit-cycle) oscillators was developed that shows how a small modulation of the signal pathway at an early stage can lead to large changes in calcium metabolism of the cell (Eichwald and Kaiser 1993). These assertions have not yet been tested by experiment.

Relationship Between Very-Low-Frequency Electric and Magnetic Fields and Bone-Healing Protocols for Osteogenesis The efficacy of exposures to electric and magnetic fields in bone healing has been observed (Falugi et al. 1987; Bassett 1990), at least when applied directly to the bone. Typical bone-healing protocols have involved pulsed 20-μsec magnetic fields. The magnetic-field strength used varies from 0 to 2,000 μT (Falugi et al. 1987). One of the mechanisms attributed to the effect of magnetic fields is that the ions in the medium between the bone ends are moved back and forth by the ac field. Ions trapped by the bone matrix become bound to the matrix, thus reducing the concentration of free ions in the medium in the vicinity of the bone. A concentration gradient is set up, and new ions diffuse to the bone, becoming available for bone repair (Parkinson and Hanks 1989b).

More recently, McLeod and Rubin (1992) demonstrated that exposure to pulsed electric fields prevented osteopenia in turkey wings, stimulating an overall 10% increase in the bone area. They used 15-, 75-, and 150-Hz sinusoidal electric fields, with an estimated peak electric field in the tissue of no more than 0.01 mV/cm. The osteogenic influence was greatest at 15 Hz stimulation.

Summary   An enormous body of work seeking a relationship between exposure to electric and magnetic fields and changes in calcium concentrations has been accumulated over the past two decades. Much of it shows some sort of positive association, albeit often requiring frequency windows, temperature windows, or power-density windows for an explanation. Some of the work, particularly the early work, is flawed by unstable preparations and uncontrolled thermal and exposure conditions. Many of the effects are difficult to observe or are only borderline significant and require pooling of data to obtain statistical significance. Many of the experiments have not been replicated adequately by others, perhaps because the exact experimental protocols have not been followed; in other cases, independent investigators were unable to replicate experiments.

Of the recent experiments summarized in Table A3-2, only three meet the exacting requirements of replication by independent laboratories, publication in peer-reviewed journals, and explicit identification of exposure strengths. Those experiments were on thymic lymphocytes in which Con-A stimulated cells showed an increase in calcium transport resulting from exposure to pulsed magnetic fields having flux densities about 10,000 times larger than those found in the average human environment (Liburdy 1992b; Walleczek and Budinger 1992; Yost and Liburdy 1993).

4

Animal and Tissue Effects

SUMMARY AND CONCLUSIONS

The published literature regarding the exposure of animals and tissues to power-frequency electric and magnetic fields is discussed in this chapter. The committee focused on three areas of principal interest: carcinogenesis, reproduction and development, and neurobehavioral and neuroendocrine responses. On the basis of an evaluation of peer-reviewed literature, the committee has made the following conclusions:

• There is no convincing evidence that exposure to power-frequency electric or magnetic fields causes cancer in animals.

A limited number of laboratory studies have been conducted to determine if any relationship between exposure to electric and magnetic fields and cancer exists. To date, no reports have been published showing demonstrable effects of electric- or magnetic-field exposures on the incidence of various types of cancer. However, some recent, as yet unreplicated laboratory evidence suggests a positive relationship between magnetic-field exposures at field strengths of approximately 100 μT (1 G) and the incidence of breast cancer in animals treated with carcinogens.

• There is no convincing evidence of adverse effects from exposure to power-frequency electric and magnetic fields on reproduction or development in animals.

Reproduction and development in animals, particularly mammals, have not been shown to be affected by exposure to very-low-frequency electric or magnetic fields.

• There is convincing evidence in animals of neurobehavioral responses to strong 60-Hz electric fields; however, adverse neurobehavioral effects of such fields have not been shown.

Laboratory evidence clearly shows that animals can detect and respond behaviorally to electric fields. Evidence of behavioral responses in animals to ac magnetic fields is much more tenuous. In either case, general adverse behavioral effects have not been shown.

• There is evidence of neuroendocrine changes associated with 60-Hz magnetic-field exposure in animals; however, alterations in neuroendocrine functions have not been shown to cause adverse health effects.

The majority of studies that investigated magnetic-field effects on pineal-gland function suggest that these fields might inhibit night-time pineal and blood melatonin concentrations; in those studies, the effective field strengths varied from 10 μT (0.1 G) to 5.2 mT (52 G). The data supporting an effect of sinusoidal electric fields on melatonin production are not compelling. Other than the observed changes in pineal function, an effect from magnetic-field exposure on other neuroendocrine or endocrine functions has not been clearly shown in the few studies reported.

Despite the observed reduction in pineal and blood melatonin concentrations in animals as a consequence of magnetic-field exposure, no evidence to date shows that melatonin concentrations in humans are affected similarly. In animals in which melatonin changes were seen, no adverse health effects have been found to be associated with electric- or magnetic-field-related depression in melatonin.

• There is evidence that pulsed magnetic-field exposures greater than about 0.5 mT (5 G) are associated with bone-healing responses in animals.

Replicable effects have been clearly shown in the bone-healing response of animals exposed to electric and magnetic fields at sufficiently high field strengths.

CRITERIA FOR CONSIDERATION OF LITERATURE

Consistent with the review guidelines established by the committee, only peer-reviewed literature is considered in this report unless otherwise noted. Results are reported only if they are exposure related and are statistically significant according to the authors' criteria unless otherwise noted. Greatest weight is given to studies that were confirmed in some manner in the peer-reviewed literature and that were blinded studies.

USE OF ANIMAL STUDIES IN EVALUATING RISK

Data from animal studies are an important component of estimating risk from nearly all agents. A gradient in the degree of an association between exposure to a toxic agent and the effects that agent can produce is called the dose-response relationship. The dose-response relationship forms the basis for the science of toxicology and health physics and allows scientists to predict toxic or adverse health effects. The dose-response relationship is expected because interactions between organisms and chemicals and energy deposition occur according to the basic laws of physics and chemistry and therefore are predictable.

Dose-response relationships are of two types: one describes the response of an *individual* to different doses of an agent, and one describes the distribution of responses of a *population* of individuals to different doses. When toxic or adverse effects are considered, individual dose-response relationships are characterized by a dose-related increase in the magnitude of the response. Interpretation of individual dose-response relationships can be confused by the multiple sites of action of most toxic agents. Each site has its own dose-response relationship. Population dose-response relationships consist of a specific end point and the dose required to produce that end point for each individual in the population.

Three assumptions are made when considering dose-response relationships:

1. The response is due to the agent administered. Although this assumption seems trivial in laboratory studies, it is not so apparent in epidemiologic studies. For example, epidemiologic studies might find an association between a response (disease) and one or more variables. Use of the term ''dose-response'' relationship in this context is always suspect until the variable is shown to be a representative factor of the putative causative agent.

2. The response is related to the measurement of the dose. The most accurate way to determine dose-response curves is to measure the dose actually reaching the site at which an effect is detected within a cell. However, measuring the dose at the site of action generally is prohibitively expensive and has been done in only a few cases. Some measurement of exposure is nearly always substituted for a true measurement of dose.

3. A quantifiable method of measuring and a precise means of expressing toxicity are available. Early in an investigation of the toxicity of an agent, the best end point for effects might not be apparent, but as more is known about the manifestations of toxicity, the dose-response relationship should become more quantifiable.

These assumptions hold true for all types of toxic agents, presumably including extremely-low-frequency electric and magnetic fields, if such fields are found to exhibit toxicity.

Types of Animal Studies Used in Descriptive Toxicology

Two main principles underlie all descriptive animal studies of toxicity (as reviewed by Klaassen and Eaton 1991). The first is that the effects produced by an agent in laboratory animals are applicable to humans. The second is that exposure of laboratory animals to toxic agents in high doses is a valid method of discovering possible hazards in humans. Toxicity tests are not designed to demonstrate that a chemical is safe but rather to characterize the toxic effects that can be produced.

The toxicologic studies that are generally used to predict adverse effects in humans are

- Acute lethality
- Skin and eye irritations
- Sensitization
- Repeated dose (sometimes referred to as ''subacute'' toxicity)
- Subchronic toxicity
- Chronic toxicity
- Mutagenicity
- Developmental and reproductive toxicity
- Other tests, including those for immunotoxicity and toxicokinetics (absorption, distribution, biotransformation, and excretion).

The most pressing issues with regard to residential electric- and magnetic-field exposure focus on carcinogenicity and possible adverse developmental and reproductive effects. Therefore, this discussion focuses on acute lethality, repeated dose, subchronic and chronic toxicity, mutagenicity, and developmental and reproductive toxicity tests, because these tests are used most often to address carcinogenicity and adverse developmental and reproductive effects.

Acute Lethality

The initial starting point for nearly all toxicologic studies is a determination of acute toxicity. The LD_{50} (the median lethal dose) and other acute toxic effects are determined for one or more routes of administration in one or more species and, in most currently used test regimes, are conducted over a 14-day period. Acute toxicity tests (1) provide a quantitative estimate of acute toxicity for comparison among substances; (2) identify target organs and other clinical signs of acute toxicity; (3) establish the reversibility of toxic responses; and (4) give guidance on dosages for other studies. The information obtained in acute toxicity studies forms the basis of the dosing regimes used in repeated-dose studies.

In animal studies on the effects of exposure to electric and magnetic fields, acute toxicity studies involve effects from high-strength current flows. The physical effects and behavioral changes present in animals receiving perceptible electric shocks do not seem appropriate for the exposure conditions under which most people are exposed to electric and magnetic fields. Moreover, human data on electrocutions are sufficient to make animal testing unnecessary.

Repeated-Dose Studies

Repeated-dose studies are performed to obtain information on adverse effects after repeated administration and as an aid to establish the dosages for subchronic toxicity studies. In most currently used test regimes, repeated-dose studies are performed after 14 days of exposure. Biologic effects reported in short-term studies using electric and magnetic fields are reviewed in this report. However, results from short-term studies often are not reproducible and are of questionable value in evaluating possible adverse health effects.

Subchronic Toxicity Studies

The principal goals of the subchronic toxicity study are to establish a no-observable-effect level (NOEL) and to further identify and characterize the organs affected by the test agent after repeated administration. Subchronic toxicity studies more precisely define the dose-response relationship of a test agent and provide the data needed to predict the appropriate dosages for chronic toxicity studies. Subchronic exposures can last for different periods of time, but in currently used test regimes, 90 days is the most common exposure duration. No subchronic toxicity studies using electric and magnetic fields have been conducted that meet the criteria necessary for defining subchronic toxicity. This deficiency is primarily due to the lack of repeatable toxic effects and the lack of a definition of dose-response relationships required from repeated-dose studies to establish dosages for a successful subchronic toxicity study.

Chronic Toxicity Studies

Dosage selection is critical to the successful completion of chronic toxicity studies. If dosages are too high, not enough animals will be alive at the end of a study to allow sufficient definition of the dose-response relationship to be useful for predicting adverse effects. If dosages are too low, not enough effects will be present to allow sufficient definition of the dose-response relationship to be useful for predicting adverse effects. Chronic exposure studies last longer than 90 days. Because humans are exposed to various types of electric and magnetic fields over their entire lifetime, exposures in chronic studies using rodents are most appropriately of 2 years duration. As is the case for subchronic toxicity studies, no chronic toxicity studies using power-frequency electric or magnetic fields have been conducted that meet the criteria necessary for defining subchronic adverse effects. The acute through subchronic dose-response relationships necessary for successful completion of chronic toxicity studies are not available.

Developmental and Reproductive Studies

Four types of animal tests are used to examine the potential of an agent to adversely affect reproduction and development—short-term, segment I, segment II, and segment III tests.

Short-term tests use whole embryos in culture, organ cultures, and cell lines.

They are not used for assessing risk directly, but they can contribute greatly to the design of developmental and reproductive studies by providing an understanding of the mechanisms by which an agent adversely affects development and reproduction.

Segment I tests are designed to address general fertility and reproductive performance. Segment I studies typically begin at an appropriate time before mating and last throughout gestation, lactation, and the first 3 weeks of life.

The potential for an agent to cause birth defects (teratogenicity) is tested in segment II studies. Segment III tests address the potential for agents to cause toxicity after birth and often include multigenerational studies.

To conduct reproductive and developmental studies properly, concentrations must be known that do not result in overt adverse effects in males and females; overt toxicity is widely known to have severe effects on reproduction and development in males and females. Thus, in the absence of good dose-response information from acute toxicity, repeated-dose, and subchronic toxicity studies, informative reproductive and developmental toxicity studies are nearly impossible to conduct.

In studies involving electric and magnetic fields, the lack of repeatable reproductive and developmental effects and the lack of a definition of reproductive and developmental dose-response relationships are not surprising given similar negative results in studies of toxicity as discussed above.

Cocarcinogenicity and Copromotion Studies of Electric and Magnetic Fields

Carcinogenesis is a multistep, multipathway process, and carcinogens probably have different potencies for each of the different steps. Experimentally, it has been difficult to identify specific steps and determine which are necessary and sufficient to cause frank malignancy. Certain systems have been developed that provide evidence of malignant transformation in vitro or malignant tumors in vivo when subjected to combinations of agents. A possible observation in these systems is the determination of whether the potency of two agents can be enhanced when they are delivered together or in a specific sequence. The term "initiator" is used for agents that are most potent when delivered first, and the term "promoter" is used for agents that are effective when delivered after initiators. Magnetic fields have been evaluated in those systems in vitro and in vivo; the data show negative and positive results. Each system is sensitive to the effect of different initiators and promoters; thus, negative data in one system do not necessarily contradict positive data in other systems. Positive results have not been replicated, but some of the data show a dose-response relationship for exposure to magnetic fields and to the interacting carcinogen. Thus, although the pattern of interaction of electric and magnetic fields with known carcinogens is not consistent, the possibility that magnetic fields in combination with some carcinogens produce transformation in these systems cannot be excluded at this time. However, these few systems cannot predict hazard to human populations living in realistic environments. The doses of carcinogens and promoters used in combination with test agents, such as electric and magnetic fields, are invariably large and represent nonphysiologic exposure. The extent to which the highly treated cells in these test systems are representative of actual potential target cells in the soma of exposed individuals is tenuous. In experimental systems in which combinations of agents are used to produce an end point, extrapolation to lower concentrations that represent actual exposure concentrations in human populations is difficult. Thus, although data in these systems are useful for the study of mechanisms and identification of possible interactions, they offer little information on the potency of lower exposure concentrations of agents in the human environment.

The data base that has been developed for initiation and promotion test systems is significant. These systems have shown positive results (i.e., enhanced carcinogenicity) for tests of copromotion and cocarcinogenicity with known and potent carcinogens, but positive results have also been observed when using other agents that are not considered potent carcinogens. For instance, acetic acid, beta-carotene, citrus oil, vitamin E, indomethacin, and putrescine have all yielded positive results in studies of copromotion or cocarcinogenicity using these in vivo test systems under certain conditions. Thus, the positive results in such tests are questionable until detailed studies have identified the underlying mechanism and the probable interaction of doses at environmental concentrations. Nevertheless, electric and magnetic fields, principally magnetic fields, have been shown to interact with carcinogens in some of these systems both in vitro and in vivo, and that fact raises some concern and deserves further attention. The committee provides suggestions for further study in this area in Chapter 7.

CARCINOGENIC AND MUTAGENIC EFFECTS

Because of epidemiologic reports of positive correlations between estimated exposures to power-frequency electromagnetic fields and cancer (see Chapter 5), considerable research interest has been generated concerning a possible connection between magnetic fields and cancer. To date, few laboratory animal studies have been published that bear directly on this question; however, an increasing number of investigations are being conducted. Studies that have been reported in the peer-reviewed literature examining the issue of magnetic-field exposure and cancer are discussed in the following pages and summarized in Appendix A, Table A4-1.

Several approaches and animal models can be used in laboratory cancer studies. The selection of a specific model depends largely on the hypothesis chosen to evaluate a particular underlying mechanism. For example, if an agent, such as an electric or magnetic field, is tested for its potential to be a complete carcinogen (an agent that by its application alone causes cancer to develop), 1.5

to 2 years of exposure of mice or rats to the agent is necessary. During that time, exposure to other possibly confounding agents must be kept to a minimum. In this regimen, the animals are observed during the major portion of their lifetimes, and the number, type, and time of development of tumors are the critical end points. This type of study should include several dosage groups and requires a relatively large number of animals, particularly if the natural incidence of a tumor type is low. Studies evaluating complete carcinogenicity are quite expensive due to the length of time and the number of animals involved.

Carcinogenesis is considered to be a multistep process; therefore, another approach is to assume the agent of interest acts either as an initiator or a promoter in which a two-phase protocol is required for testing. Initiation is defined as a genotoxic event in which the carcinogen interacts with the organism to affect the DNA directly. Promotion is operationally defined as an experimental protocol in which the promoting agent is applied subsequent to initiation and generally over a protracted time. Promotion is associated with a number of subcellular events that are generally nongenotoxic and is responsible for the conversion of initiated cells to cancerous cells. To evaluate electric and magnetic fields as an initiator, one high-dose exposure would be given followed by repeated exposures to a model promoter (e.g., 12-O-tetradecanoylphorbol-13-acetate, TPA) over a long-term period. If electric or magnetic fields were to be investigated for possible promotional effects, the animals would be exposed to a known initiator (e.g., 7,12-dimethylbenz[a]anthracene, DMBA) and subsequently exposed to electric or magnetic fields over a long-term period (e.g., months). The initiation and promotion approaches have the advantages of using fewer animals, less time, and less cost. However, a given model is usually limited to an evaluation of a specific type of cancer and might provide only general information on possible biologic mechanisms of the agent of interest and cancer development. Initiation and promotion studies use initiating or promoting agents, such as DMBA and TPA, respectively, at exposure concentrations that far exceed any possible comparable exposure concentrations in humans. Interpretation of such studies is for identification of possible toxic mechanisms, not for direct extrapolation to human risk.

Complete Carcinogen Studies

Few life-long animal studies examining power-frequency electric or magnetic fields as a complete carcinogen have been completed, although several are under way in the United States, Italy, Japan, and Canada. Several studies designed to evaluate magnetic fields as a promoter of cancer contained control groups that were exposed to magnetic fields without being exposed to a chemical initiator. These studies include a mammary tumor-promotion study in rats (Beniashvili et al. 1991), a lymphoma study in mice (Svedenstal and Holmberg 1993), and a mouse skin-tumor-promotion study (Rannug et al. 1993a). A major deficiency of using such studies to evaluate complete carcinogenicity is the small size of groups involved. The Beniashvili et al. (1991) study found an increase in mammary gland tumors in rats exposed to magnetic fields at 20 μT for 3 hr per day as compared with unexposed animals. The other two studies (Rannug et al. 1993a; Svedenstal and Holmberg 1993) reported no increase in tumors with long-term exposure to magnetic fields at strengths of 500, 50, and 15 μT.

Tumor-Initiation Studies

No tumor-initiation studies of exposures to power-frequency electric or magnetic fields have been reported in the literature. Very little motivation exists for such studies because the energies involved are too weak to break chemical bonds. Furthermore, in vitro studies have not provided evidence that DNA molecules can be damaged directly by exposure to 50- or 60-Hz electric or magnetic fields.

Tumor-Promotion Studies

Despite the obvious need for promotion studies because of the suggested association between indirect measurements of exposure to electric and magnetic fields and cancer observed in epidemiologic investigations, few animal experiments have been completed. Skin-tumor promotion, after initiation with DMBA, was examined in mice exposed continuously to a 60-Hz magnetic field at 2 mT, 6 hr per day, 5 days per week, for up to 21-23 weeks (McLean et al. 1991). None of the exposed or sham-exposed mice developed papillomas. When magnetic-field exposure was combined with application of TPA, a slightly earlier development of tumors was observed in the field-exposed animals (Stuchly et al. 1992).

Rannug and co-workers (1993a,b,c) conducted skin-tumor and liver-foci studies in Sweden. In the 2-year skin-tumor-promotion study, mice were initiated with DMBA, then exposed to 50-Hz magnetic fields at either 0.5 mT or 50 μT for 19-21 hr per day. No evidence of a field-exposure effect was observed either in the development of systemic or skin tumors or in skin hyperplasia. In the liver-foci study, rats were exposed to similar magnetic-field strengths over a 12-week period. The exposed animals showed no differences in foci development from the sham-exposed rats. In animals exposed to chemical promoter (phenobarbital) and the magnetic field, foci formation was slightly inhibited when compared with initiated-only animals.

In a series of four experiments, rats were exposed for 91 days to 50-Hz magnetic fields at 30 mT (Mevissen et al. 1993). Initiation was accomplished with repeated oral doses of DMBA, and mammary tumors developed subsequently. In one experiment, the number of tumors per tumor-bearing animal was increased in animals exposed to the magnetic field. In a repeat of that experiment, however, no difference between exposed and sham-exposed animals was observed. This study is handicapped by the small number of animals in each group.

Before the Mevissen et al. (1993) study, a group in Georgia examined mammary carcinogenesis in magnetic-field exposed animals that were initiated with N-nitroso-N-methylurea (NMU) (Beniashvili et al. 1991). In the groups of animals exposed to a 60-Hz magnetic field at 20 μT for 3 hr per day for the lifetime of the animals, the incidence of NMU-induced mammary tumors increased over that in sham-exposed animals or in animals exposed for only 0.5 hr per day.

An additional mammary carcinogenesis study was performed in which DMBA was used to initiate mammary tumors in rats. Löscher and co-workers (1993) reported a significant increase in mammary-tumor induction in the rats exposed to a magnetic field. All rats received four weekly doses of 5-mg DMBA beginning at 52 days of age. After DMBA administration, 99 rats were exposed to 50-Hz magnetic fields at a flux density of 0.1 mT for 24 hr per day for 3 months. Another 99 rats were sham exposed. After 3 months of exposure, mammary-tumor incidence was about 50% higher in the exposed group (51 tumors) than in the sham-exposed group (34 tumors). The difference was statistically significant ($p < 0.05$). The tumors were also larger in the exposed group ($p = 0.0134$), but a difference was not found in the number of tumors per tumor-bearing rat. Note that this exposure is about 1,000 times that of the usual residential field strengths.

REPRODUCTIVE AND DEVELOPMENTAL EFFECTS

This section deals with in vitro and in vivo reproductive and developmental biologic effects of electric and magnetic fields at frequencies of 50 or 60 Hz in exposures that are relevant to those associated with power transmission and use. It is divided into considerations of effects of electric fields and magnetic fields. This division is somewhat artificial because all time-varying electric fields have an associated magnetic field; however, at these low frequencies, the fields can be considered independently to a high degree of accuracy. Nonmammalian and mammalian studies are also considered separately. The studies are summarized in Appendix A, Table A4-2.

Nonmammalian Studies of 50- or 60-Hz Electric Fields

Fish

Embryonic effects of concurrent exposure to power-frequency electric and magnetic fields have been studied in Medaka fish by Cameron et al. (1985). Two- to four-cell-stage embryos were exposed for 48 hr either to 60-Hz electric fields that produced a current density of 300 mA/m², to a magnetic field of 100 μT (1.0 G) root mean square (rms), or to combined fields. No significant developmental delays were reported immediately after exposure. Delays averaging 18 hr were detected 36-73 hr after removal from the magnetic field and the combined field exposure. Developmental delays did not result in abnormal development or decreases in survival through hatching.

Chicken

The chicken embryo has been used to study potential effects of electric fields. Blackman et al. (1988a,b) studied brain tissue from embryos in chicken eggs exposed to 50- or 60-Hz fields at 10 V/m rms. The associated magnetic field was less than 70 nT (1 nT = 10^{-9} T) rms. Brain tissue was removed 1.5 days after hatching. The tissue was placed in a physiologic salt solution containing radioactive calcium and then placed in the same solution with no radioactive calcium and exposed to 50- or 60-Hz fields at 15.9 V/m rms and 73 nT rms for 20 min. The calcium efflux from the brain tissue of chicks exposed as embryos to 60-Hz fields was affected (see the description and analysis of these experiments in Chapter 3). The same phenomenon was not observed with embryos exposed to 50-Hz fields. Three replicates of the Blackman study by other laboratories have not produced consistent results.

Mammalian Studies of 50- or 60-Hz Electric Fields

Mice

Male and female mice were exposed to either horizontal or vertical electric fields in two studies by Marino et al. (1976, 1980). In the first study, mice were exposed to electric fields at 10 and 15 kV/m that led to effects attributed by the authors to microshocks. The second study involved three generations of mice. Although the postnatal-weight gains were similar in exposed and unexposed mice, a higher mortality was observed in the exposed mice. This is the only report of that phenomenon, and the results have not been supported by data from studies conducted at other laboratories.

Unlike the work of Marino and co-workers, Fam (1980) was unable to identify an exposure-related change in mortality of the progeny of mice exposed to 60-Hz electric fields at 240 kV/m. In this study, mice were exposed throughout gestation, the offspring were bred, and their litters were monitored for growth, blood histologic and biochemical changes, and histologic changes of major organs. In agreement with the results of Marino et al. (1980) except those on mortality, no changes were observed in growth or in any of the other measurements as a result of exposure.

Kowalczuk and Saunders (1990) were unable to detect any exposure-related dominant lethal mutations in male mice exposed to 50-Hz electric fields at 20 kV/m. Males were exposed for 2 weeks before breeding, and no exposure-related effects in offspring were detected in in utero death, litter size, or viability of offspring. Females were not exposed.

Zusman et al. (1990) found in vitro effects of electric fields in embryos of rats and mice, but they were unable to detect effects in fetuses exposed in vivo. Field frequencies used on mice in vitro and in vivo were 1, 20, 50, 70, or 100 Hz at 0.6 V/m with a pulse duration of 10 msec. Preimplantation mouse embryos were exposed and monitored through the blastocyst stage; 10.5-day-cultured rat embryos were exposed to the same fields. Cultured rat embryos showed abnormal limb development, and mouse embryos showed retarded development at some frequencies. When rats were exposed to the same fields and the offspring were examined at term, malformations did not increase. This study is greatly weakened because no indication is given that evaluators were blinded to the exposure group, and no correction is given for the use of multiple t-test.

Rats

Andrienko (1977) reported a study in rats exposed to 50-Hz electric fields at 5 kV/m in which they claimed to find exposure-related effects on several reproductive and developmental end points. These results are not considered further in this report because of the lack of details on experimental and statistical design furnished in the text.

Free et al. (1981) were unable to detect exposure-related effects of 60-Hz electric fields on neonatal rats. They exposed rats to 64-kV/m electric fields for 7 weeks beginning at 20 days of age and measured a spectrum of hormones that are part of the reproductive cycle (testosterone, follicle-stimulating hormone, luteinizing hormone, corticosterone, prolactin, thyroid-stimulating hormone, reduced glutathione, and thyroxin).

Burack et al. (1984) reported some effects in prenatally exposed rats; the effects appeared to be related to general stress rather than specific effects from electric fields. In this study, 17 pregnant female rats were exposed to 60-Hz electric fields at 80 kV/m on days 14-21 of gestation. Twelve pregnant females served as controls. After birth, litters were examined for viability, body weight (representing growth and maturation), and developmental landmarks (ear flap separation, eye opening, anogenital distance, age of testes descent, age of vaginal opening, and male sexual response in testosterone-treated gonadectomized animals). No exposure-related effects were detected except that exposed males displayed reduced copulatory behavior when compared with controls. As discussed in a review by Chernoff et al. (1992), the behavioral changes reported in that study match those expected from general stress.

According to Chernoff et al. (1992), Margonato and Viola (1982) were unable to detect any exposure-related effects in the offspring of male rats exposed to 50-Hz electric fields at 100 kV/m. Male rats were exposed for 30 min per day or 8 hr per day for up to 48 days. No exposure-related changes were detected in fertility, sperm viability, or morphology of exposed males or in the number of implantations, percent live per litter, or incidence of malformation in offspring.

Seto et al. (1984) was unable to find any exposure-related effects in the fetuses of rats exposed to 60-Hz electric fields at 80 kV/m for 21 hr per day for four generations. No significant effects were detected in fertility measures or postnatal growth for the first three generations, and no intrauterine effects on frequency of placental resorptions, fetal deaths, or fetal malformations were seen in the fourth generation.

Sikov et al. (1984) were unable to detect any exposure-related effects on the perinatal development of rats exposed to 60-Hz electric fields at 65 kV/m (effective field strength). Measurements included reproductive performance of adults, prenatal development of offspring, and perinatal effects of exposure through 25 days of age.

In a study designed as a follow-up to experiments done on swine (Sikov et al. 1987), Rommereim et al. (1987) exposed female rats to 60-Hz electric fields at 65 kV/m for 19-20 hr per day, beginning at 3 months of age and continuing through two breedings. Female offspring of the first breeding were exposed under the same exposure regimen of Sikov et al., bred at 3 months of age from selected animals, and killed for teratologic examination. Teratologic examinations of fetuses followed the first and second breedings. The study was conducted in two full replicates. No exposure-related effects were both replicated and statistically significant.

In the study by Rommereim et al. (1987), several effects were reported that were not replicated or consistent. In the second pregnancy of the first generation of animals, the percentage of litters with placental resorptions among exposed litters decreased significantly. Prenatal mortality decreased in litters of exposed female offspring. Neither effect was repeated during the replicate study. Significant divergent sex ratio occurred in the second pregnancy of the first generation of females, but that effect was not replicated. An increase in the degree of ossification of the skull was detected in exposed litters of the first pregnancy of the first generation, but that effect was not seen in the subsequent generation or the replicate. In the exposed group of the second pregnancy of the first generation, the incidence of reduced ossification (formation of bone) of the sternebra (primordial sternum of the embryo) was increased. The incidence of litters with reduced ossification of the phalanges was decreased in exposed litters of the second generation of females. If $p < 0.12$ is accepted as significant, incidence of malformations of all types (minor and major combined) was found in exposed animals in the second breeding of the first generation, but that effect was not detected in the first breedings of any experiment or in the second breeding of the replicate study. Growth and viability did not differ between exposed and control groups. The authors suggested that the pattern of significant results and the failure to replicate effects might have been due to the presence of a 65-kV/m electric-field threshold for the developing rat.

A second set of studies was conducted by Rommereim et al. (1989) to test the threshold hypothesis described above. Male and female rats were exposed to 60-Hz electric fields at 112 or 150 kV/m. Exposures were for 19 hr per day beginning 1 month before and continuing through breeding, completion of gestation, and rearing of offspring through weaning. No differences were detected in breeding success, pregnancy rate, litter size, or postnatal growth and development.

In a third follow-up study (Rommereim et al. 1990) to the swine study conducted by Sikov et al. (1987), rats were exposed to 60-Hz electric fields at 10, 65, or 130 kV/m. Exposed males and females were bred and allowed to litter; the offspring were also exposed during breeding and subsequent pregnancies. Teratologic evaluation of over 7,000 fetuses was done at the termination of the study. No effects of exposure at any of the three field strengths were detected in reproductive outcomes, including course of pregnancy, pup weight at birth, and postnatal growth and development. No exposure-related increase in malformations was detected. The lack of reproductive effects reported in this study was even further strengthened by field-strength-related increases in chromodacryorrhea (a stress response consisting of release from the eye of a porphyrin-based material secreted by a gland behind the eye) in dams and their offspring. That effect indicated that the field strengths used were capable of producing biologic effects.

As mentioned in the previous section on mice, Zusman et al. (1990) found in vitro effects of electric fields in embryos of rats and mice, but they were unable to detect effects in fetuses exposed in vivo.

Swine

Sikov et al. (1987) reported an extensive study in which miniature swine were exposed to 60-Hz electric fields at 30 kV/m for 20 hr per day, 7 days per week for 4-18 months. Taken as a whole, this study did not show consistent exposure-related effects. Effects in some biologic results that were statistically significant were inconsistent throughout the study, and in follow-up studies in rats (Rommereim et al. 1987, 1989, 1990), no exposure-related effects were detected.

In the first generation of the study by Sikov et al. (1987), sows were exposed to electric fields for 4 months before breeding; pregnant animals were either killed before term for examination of fetuses or allowed to farrow. The mean number of live fetuses per litter increased, fetal deaths decreased, and fetal malformations decreased in offspring of exposed sows killed before term. Exposed sows that were allowed to farrow were bred again and killed before term for examination of fetuses. No differences occurred in fertility, fetal weight, or perinatal mortality in offspring from exposed animals. Exposed fetuses were smaller (including fetal mass, crown-rump length, maximal skull width, and intraorbital distance), and no exposure-related change was observed in malformation rate. Offspring from the first group of sows allowed to farrow were exposed to electric fields for 18 months and then bred. This second group of animals were allowed to farrow, subsequently bred again, and killed before term for examination of fetuses. The increase in the number of malformations in the fetuses was significant. Offspring from the second group allowed to farrow were killed after 10 months of exposure. No significant adverse effects were detected in those animals. These studies were complicated by a disease outbreak during the course of the second breeding of the first generation of sows. The presence of the disease and the associated exposure to electric fields make interpretation of the increase in the number of malformations detected in those animals very difficult.

Cattle

Algers and Hultgren (1987) exposed pregnant cattle to 50-Hz electric fields at 4 kV/m and magnetic fields at 2.0 μT (20 mG) by keeping the animals beneath 400-kV power lines. The animals were exposed continuously for 4 months. No changes were detected in fertility, estrous cycle, progesterone levels, intensity of estrous, viability of offspring, or incidence of malformations.

Nonmammalian Studies of Time-Varying Magnetic Fields

Chicken

Numerous studies have been done in which chicken eggs were exposed to magnetic fields. For a comprehensive review of these studies, the reader is referred to Chernoff et al. (1992). Because of the plethora of conflicting results reported in the literature and the lack of replication of studies by different laboratories, a multilaboratory definitive study was conducted; nearly identical protocols in six laboratories in six locations were involved (Berman et al. 1990).

Berman et al. (1990) exposed eggs to 100-Hz pulsed magnetic fields at 1-μT amplitude, 0.5-msec duration, and 2-msec rise and fall time. Exposure occurred for the first 48 hr of incubation. Embryos were then examined for fertility, developmental stage, and morphology. Two of six laboratories detected a decrease in the percentage of normal embryos as a function of the number of fertile eggs and live embryos. That effect was significant when the results of all laboratories were pooled. Interestingly, the results of this study reflect the range of previous studies in that the strongest effect was apparently the laboratory site. Unfortunately, because the strongest effect is more related to the laboratory site than to the fields used, the significance of the results of this study is difficult to determine, and the results do not give confidence that the study shows a true biologic effect of magnetic-field exposure.

As a follow-up to the study of Berman et al. (1990), Martin (1992) exposed eggs to 60-Hz peak-to-peak fields at 3 μT for the first 48 hr of incubation. In three subsets, the embryos were evaluated immediately after incubation; in one, the embryos were allowed to develop for another 72 hr. Despite earlier reports of positive effects by this laboratory, no exposure-related effects were detected.

Cox et al. (1993) attempted to confirm earlier studies by closely replicating appropriate exposure conditions. They exposed chicken embryos to 50-Hz fields at 10 μT with a superimposed field at 17 μT for 52 hr and allowed the embryos to develop for an additional 68 hr. No exposure-related increases in abnormal development were detected.

Mammalian Studies of Time-Varying Magnetic Fields

Mice

McRobbie and Foster (1985) exposed mice to pulsed magnetic fields ranging from 3.5 to 12 μT with pulse periods ranging from 0.33 to 0.56 msec. No exposure-related effects were detected. This study did not follow scientifically accepted test guidelines, and the data are of little value in evaluating biologic effects of magnetic fields.

Wiley et al. (1992) were unable to detect any exposure-related effects in mice exposed to fields designed to be relevant to the magnetic fields generated by video-display terminals. Mice were exposed to a magnetic field with amplitude varying in a saw-tooth shape with a repetition rate of 20 kHz and field strengths of 3.6, 17, or 200 μT from day 1 to day 19 of pregnancy. This study was unusual in that large numbers of animals were used (185 controls and three groups of 186 pregnant females). Dams were killed on day 18 of pregnancy and evaluated for implantations; litters were evaluated for fetal deaths, placental resorptions, body weights, and gross external, visceral, and skeletal malformations.

Frolen et al. (1993) exposed pregnant mice to pulsed magnetic fields (saw-toothed, linear rise time of 45 msec and a 5-msec decay time, 15-μT peak field strength, and a frequency of 20 Hz). No change in the rate of exposure-related malformations occurred. Exposures were begun on day 1 of gestation in two experiments and days 2, 5, and 7 of gestation in three additional experiments, respectively. All exposures were continued until day 19 of gestation. The number of implantations, placental resorptions, living and dead fetuses, and malformations and the length and weight of live fetuses were recorded. An increased rate of placental resorption was detected in exposed mice in all experiments except the one in which fetuses were exposed on day 7 of gestation. None of the increases in placental resorption rates was reflected in reductions in litter size. Body mass and length of exposed fetuses were reduced in the experiment in which fetuses were exposed on day 7 of gestation. The lack of correlation between increases in rates of resorptions and litter size makes it unlikely that the detected increase is of biologic significance.

Rats

Persinger et al. (1978) exposed pregnant rats to 0.5-Hz rotating fields at either 5, 100, or 1,000 μT from day 19 of gestation to 3 days after birth. Effects appeared to be unrelated to exposure.

Stuchly et al. (1988) exposed rats to 18-kHz saw-toothed waveform (44-μsec rise time, 12-μsec fall time) magnetic fields at 5.7-, 23-, or 66-μT peak-to-peak strengths. The study appeared to have largely negative results. Rats were exposed for 2 weeks before mating and throughout gestation for 7 hr per day. No exposure-related differences were detected in maternal measurements, fetal weight, or fetal malformations. A significant decrease in the incidence of bipartite or semipartite thoracic centra (primordial ossification points within the thoracic vertebra) was detected in the 2.9- and 33-μT exposure groups. A significant increase in the incidence of minor skeletal anomalies was detected in fetuses (but not litters) in the 33-μT exposure group. Because minor skeletal changes in the absence of terata (abnormalities in the developing or newborn fetus) are not likely to indicate serious adverse effects, the significance of the skeletal changes observed in this study with regard to biologic effects is difficult to assess, but the results indicate no abnormal effects on development.

McGivern et al. (1990) exposed rats to a 15-Hz pulsed magnetic field of 0.3-msec duration, 330-msec rise time, and peak strength of 800 μT. Pregnant animals were exposed for two 15-min periods on days 15 to 20 of gestation. At birth, no exposure-related effects on offspring were detected for viability, average weight, or anogenital distance. At 120 days after birth, no exposure-related effects were detected in circulating concentrations of testosterone, luteinizing hormone, and follicle-stimulating hormone. Increases in accessory-sex-organ weights and reductions in scent-marking behavior were detected in offspring of exposed dams. The committee is not aware of any attempts to replicate these results by any other study or laboratory.

Huuskonen et al. (1993) exposed mated female rats to 50-Hz (sine wave, peak-to-peak) magnetic fields at 35.6 μT or 20-kHz saw-toothed magnetic fields at 15.0 μT. No increases in malformation rates or placental resorptions were detected in the study. The mean number of implantations and living fetuses per litter was increased in rats exposed to the 50-Hz field. The increase was most likely an artifact due to the high number of resorptions in the control group. The incidence of fetuses with minor skeletal anomalies increased in both exposure groups similar to that reported by Stuchly et al. (1988). Nevertheless, such skeletal anomalies are common in teratologic studies and generally are not considered by most teratologists as indicating abnormal development.

Summary of Reproductive and Developmental Effects

The peer-reviewed literature appears to offer very little evidence of adverse effects on animals from power-frequency electric and magnetic fields. Some in vitro biologic effects might occur, but evidence of in vivo effects from either electric or magnetic fields has very little support at strengths below those perceived (see following section) by animals. Experiments have also failed to support any mechanism for in vivo effects on reproduction or development.

NEUROBEHAVIORAL EFFECTS

A survey of the literature on the neurobehavioral effects of extremely-low-frequency electric- and magnetic-field exposure revealed that this literature has been reviewed many times. For the purposes of assessment, neurobehavioral effects considered are behavioral, anatomic, and physiologic alterations and chemical changes that may be taken as correlates of behavioral effects. Only those reports published in peer-reviewed journals and with methods adequately described to allow for replication were included in the final evaluation; those reports are summarized in Tables A4-3 through A4-6. Some reports fulfilled these requirements, but others used inappropriate controls or inadequate exposure apparatus. All studies that met the committee's basic requirements are included in the tables; however, only those reports that were repeatable and reliable are discussed herein.

This section is divided into discussions of studies using electric fields and those using magnetic fields or combined electric and magnetic fields. Simple and complex responses are also discussed separately. Simple responses include detection threshold levels (behavioral or physiologic responses) and general activity levels. Complex responses include aversion, avoidance, social behavior, learning, and analgesia.

Electric Fields

Over the past 15 years, several studies using a variety of subjects proved that mammals can detect 60-Hz electric fields as a sensory stimulus. An example that established detection and also determined the approximate threshold level was published by Sagan et al. (1987). Two operant behavioral techniques were used to estimate the minimal field strength necessary for rats to detect the electric field. The investigators found that not only did the rats respond in a way that indicated they detected the fields but also that the rats' performance was correlated accurately with the magnitude of the field. The two behavioral protocols yielded average threshold estimates of 13.3 and 7.9 kV/m rms, which were similar to the thresholds produced by other investigators (between 4 and 10 kV/m) using different behavioral protocols. Although the results clearly showed that rats can detect electric fields, these investigations did not determine the positive or negative effects of electric fields on behavior. Stern and Laties (1989) tested whether 60-Hz electric fields at 90 or 100 kV/m were perceived as an aversive stimulus to rats. In this study, the rats were given the opportunity to turn off the electric fields that they were exposed to chronically, and when the rats turned off the electric fields, they were given the opportunity to turn it on again. None of the rats performed differently in the presence of electric fields or in control conditions where electric fields were never present. As a control for the protocol, illumination from an incandescent light was used instead of electric fields. The incandescent light served as an aversive stimulus, and the rats turned the light off at a rate dependent on the intensity of the light. Results from these studies showed that 60-Hz electric fields at 100 kV/m are not a detectably aversive stimulus to rats. However, the mechanism through which the electric field acts is not known.

In one study, investigators attempted to determine whether the electric field could be exerting its effect through stimulation of the hair follicles or the skin rather than through a direct action on neuronal membranes (Weigel et al. 1987). Using the exposed surface of an anesthetized cat's paw, 60-Hz electric fields at up to 600 kV/m were applied while simultaneously recording from the sensory dorsal root fibers, which transduce afferent impulses that originate from various receptors in the exposed paw. The results clearly showed that electric fields can elicit activation of the cutaneous mechanoreceptors with persistent duration lasting up to 90 min in some cases without fatigue. The mechanism for that response could be through the vibration of the hair follicles or through displacement of the skin by the force of the field that stimulates the receptors. Those two external mechanisms are separate from the possibility of a direct interaction of the induced currents produced in the skin with the neuronal membranes that stimulate the receptor to fire. By shaving the hair off the paw and applying mineral oil to the paw, a significant reduction in firing rate to stimulation was recorded, suggesting that the major part, but not necessarily all, of the mechanism for electric-field detection is through vibrations of the skin and hair.

Magnetic Fields

Although signal detection methods have provided evidence of the ability of mammals to detect electric fields, such evidence is not available for magnetic fields except at very high field strengths (i.e., magnetic excitation of endogenous phosphenes). A comprehensive series of studies examined the effects of chronic exposure of nonhuman primates to 60-Hz electric and magnetic fields on general health and behavioral performance, chemistry, and neurophysiology. In the first study, Wolpaw et al. (1989) exposed pigtail macaque primates to electric and magnetic fields at 3 kV/m and 10 μT, 10 kV/m and 30 μT, and 30 kV/m and 90 μT, respectively, for three 21-day periods; 21-day sham exposures preceded and followed the experimental period. General health examinations, including weight, blood chemistry, blood-cell counts, performance on a simple motor task, and postmortem examinations, were conducted on the animals. No detectable effects of electric and magnetic fields were discernible between sham exposures and experimental periods.

A companion paper (Seegal et al., 1989) reported the effects of twice-weekly-evoked potentials during the daily 6-hr field-off period. No effects of field exposure were detected on the auditory-, visual-, or somatosensory-evoked potentials of the early or mid-latency components of the response. A significant decrease in the amplitude of the late components of the somatosensory-evoked potentials was detected during two high-strength field exposures. The discussion of the results suggested that these changes might be due to opiate antagonistic effects of exposure to electric and magnetic fields. The metabolites serotonin and dopamine were changed in the monkeys exposed to electric and magnetic fields. Substantial data relates the endogenous serotonin system with analgesia; however, the mechanism is not clear through which electric and magnetic fields have an influence on serotonin and its effect on somatosensory-evoked potentials.

In a related study in rats, Ossenkopp and Cain (1988) showed that 1-hr exposures to 60-Hz magnetic fields at 100 μT (1 G) resulted in a shorter duration of fully developed seizures. These investigators also linked their results to the substantial evidence that magnetic fields inhibit the nocturnal analgesic effects of morphine in a field-strength-dependent manner (Ossenkopp and Kavaliers 1987). The mechanism of this effect is not known; however, studies using calcium-channel agonists and antagonists administered with morphine demonstrate that calcium channel antagonists inhibit and agonists enhance the analgesic effects of morphine in the presence of magnetic fields. The authors of those studies proposed that the effect of magnetic fields on analgesia is mediated through the calcium channels and cited the in vitro results of magnetic fields on calcium channels as evidence. However, direct evidence for the mechanisms of action remains undetermined.

Several studies examined the effects of magnetic fields on learning and performance in simple and complex behavioral tasks. Examples from even the best studies show mixed results. Hong et al. (1988) exposed infant rats to a static magnetic field at 0.5 T for 14 postnatal days. After a 1-month rest period, exposed and sham-exposed rats were trained to reverse a position habit in an enclosed T-maze four times. Although exposed and unexposed male and female rats differed, no differences were detected for total errors committed over the four reversal problems.

In contrast to this static-field report, Salzinger et al. (1990) reported results from rats exposed perinatally to 60-Hz electric fields at 30 kV/m and 100-μT magnetic fields for 22 days in utero and for 20 hr per day during the first 8 days postpartum. As adult rats, they were trained to emit responses for food on a random-interval schedule. When the rats were tested as adults, the exposed rats consistently responded at lower rates than the sham-exposed rats. In addition, the decrease in response was not eliminated by extinction procedures or by an additional month of testing. These results do not necessarily imply a deleterious effect of perinatal exposure to magnetic fields, but they do appear to indicate an effect was produced.

In support of an effect, Thomas et al. (1986) and Liboff et al. (1989) reported a temporary loss of stable baseline performance on a component of the multiple fixed-ratio-differential low-rate schedule dealing with differential reinforcement of low rates of responding. This loss followed a 30-min exposure to a combination of a static magnetic field at 26.1 μT and a 60-Hz linearly polarized magnetic field at 0.139 μT. Once again a decrease in performance accuracy on this task does not imply a deleterious effect of magnetic-field exposure and might be more in line with a detection or perception of the field.

To assess the potential aversion quality of 60-Hz magnetic fields, Lovely et al. (1992) tested the preference or aversion to 60-Hz magnetic fields at 3.03 mT in a shuttle box. In two sequential studies using the appropriate control and sham conditions, animals did not prefer or avoid the exposed chamber. The authors discussed their results in relation to significant responses observed with large 60-Hz electric fields. They suggested that the lack of aversion in these magnetic-field experiments indicates that aversive behavior produced by electric fields might be associated with body-surface interactions rather than internal-body currents resulting from electric-field exposure.

Summary of Neurobehavioral Effects

Mammals clearly can detect 60-Hz electric fields at relatively modest field strengths (above a few kilovolts per meter). However, the effect of electric fields, even at field strengths an order of magnitude higher, is not perceived as aversive. Further, the action of the field appears to be mediated primarily through the stimulation of the receptors and the skin through hair movement or vibration rather than through the direct interaction with neuronal membranes.

Even though little evidence exists showing that 60-Hz magnetic fields can be detected by animals, at the highest field strengths where rats appear to detect such fields (3 mT at 60 Hz), they do not produce an avoidance behavior. In addition, no general adverse health effects are detectable for field exposures, as measured behaviorally, chemically, or pathologically. However, repeated studies have reported behavioral, chemical, and electrophysiologic effects of long-term and short-term exposure to 60-Hz magnetic fields. These effects include a decrease in stable baseline performance on multiple-operant schedules dealing with reinforced behavior, on the one hand, and a suppression or decrease in induced-seizure duration, on the other hand. Both of those effects could be linked hypothetically by reports that 60-Hz magnetic fields inhibit endogenous opiate activity. A decrease in opiate activity could decrease the reinforcing properties of stimuli and exogenous opiates are known to enhance seizures. Thus, a decrease in endogenous opiates might inhibit seizures.

The underlying biologic mechanisms that mediate these effects are not known, but the results present interesting biologic questions that might or might not be construed as being health related. No link has been made between the effects observed in animals and those observed in studies of the cellular effects of exposure to electric and magnetic fields. Those studies are discussed in Chapter 3 of this report.

IN VIVO NEUROCHEMICAL AND NEUROENDOCRINE EFFECTS

Neurochemical Effects

A variety of chemical transmitters in the brain mediate interactions between neurons. These chemical agents are released from the terminals of one neuron near the limiting membrane of another neuron, where they typically interact with specific receptors on the postsynaptic cell. The synthesis of these neurotransmitters and their release into the synaptic cleft between cells is an important aspect of cell-to-cell communication with the brain.

Relatively few studies have examined changes in brain neurotransmitter metabolism as a consequence of exposure to electric and magnetic fields. According to Vasquez et al. (1988), the chronic exposure of rats to a 60-Hz electric field at 39 kV/m changes the metabolism of brain monoamines in rats as reflected by alterations in the circadian rhythms of these chemicals. Hypothalamic and striatal norepinephrine, serotonin, dopamine, and 5-hydroxyindole acetic acid, as well as the dopamine metabolite dihydroxyphenyl acetic acid were measured following the exposure of rats to an electric field for 20 hr per day for 30 days; the measurements were made at six time points (three during the night and three during the day) throughout a 24-hr period. In the hypothalamus, the rhythms of norepinephrine, dopamine, and 5-hydroxyindole acetic acid in exposed rats differed from those in sham-exposed controls. The differences were in terms of the phasing of the rhythms rather than in their amplitude. In the striatum, only the dihydroxyphenyl acetic acid rhythm was changed by the electric-field exposure. Changes such as those, particularly in the hypothalamus, could be related to the hormonal alterations of the neuroendocrine axis that have been reported. On the other hand, static measurements, such as those reported by Vasquez and colleagues (1988), are not informative in terms of the synthesis or release of the specific chemicals in question. In addition, the neural concentrations and rhythms of the chemicals measured are highly labile (Morgan et al. 1973; Kempf et al. 1982); thus, the importance of the reported changes in terms of the physiology of the organism remains unknown.

The work of Seegal et al. (1989) also indicates that brain monoamine metabolism changes as a consequence of electric-field exposure. They observed that when macaque monkeys were exposed to a 3- to 30-kV/m electric field for 20 days, homovanillic acid, a dopamine metabolite, and 5-hydroxyindole acetic acid, a serotonin metabolite, were depressed in cerebrospinal fluid. Concentrations of these metabolites in the cerebrospinal fluid are generally reflective of brain neurotransmitter metabolism.

Amino-acid neurotransmitter concentrations have also been measured in the striatum of rats exposed to a 50-Hz electric field that ranged in strength from 20 to 180 kV/m; the exposures continued for either 14 or 58 days, and sham-exposed rats were used as controls (Vasquez et al. 1988). The neurotransmitters measured were taurine, glycine, aspartate, glutamate, gamma aminobutyric acid, and alanine. Following exposure of the rats to electric fields for 14 days, a generalized increase in striatal concentrations of all the neurotransmitters was observed. On the other hand, after 56 days of exposure, the concentrations of these neurotransmitters in the striatum were depressed. The authors pointed out that the changes were minor, and although the differences between the exposed rats and the controls were statistically significant, the mean values for the neurotransmitter concentrations were within normal limits of variation.

No data are available on the potential effects of sinusoidal magnetic fields on neurotransmitter metabolism in the central nervous system.

Melatonin Effects

Melatonin is a ubiquitously acting hormone possibly produced in all animals, including humans. In mammals, a major site of production is the pineal gland. The pineal gland is an end-organ of the visual system and is innervated by postganglionic neurons whose activity is determined by light perception at the retinas (see Figure 4-1).

Melatonin is an aminoindole and the product of the metabolism of tryptophan. Tryptophan is taken from the circulation into the pinealocytes, the hormone producing cells of the pineal gland where it is converted to serotonin (Figure 4-1). Serotonin is metabolized to melatonin in a two-step process; initially it is *N*-acetylated by the enzyme *N*-acetyltransferase to *N*-acetylserotonin. This product is then *O*-methylated by the enzyme hydroxyindole-*O*-methyltransferase, and melatonin (*N*-acetyl-5-methoxytryptamine) is formed (Reiter 1991). Melatonin production is higher at nighttime (in darkness) than daytime. After it is produced, melatonin is quickly released into the systemic circulation, causing concentrations of melatonin to be higher during the night than the day. Once in the circulation, melatonin readily enters cells to exert its effects because it is highly lipophilic.

The synthesis of melatonin is controlled by exposure to electromagnetic radiation of wavelengths in the visible regions (Reiter 1985). Similarly to visible light, certain ultraviolet wavelengths and infrared wavelengths also alter pineal melatonin production (Brainard et al. 1993; Reiter 1993a). No retinal photoreceptors are known to be activated by either the ultraviolet or the infrared wavelengths used in these studies. This fact implies that wavelengths of electromagnetic radiation outside the visible range change the ability of the pineal gland to produce melatonin; furthermore, the result is achieved by something other than the classical photoreceptor mechanisms at the level of the retinas. This finding might be germane to the subsequent discussion only in the sense that the electric and

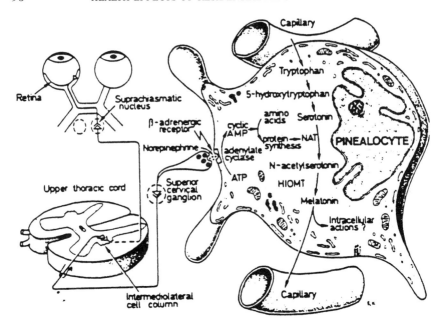

FIGURE 4-1   Illustration of the association of the retina with the pineal gland, in this case represented by a single pinealocyte. The pineal hormone melatonin is synthesized from the amino acid tryptophan; serotonin is an intermediate. The conversion of serotonin to melatonin requires two enzymes—*N*-acetyltransferase (NAT) and hydroxyindole-*O*-methyltransferase (HIOMT). The NAT-serotonin reaction is the rate-limiting step in melatonin production. SOURCE: Modified from Reiter 1991.

magnetic fields (extremely low frequencies and long wavelengths) that reportedly alter pineal melatonin synthesis are well out of the visible range. It should be emphasized, however, that the extremely low frequencies associated with power transmission differ by a factor of 10^{12} relative to visible and infrared radiation. Thus, the mechanisms of physical interactions of these different wavelengths might vary radically.

Effects of Electric Fields on Animals

The peer-reviewed reports that investigated the effects of sinusoidal electric fields on pineal serotonin metabolism and melatonin production and secretion are summarized in Table A4-7. Over a decade ago, Wilson and colleagues (1981) described a marked suppression of pineal melatonin production in rats exposed for 4 weeks to a 60-Hz electric field at 39 kV for 20 hr per day for 30 days. According to the experimental protocol, the authors took numerous precautions to ensure the health of the rats. After a 2-week period of isolation from other animals, the rats were randomly divided into groups that were subsequently either exposed or sham exposed to the electric field. They were 56 days of age at the onset of the study.

The exposure apparatus was state of the art and is described in detail in a separate report (Hilton and Phillips 1980). Although the rats were subjected to an unperturbed field strength of 65 kV/m, the effective field strength was calculated to be about 35% lower because of mutual shielding, namely, 39 kV/m (Kaune 1981a). The control rats were placed in an identical exposure facility, but the coils were not energized; thus, they were true sham controls. The report does not indicate whether the investigators knew which of the two coil systems was energized during the experiment.

At the conclusion of the study, the animals were killed during either daytime (at 0800 hr or 1400 hr) or nighttime (at 2200 hr or 0200 hr, under dim red light). Pineal glands were collected and assayed for their contents of two pineal constituents, melatonin and 5-methoxytryptophol, using a gas chromatography-mass spectroscopy method developed by one of the authors of the report (Wilson et al. 1977). Additionally, pineal *N*-acetyltransferase (NAT) activity, which limits the rate of melatonin production, was measured by a standard radioenzymatic assay (Deguchi and Axelrod 1972).

Wilson et al. (1981) performed two similar experiments. In the first experiment, a reduced concentration of melatonin was observed 6 hr after onset of darkness in rats that had been exposed to 60-Hz electric fields for 1 month; however, a change in the 5-methoxytryptophol concentration after exposure was not significant. In the second experiment using comparable exposure conditions, pineal melatonin concentrations and NAT activity were estimated. Melatonin was found to be significantly depressed, but NAT activity remained unchanged as a consequence of exposure. The dichotomous response of the activity of NAT and the pineal content of melatonin is somewhat unusual inasmuch as the *N*-acetylation of serotonin is widely accepted as the rate-limiting step in melatonin synthesis (Reiter 1991). The changes observed were theorized to be a possible consequence of the reduction in the firing rate of the sympathetic neurons that terminate in the pineal gland, because earlier work (Jaffe et al. 1980) found that 60-Hz electric-field exposure reduced the frequency of action potentials in the superior cervical ganglia.

Wilson et al. (1981) gave the impression that effective electric-field strengths of 39 kV/m reduced the synthesis of melatonin in the pineal gland of rats. The results were seemingly compelling in terms of the magnitude of the inhibition; however, the field strength (39 kV/m) was high. In an erratum published 2 years later, Wilson et al. (1983) reported that the animals were actually exposed to field strengths of 1.7 to 1.9 kV/m, the difference being due to a malfunction of a transformer at the time the study was performed. In the 1-page erratum, the authors reported that the experimental results were duplicated in a study using an effective field strength of 65 kV/m, but no actual data are shown. The implication of those findings, and one espoused by the authors, is that a wide range of field strengths (1.7-65 kV/m) reduced nocturnal pineal melatonin production in rats.

In 1986, Wilson and colleagues confirmed their original findings in studies in which rats were exposed to a 39-kV/m electric field for either 1, 2, 3, or 4 weeks (Wilson et al. 1986). After 3 and 4 weeks, the drops in nocturnal pineal melatonin concentrations were significant compared with those in sham-exposed controls. Likewise, NAT values were depressed at night in the pineal glands of rats exposed to the fields for either 3 or 4 weeks (see Table A4-9). In the same study, Wilson et al. (1986) found that withdrawal of the fields after 4 weeks caused a quick return (within 3 days) to the day-night melatonin rhythm.

As part of a larger study related to the potential consequences of electric-field exposure on fetal development, Reiter et al. (1988) exposed pregnant female rats in utero and the dams and newborns for 23 days after birth to field strengths of 10, 65, or 130 kV/m. At 23 days of age, pineal glands were collected from the young rats during the day or the night for pineal melatonin measurements. The assays for melatonin were performed blind in an independent laboratory that had not conducted electric- and magnetic-field research; a radioimmunoassay was used for measuring melatonin (Rollag and Niswender 1976; Champney et al. 1984). In the Reiter et al. (1988) study, each of the field strengths used (10, 65, and 130 kV/m) caused a slight but significant reduction in nocturnal pineal melatonin concentrations at one time point only. (Melatonin was measured at three different nighttime points.) No dose-response relationship was apparent. The nocturnal rise in pineal melatonin concentrations was also calculated to be delayed by 1.4 hr in each of the exposed groups of rats. This study used the same exposure facility as Wilson et al. (1981, 1986), but the experiments differed in that the initial studies used adult rats and the later study used fetuses and newborns. The melatonin assay procedures were different also (mass spectrometry versus radioimmunoassay).

In a study specially designed to test the effects of electric-field exposure on the melatonin synthetic activity of the pineal gland in rats, Grota and colleagues (1994) used a protocol similar to that used by Wilson and colleagues (1981, 1986). The exposure facility was state of the art, and the project was carefully supervised by a group of scientific advisors. The end points included pineal and blood melatonin concentrations as well as the activities of the two enzymes, NAT and hydroxyindole-O-methyltransferase (HIOMT) (Figure 4-1), required to convert serotonin to melatonin. Adult Sprague-Dawley rats (56 days of age at exposure onset) were exposed to a 65-kV/m 60-Hz electric field for 30 days. The control animals were sham exposed, and the pineal assays were conducted blind in an independent laboratory. After an exposure regimen of 30 days, neither daytime or nighttime pineal NAT activity, pineal HIOMT activity, nor pineal melatonin concentrations differed between the exposed and the sham-exposed animals. Serum melatonin concentrations were reported to be significantly

(p < 0.05) lower in exposed versus sham-exposed controls at one time point. In view of the failure of either pineal NAT activity, pineal HIOMT activity, or pineal melatonin concentrations to change, Grota et al. (1994) expressed concern that the measured depression in serum melatonin might have been due to chance rather than being specifically related to the field exposures. In addition, the statistical methods used in this study did not take into account the multiple hypotheses that were being tested simultaneously; such a consideration would render the differences between sham-exposed and exposed statistically insignificant.

In general, although the early studies on the suppression of melatonin synthesis by electric-field exposure were somewhat convincing, recent studies have failed to confirm a marked effect of such fields on the ability of the pineal gland to convert serotonin to melatonin. The current evidence is not convincing that electric-field exposure significantly impairs the melatonin-producing ability of the pineal gland.

Effects of Magnetic Fields on Animals

Interest in the potential neuroendocrine consequences of sinusoidal magnetic-field exposure has increased in recent years, and several approaches have been used to either directly or indirectly assess the effects of such fields on the physiologic integrity of the pineal gland.

Four studies have reported morphologic changes in the pineal gland following exposure of rats to 50-Hz magnetic fields (Table A4-8). The first study in a series investigated the ultrastructural appearance of the pineal gland of rats after their exposure to a very high magnetic-field strength of 0.7 T for 20 min daily for 2 weeks (Milin et al. 1988). Although the authors reported changes that implied significant alterations in the secretion of peptides by the pineal gland, the interpretation of the findings is greatly confounded by the restraints placed on the animals (a severe stress for rats) during the exposures. Also, no proof has been given that the pineal gland secretes any peptides. Two other studies also used high field strengths (5.2 mT), and the outcomes were inconsistent (Gimenez-Gonzales et al. 1991; Martinez-Soriano et al. 1992). A fourth study by Matsushima et al (1993) was well controlled (exposed versus sham-exposed animals) and used a 50-Hz circularly polarized magnetic field at 5 mT, which had been claimed to influence pineal melatonin synthesis (Kato et al. 1993). Although some changes in pinealocyte size occurred in the exposed animals, they varied according to location in the gland and according to the time of year the study was conducted. How or whether these changes are significant in terms of pineal-gland function remains unknown.

An extensive number of earlier reports illustrated the suppressive effects of pulsed or perturbed static magnetic fields on pineal melatonin production (Olcese and Reuss 1986; Wilson et al. 1989; Villa et al. 1991; Reiter and Richardson

1992; Reiter 1992, 1993b,c). However, recent reports focused on the effects of sinusoidal magnetic fields on the melatonin-producing ability of the pineal gland (Table A4-9). The first report of a change in circulating melatonin concentrations associated with exposure of rats to sinusoidal magnetic fields (Martinez-Soriano et al. 1992) did not provide a complete description of the methods used; thus, their observations are of questionable significance. As in later publications by other workers, however, they did observe a reduction in blood melatonin concentrations 15 days after an intermittent exposure of the animals to a 5.2-mT (52-G) sinusoidal magnetic field.

Three subsequent studies were carried out in Japan by Kato et al. (1993, 1994a,b) who used exposed and properly sham-exposed rats. The first of these reports is comprehensive, and the methods are well described (Kato et al. 1993). In that study, rats were exposed to a 50-Hz circularly polarized magnetic-field strength of 0 (control), 0.02, 0.1, 1, 50, or 250 μT continuously for 42 days. In repetitive studies, Kato and colleagues showed that pineal and blood melatonin concentrations during daytime and nighttime were reduced by the exposures used. The main point of this report is that magnetic-field strengths of 1 μT and higher reproducibly reduced both pineal and blood melatonin concentrations at night. The degree of reduction was typically on the order of 25-30% and had p values of 0.05 to 0.01. The only perplexing and seemingly contradictory finding in the Kato et al. (1993) report is the modest rise in daytime pineal melatonin concentrations in one study in which rats were exposed to field strengths of either 1, 5, or 50 μT. Among all the reports in which pineal melatonin synthesis was studied in relation to either static or sinusoidal magnetic-field exposure, the Kato et al. (1993) result is the only hint that such fields might do something other than suppress the melatonin-producing ability of the pineal gland. In every other report, using a wide variety of exposure conditions, field exposures of all types have been reported only to suppress melatonin production (when a change in melatonin concentration was observed) (Olcese and Reuss 1986; Wilson et al. 1989; Villa et al. 1991; Reiter and Richardson 1992; Reiter 1992, 1993b,c).

In confirming the efficacy of rotating magnetic fields in suppressing pineal melatonin production, Kato et al. (1994a) made similar observations using pigmented (Long-Evans) rats and the same exposure conditions used in their original report. A comparison of the responsiveness of albino and pigmented animals to 1-μT sinusoidal magnetic fields was made because of an earlier publication in which it was claimed that the ability of static magnetic-field exposure to suppress melatonin synthesis is related to cutaneous pigmentation (Olcese and Reuss 1986).

In a series of three experiments, Kato et al. (1994b) claimed that the exposure of rats to either a horizontally or vertically oriented (as opposed to a rotating vector) 50-Hz magnetic field at 1 mT for 6 weeks was not associated with a significant alteration of either daytime or nighttime blood or pineal melatonin concentrations when compared with the controls sham exposed at about 0.02 mT. The authors speculate that the complexity of multidimensional magnetic fields, which are a consequence of the circularly polarized fields (as opposed to the one-dimensional horizontal or vertical fields), might be operative, possibly by the induction of eddy currents, in suppressing the conversion of serotonin to melatonin in the pineal gland. These provocative findings, however, are near the limit of plausibility and must be replicated before the results can be accepted.

Until this point, the tacit assumption has been made that, for either electric- or magnetic-field exposures to suppress nocturnal melatonin production, at least part of the exposure must occur during the night when the synthesis of melatonin is increased. The studies of Yellon (1994) indicate otherwise. In successive studies performed during various seasons over several years, Yellon claimed that brief daytime exposures to unusual magnetic-field environments alter the underlying biologic clock mechanisms of the organism, thereby changing nocturnal melatonin synthesis by the pineal gland. In these studies, male and female Djungarian hamsters were exposed to a 60-Hz horizontal magnetic field at 100 μT (1 G) for 15 min beginning (at 18:00) 2 hr before onset of darkness (at 20:00). Sham-exposed hamsters served as controls. Immediately before darkness onset and during the night at 0.5 to 2-hr intervals, blood samples and pineal glands were collected from experimental and control animals. In two of three studies, the brief daytime exposure to magnetic fields significantly delayed and reduced the nighttime rise in melatonin; in the final experiment, no such observation was made. The differences are difficult to explain, but the author believes they might be related to the heterogeneity of the experimental animals. The animals were from a breeding colony maintained by the researcher, and, although all were adults, they ranged widely in age and included males as well as females, which were in different phases of their estrous cycles. Whether any of those factors confounded the outcome of the Yellon studies remains unknown. The findings point out, however, that daytime exposure to certain magnetic fields might affect the ability of the pineal gland to either synthesize properly or release melatonin on the subsequent night.

The papers summarized above represent all those published that satisfy the committee's criteria for inclusion in this report.

Effects of Combined Electric and Magnetic Fields on Animals

Lee et al. (1993) used animals in a natural setting to examine the effects of combined 60-Hz electric and magnetic fields on the circadian melatonin cycle (Table A4-10). In these complete and carefully supervised experiments, female lambs were maintained under a 500-kV high-voltage transmission line (mean electric field at 6 kV/m, mean magnetic field at 4 μT) for 10 months. Control animals were caged 229 m from the transmission line (mean electric field at < 10 V/m and mean magnetic field at < 0.02 μT). Blood samples were collected over 48-hr intervals during the 10-month period, and blood melatonin concentrations were estimated with a radioimmunoassay. Through the 10-month period, the transmission-line-exposed and control animals (10 each) exhibited essentially indistinguishable 24-hr melatonin rhythms. The cycles were compared during different seasons and under different environmental conditions and temperatures.

Combined electric and magnetic fields were also used by Rogers et al. (1995) who examined the effects of such exposures on blood melatonin concentrations in baboons. When two baboons were subjected to a 60-Hz field at strengths of 30 kV/m and 100 μT (1.0 G) that was intermittent, irregularly scheduled, and switched rapidly on and off for 10 days, low nighttime concentrations (equivalent to those measured during the day) of melatonin were reported. The animals served as their own controls and were sham exposed several weeks before exposure to electric and magnetic fields and were found to have a normal nighttime increase in blood melatonin. One strength of the study is that the animals served as their own controls; that strength, however, also led to a major shortcoming—the control and experimental periods obviously were not simultaneous.

Effects of Electric and Magnetic Fields on Humans

Although the melatonin rhythm has been frequently studied in humans and found to be similar to that in other mammals, only on a couple of occasions has circulating melatonin been examined after the exposure of individuals to experimental electric- and magnetic-field environments (Table A4-11). In two reports, adult males were exposed to the complex electromagnetic environment used for magnetic resonance imaging (MRI) for up to 60 min. In neither male was a significant change in blood melatonin concentrations measured (Prato et al. 1988-1989; Schiffman et al. 1994). The subjects served as their own controls and were sham exposed on different nights. Although no significant alterations in blood melatonin concentrations were noted, interpretation of the data is confounded by the observation of Schiffman et al. (1994) that even bright-light exposure had relatively little effect on circulating melatonin concentrations. A single report claims that the exposure of rats to MRI fields also is without effect on pineal serotonin metabolism (LaPorte et al. 1990).

In one report of men and women using electric blankets at night, melatonin was indirectly assessed by examining the urinary excretion of 6-hydroxy melatonin sulfate (Wilson et al. 1990a). In about 25% of the individuals, some change occurred in the concentrations of the urinary metabolite when the blanket was used and when its use was discontinued. Whether these reported changes are at all related to the electric- or magnetic-field features of the blankets remains unknown. A claim has been made that humans vary in their sensitivity to electric and magnetic fields (Rea et al. 1991). That claim, albeit of potential interest and importance, requires confirmation under conditions in which the experimenter is assured that the subjects did not receive other clues that alerted them to the onset and the cutoff of the field exposures.

Neuroendocrine Effects

Compared with the amount of information related to the effects of electric and magnetic fields on pineal morphology and physiology, the data on the impact of such fields on the hypothalamus-pituitary target-organ axis are sparse. The potential activation of the pituitary-adrenal system in animals exposed to electric and magnetic fields is of particular interest. The goal of these studies was to determine whether exposures to such fields constitute a stress to the animals. The typical stress hormones are pituitary adrenocorticotropin (ACTH), adrenocortical steroids (e.g., corticosterone and cortisol), and catecholamines (e.g., norepinephrine and epinephrine), which are released during stress into the blood from a variety of sites but particularly from the adrenal medulla.

Earlier studies by Dumanskii et al. (1977) and Marino et al. (1977) concluded that electric-field exposure induces a mild stress response in animals, although the evidence presented is generally unconvincing. Unfortunately, on the basis of seemingly meager data, Marino and Becker (1977) concluded that such field exposures might have implications for human health.

In a series of 10 studies, Marino et al. (1977) exposed young rats to a 60-Hz electric field at 15 kV/m for 10 months. At the conclusion of the exposure interval, 6 of the 10 experiments found a modest suppression of blood adrenocorticoid concentrations. The suppressions, which average about 30%, were in some cases statistically significant and in others not statistically verified. In the other four studies, blood adrenosteroid concentrations did not differ in exposed and nonexposed animals. Despite the fact that stressors of any type increase, rather than decrease, the secretion of steroids from the adrenal medulla, Marino et al. (1977) concluded from their findings that electric-field exposure constitutes a stress.

In a more carefully conducted study with mice, Hackman and Graves (1981) determined that, if electric-field exposure is a stress to animals, it is a very weak one. These authors exposed groups of mice to a 60-Hz electric field at a field strength of either 25 or 50 kV/m. The onset of the fields was associated with a transient and low-amplitude rise in circulating corticosterone concentrations; that short-term peak lasted less than 15 min. Thereafter (up to 20 min), basal (unstressed) concentrations of corticosterone were measured in the exposed mice. That effect contrasts with that in control mice that were exposed to sound stress for the same duration; in the controls, circulating corticosterone concentrations were high and were maintained throughout the 120-min experiment.

The findings of Hackman and Graves (1981) are essentially consistent with those of another report that appeared at about the same time (Free et al. 1981). Free et al. exposed adult male rats to a 60-Hz electric field at 64 kV/m for 120 days, after which circulating corticosterone concentrations were measured. The outcome of the study was inconsistent in that corticosterone values in all groups of animals were highly variable. When young rats were exposed to a 60-Hz electric field at 8 kV/m from 20 to 56 days of age, the blood corticosterone rhythm seemed to be slightly shifted.

In the seemingly carefully controlled study on the potential stress from electric-field exposure, Quinlan et al. (1985) concluded that such exposures did not evoke an activation of the pituitary-adrenocortical axis. They based their conclusion on the observation that the acute exposure of adult Long-Evans rats to a 60-Hz electric field at 100 kV/m for either 1 or 3 hr did not change blood or adrenocortical concentrations of corticosterone compared with those measured in sham-exposed controls. A particular strength of the Quinlan et al. (1985) study was the exposure facility used. The details of the facility have been published (Stern et al. 1983), and the field measurements were made by an individual from the National Bureau of Standards. Additionally, the authors took numerous precautions to ensure that the animals were isolated from known stressors.

Although Quinlan et al. (1985) reported no change in adrenocortical function as a consequence of electric-field exposure, growth hormone (GH) concentrations were raised in the blood after intermittent (16-sec on/off) electric-field exposure for 3 hr. The release of GH from the pituitary has been interpreted by some investigators to be a response to stress. Other adenohypophyseal hormones (thyrotropin and prolactin) were not changed by either continuous or intermittent electric-field exposure.

The final study on the neuroendocrine physiology of animals exposed to electric fields was published by Portet and Cabanes (1988). Following the exposure of rats to a 50-Hz, 50-kV/m electric field for 8 hr daily for 4 weeks, they observed no differences in blood corticosterone, ACTH, thyrotropin, triiodothyronine, and thyroxine concentrations or in adrenocortical corticosterone concentrations; comparisons were made with sham-exposed controls. At the conclusion of the study, the adrenal-gland weights were compared between exposed and sham- exposed rats, and the mean weights did not differ significantly; likewise, all the glands were histologically similar.

As another potential index of stress, Leung et al. (1990) quantified the degree of chromodacryorrhea around the eyes of rats exposed to a 60-Hz electric field with strengths of either 65 or 130 kV/m from day 10 of gestation until adulthood. This brown discoloration of the fur around the eyes is a consequence of secretory products from the intraorbital harderian glands and its presence, in excess, might be an indication of stress to the animals (Harkness and Ridgway 1980, Sakai 1981, Leung et al. 1990). Compared with sham-exposed control rats, both the 65- and 130-kV/m fields significantly increased the degree of chromodacryorrhea. No pituitary-adrenal function measurements were included to verify whether the field-exposed rats were actually stressed.

In general, the reports on the effects of electric fields on pituitary-adrenal function in rats suggest that this system is not significantly changed during field exposure. Likewise, no other pituitary hormones have been found to be substantially modified by electric-field exposures. Despite the fact that the emphasis of much research has shifted to the potential biologic consequences of magnetic-field exposures, no data are available on any pituitary or target-organ hormones in animals exposed to magnetic fields under controlled conditions. However, Lee et al. (1993) reported no change in circulating progesterone in sheep maintained under a 500-kV transmission line.

Consistency and Plausibility of Results

Surprisingly few studies have examined the potential neurochemical changes associated with the exposure of animals to very-low-frequency sinusoidal electric or magnetic fields. In fact, despite the current widespread interest in the biologic consequences of time-varying magnetic fields in this area of neurobiology (exclusive of those on the pineal gland), there is a dearth of information in the peer-reviewed literature. Because it is usually intuitively assumed that the brain or its appendages would be likely to detect changes in the electric-field environment, this dearth of information is remarkable. Among many potentially important molecular biologic assessments, studies of gene expression in the central nervous system of electric- and magnetic-field-exposed animals would seem appropriate.

The effects of exposure to sinusoidal electric and magnetic fields on pineal synthesis and secretion of melatonin have generated a great deal of interest. Electric-field-induced reduction of melatonin was initially reported to be profound; however, recent studies were unable to confirm those findings. A reasonable explanation has not been presented for the apparent disappearance of the effect that was so apparent in the early studies.

Because of the failure to confirm the findings on electric-field exposures and because of the shift in emphasis of much of the work to magnetic fields, studies of melatonin suppression by changes in the magnetic field are increasing steadily. Although the publications are beyond the scope of this report, a substantial number of published peer-reviewed reports showed that exposure of mammals to perturbations of pulsed static magnetic fields reduces the nocturnal production and secretion of melatonin (Reiter 1993b). Likewise, recent studies on sinusoidal magnetic-field exposure showed that these fields also inhibit the ability of the pineal gland to transform serotonin into melatonin. Collectively, the changes in melatonin concentrations as a consequence of such exposures are shown to be highly consistent as follows: (1) When a change in melatonin concentrations has been reported, the change has always been a suppression (never a stimulation); if melatonin production were unrelated to field exposure (i.e., just a random occurrence), the field-exposed animals would have higher nighttime melatonin concentrations than controls on some occasions; such a change has never been seen. (2) When other measurements of melatonin synthesis are estimated, they change in predictable ways consistent with the reduction in melatonin production. (3) The suppression of melatonin due to exposure to sinusoidal electric and magnetic fields has been reported in three mammals, the rat (two strains), the Djungarian hamster, and the baboon.

Another reason for the interest in the reported melatonin changes is their potential relationship to the higher incidence of cancer reported in some epidemiologic studies. Two biologically plausible, although unproved, mechanisms theoretically describe a link between reduced melatonin concentrations and cancer initiation, promotion, and progression (Figure 4-2). According to Stevens (1987a,b), the increased secretion of prolactin and gonadal steroids, which are natural consequences of melatonin suppression (Reiter 1980), could lead to excessive proliferation of stem cells in the endocrine-system organs, thereby increasing the likelihood of tumor growth in these organs (e.g., breast and prostate tissue). Another theory relies on the observations related to the intracellular action of melatonin. In recent studies, melatonin was shown to be a potent hydroxyl radical scavenger (Tan et al. 1993a) and to prevent carcinogen-induced damage to nuclear DNA (Tan et al. 1993b, 1994). Thus, reduction in melatonin due to any cause might increase the likelihood of DNA damage and cancer initiation. Considering the observations that magnetic-field exposure might induce or prolong the half-life of free radicals (Grundeler et al. 1992; Nossol et al. 1993; Harkins and Grissom 1994), which are known to be scavenged by melatonin (Tan et al. 1993a), additional justification for an association between reduced melatonin production and cancer initiation is suggested. Furthermore, melatonin has been shown in a variety of test systems to reduce the growth of already initiated cancer cells (Blask 1993), so its reduction might also promote tumor growth.

Despite the suggested associations between exposure to electric and magnetic fields, reduction in melatonin, and development of cancer, no direct experimental evidence links the field-induced reductions in melatonin to increased cancer risk. Thus, even though the explanations are biologically plausible and some experimental evidence supports the connection (Löscher et al. 1993), studies investigating the potential link need to be performed.

Other than the melatonin-production effects, other neuroendocrine and hormonal effects of electric- or magnetic-field exposures seem to be minimal, although only a few studies have examined these interactions. The field exposures used to date seem not to constitute a significant stress to the animals.

BONE HEALING AND STIMULATED CELL GROWTH

Evidence showing that exposure to electric and magnetic fields influences both the normal functions and the healing processes in bone is considerable. Bone turnover and fracture healing have been reported to respond to electric and magnetic fields under a variety of circumstances. Bone-fracture healing, in particular, represents the most thoroughly documented example of effects of low-frequency, relatively low-strength electromagnetic energy on human tissues. The effects of exposure to electric and magnetic fields on bone tissue have been studied

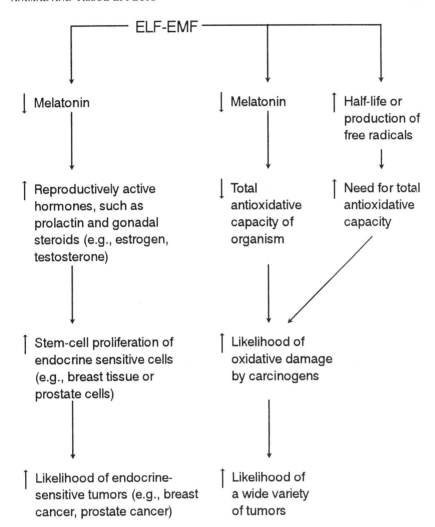

FIGURE 4-2   The two theories that have been proposed to explain the potential association of reduced melatonin production with the alleged increase in cancer after exposure to electric and magnetic fields. On the left, reduced melatonin concentrations lead to an increased secretion of prolactin and gonadal steroids. That increase causes proliferation of cell division in breast or prostate tissue and stimulates growth of initiated cancer cells. On the right, melatonin suppression reduced the total antioxidative potential of the organism, thereby increasing the likelihood of damage by a carcinogen to the DNA of any cell. DNA damage can increase the risk of cancer particularly if electric- and magnetic-field exposure also increases the half-life of production of free radicals.

in vivo in experimental animals and in humans, and the molecular mechanisms of the effects have been studied in in vitro systems. The evidence shows that electric and magnetic fields affect signal-transduction processes in bone cells, principally osteoblasts. The effects of magnetic-field exposure on bone have been observed almost entirely at field strengths of 0.1-15 mT (1-150 G) for magnetic fields and 1-100 mA/cm^2 for current density (which is proportional to the electric field); those strengths are orders of magnitude higher than the baseline values associated with household exposures. However, the field strengths reported to produce significant effects on bone overlap field strengths that can occur during intermittent exposures to household appliances or occupational electric equipment (Wilson et al. 1994). Little evidence can be found for effects of magnetic or electric fields on bone at magnetic-field strengths below 100 μT (1 G) or at current densities below 1 mA/cm^2.

Regulation and Cell Biology of Bone

Bone turnover is controlled by a number of hormones, principally parathyroid hormone (PTH) and 1,25-dihydroxyvitamin D_3. Those hormones act in bone largely by regulating activities of osteoblasts (cells that synthesize and calcify bone matrix) and osteoclasts (cells that resorb bone mineral and matrix); both hormones also maintain the balance of calcium and phosphate ions in the kidney and intestine. PTH and 1,25-dihydroxyvitamin D_3 cause increased resorption and decreased formation of bone when their concentrations are acutely raised (Auerbach et al. 1985), but both agents also promote bone formation at lower concentrations or over longer periods (Tam et al. 1982). Both agents probably exert their long-term actions directly on the cellular differentiation of osteoblasts and osteoclasts (Raisz 1977). Other hormones and cytokines also carry out specialized functions or play pathologic roles in bone metabolism (e.g., calcitonin, transforming growth factor β (TGF-β), interleukin 1, and PTH-like peptide) (Manolagas and Jilka 1995). The central role of the osteoblast in regulation of bone metabolism is emphasized by findings that the osteoblast is probably the primary target cell for PTH (Rodan and Martin 1981), which passes the hormonal regulatory message to other cell types by paracrine mechanisms.

Bone fracture is invariably accompanied by trauma and hemorrhage. Subsequent to a fracture, a specialized remodeling structure called callus forms around the fracture site. Extensive proliferation, differentiation, and tissue turnover involving both osteoblasts and osteoclasts take place over the ensuing healing period, leading eventually to resorption of the callus tissue and strengthening of the new bone bridging the fracture (Bassett 1989). In some cases, callus formation and subsequent remodeling fail for various reasons, such as necrosis, failure to vascularize, or infection. These failures to heal might be persistent or even permanent in some cases. Resistance to healing (nonunion fractures) is the primary condition for which therapeutic electric and magnetic fields have been applied most often.

Other uses of electric and magnetic fields have been to promote bridging of congenital gaps (pseudoarthroses) in bone and to enhance the density of bone in cases of osteoporosis (Polk 1993).

Endogenous Electromagnetic Properties of Bone

It is important to recognize that significant electric fields are a normal property of bone in living organisms. The modern era of research on this topic was initiated by a report showing that bone exhibited piezoelectric properties (Fukuda and Yasuda 1957). That finding was confirmed and extended to hydrated bone tissue by numerous investigators (Bassett and Becker 1962; Friedenberg and Brighton 1966; Cochran et al. 1968). Repetitive pulses of current density in the range of 10-100 mA/cm^2, with electric fields of 20-200 mV/cm, are generated in bone during normal movement because of mechanical stresses on the bone (Bassett and Becker 1962; Pienkowski and Pollack 1983). Electric fields in normal bone are from two primary sources: (1) piezoelectric responses of the calcified bone matrix to mechanical loading (Anderson and Eriksson 1970), and (2) streaming potentials due to dynamic charge separation between the essentially static charges in the collagen fibers and the ionic charges in the surrounding mobile fluids, which stream during loading and relaxation of bone (Borgens 1984). The mineral component of bone (hydroxyapatite) apparently contributes to the piezoelectric process mainly as an insulator that limits dispersion of charges produced by compaction of collagen fibers (Pollack 1984). Another postulated source of endogenous electric fields in bone is the electric processes of living bone cells, which contribute significantly to the higher current densities detected in living bone as opposed to dead bone (Friedenberg et al. 1973; Bassett 1989).

It is noteworthy in the context of the overall mission of this report that no report has been made of the magnetic component of the endogenous fields produced by bone. Moreover, in studies of the effects of various electric- and magnetic-field exposures on osteogenesis, Rubin et al. (1989) concluded that regardless of the magnetic characteristics of the applied field, the only factors relevant to biologic function in bone are those associated with the induced electric field. That conclusion has been supported by many studies and is generally accepted by most investigators in the area of research on electric- and magnetic-field effects on bone tissue (Brighton and McCluskey 1986; Bassett 1989; Polk 1993).

For over a century, bone growth, remodeling, and turnover in normal organisms have been hypothesized to be subject to the influence of endogenously generated electric fields, and application of externally generated electric fields has been hypothesized to be therapeutically useful in treatment of fractures or defects in osteogenesis (reviewed in Brighton and McCluskey 1986). Fracture of bone dramatically enhances the generation of charges and the flow of current in the area around the fracture site, especially during the first few minutes or hours after injury, and the magnitude of the currents surrounding the fracture is directly related to the healing of the fracture in subsequent processes (Friedenberg and Brighton 1966; Borgens 1984). Both piezoelectric and cellular electric processes are believed to contribute to fracture currents.

Locally generated electric phenomena in normal bone remodeling and fracture healing are believed by many researchers to be involved in a process in which areas of bone that accumulate negative charge are subject to increased deposition of bone matrix, and areas of positive charge are subject to increased resorption of existing bone matrix (Dealler 1981). That hypothesis is based on observations that, during chronic flexure of living bone, the areas of bone undergoing compression are the sites of increased bone formation ("Wolff's law"; Wolff 1892) and increased negative charge (Fukuda and Yasuda 1957; Bassett and Becker 1962), and the areas undergoing tension are the sites of increased bone resorption and increased positive charge. Moreover, numerous experimental and clinical studies (e.g., Brighton et al. 1979) have confirmed that placing dc electrodes in bone produces increased bone formation in the immediate area of the negative electrode and increased bone resorption in the area of the positive electrode. A range of current densities, roughly 10-100 mA/cm² has been reported to be optimal for observation of these effects; no effect has been observed at lower current densities, and cell death occurs at higher current densities (Friedenberg et al. 1970, Brighton and McCluskey 1986).

Clinical Stimulation of Bone Healing with Electric and Magnetic Fields

In 1964, Bassett and colleagues reported stimulation of bone growth in vivo with the use of implanted electrodes in unfractured dog bone (Bassett et al. 1964). That report led to a number of studies using various apparatus and having widely varying results (summarized in Hassler et al. 1977; Brighton and McCluskey 1986; Polk 1993). The first clearly documented successful studies of fracture healing using implanted electrodes were those reported by Friedenberg et al. (1971a,b), in which fibular fractures in rabbits were observed to heal much faster than those in sham-treated controls. Friedenberg et al. (1971a) applied 10 mA of dc field with the negative electrode implanted directly in the fracture site. In a single human case report, a nonunion fracture was healed by the application of a similar apparatus (Friedenberg et al. 1971b). Subsequent case reports and large clinical studies have convincingly documented that nonunion fractures and congenital bone defects (pseudoarthroses and failed arthrodeses) can be healed by means of implanted dc electrodes (Brighton et al. 1979).

Pulsed fields have been used more widely than dc fields for clinical bone-healing devices, at least partly because devices producing pulsed fields can be made noninvasive. During early studies of the electromagnetic properties of bone, it was found that specific time-varying current pulses could be detected in bone undergoing stresses similar to those involved in locomotion (Bassett and Becker

1962). Evidence also suggested that pulsed current delivered by implanted electrodes would decrease the amount of tissue damage due to electrolysis at the electrode surface (Levy and Rubin 1972). Bassett et al. (1974a) set out to influence osteogenesis by reproducing those pulses of current with noninvasive means as a way of avoiding the complications encountered with invasively implanted electrodes. Early attempts were made to produce pulsing currents in bone by placing the subject between electrostatically charged plates whose charge was altered rapidly (Bassett et al. 1974a). Subsequently, the Bassett group developed the strategy of using Helmholz induction to produce intratissue current flows by means of copper-wire induction coils placed noninvasively adjacent to the tissue (Bassett et al. 1974b). Electric pulses of 10-30 V applied to the induction coils were found to produce coupled pulses of about 1 mV in adjacent tissue at current densities estimated to be about 10 mA/cm^2 in tissue. Although the amplitude and general waveform of the pulses produced by this device resembled those found in living bone, the time scale was abbreviated considerably (microseconds as opposed to fractional seconds in normal locomotion) for electronic design reasons (Bassett et al. 1977). In an initial series of clinical studies with this type of device, a pulsed waveform with a single 300-μsec positive voltage pulse was used and repeated 72 times per second. At least 70% of resistant nonunion fractures and pseudoarthroses were healed by treatment with that device (Bassett et al. 1977). Subsequently, larger clinical studies reported success rates for pulsed electric-field treatment of over 80% (Bassett 1989).

Further developments in the induced pulsed electric signal involved use of a burst of about 20 200-μsec pulses, repeated 15 times per second, with a slightly improved success rate in fracture treatment (Bassett et al. 1982). The older single-pulse signal is apparently more effective in the treatment of osteonecrosis and disuse osteoporosis (Martin and Gutman 1978; Bassett 1983). A variety of other externally induced electric- and magnetic-field exposure have been used in animal and human studies (McClanahan and Phillips 1983). Success has varied. The device designed by Bassett's group and manufactured by Electro-Biology (Fairfield, N.J.) is approved by the U.S. Food and Drug Administration for clinical treatment of resistant fractures and pseudoarthroses. Side effects have been reported to be minimal. No evidence of increases in cancer or other diseases has been found despite the high field strengths used in comparison with environmental field strengths (Compere 1982; Bassett 1989). A small number of other clinical devices are also approved for use in stimulating bone healing. Most of these are based on either pulsed inductive fields or implanted dc electrodes. Some devices also use sine-wave extremely-low-frequency fields at a variety of frequencies (Polk 1993).

Despite the strong evidence that healing of nonunion fractures and pseudoarthroses is accelerated by electric and magnetic fields, there is no convincing evidence that the treatments have any influence on the healing of uncomplicated fractures. Uncomplicated fractures have not been widely treated with electric-

or magnetic-field procedures, however, for the simple reason that healing begins almost immediately in normal patients. Bassett (1982, 1983) suggests that once the final repair phases of the healing process have been triggered, whether by normal events or by exposure to electric and magnetic fields, treatment with the fields might only marginally accelerate the remaining events of fracture healing. On the other hand, such fields have been used to improve incorporation of bone grafts, to facilitate spinal fusions, and to improve certain types of osteoporosis (Friedenberg and Brighton 1981; Bassett 1983).

Potential Mechanisms of Electric- and Magnetic-Field Effects on Bone

The mechanistic bases have not been clearly established for the effects of either dc or pulsed electric fields on bone healing. Most evidence suggests that changes in osteoblast activities are the major functions responsible for bone responses to electric- and magnetic-field exposure (Watson and Downes 1979; Dealler 1981; Friedenberg and Brighton 1981; Bassett 1983). Although agreement is not explicit, different mechanisms might exist for the effects on osteoblasts by dc fields and by pulsed electric and magnetic fields (Polk 1993). Direct-current fields morphologically appear to stimulate osteogenesis mainly by stimulating the proliferation and differentiation of preosteogenic cells in the fibrocartilage matrix that fills the fracture gap in nonunion fractures (Friedenberg et al. 1974). Those cells then form new bone as if they had gone through an uninterrupted differentiation induced by the fracture process itself. Brighton and Friedenberg (1974) suggested that lowered oxygen tension in the area of the cathode might play an important role in triggering differentiation. Other possibilities are local changes in ionic concentrations or pH (Jahn 1968), stimulation of local nerves or blood vessels (Becker 1974), or direct membrane effects on cells by dc (Cone 1971). Bassett (1983), on the other hand, stressed the effects of pulsed electric and magnetic fields on functions of already differentiated bone cells rather than on precursors, suggesting that pulsed fields are less effective on osteogenesis as a proliferative process per se than it is on stimulating the function of existing bone cells at or near the fracture site. Bassett (1982, 1983) classified the demonstrated tissue effects of pulsed electric- and magnetic-field exposure as (1) a major and primary effect of reducing bone destruction, possibly by decreasing the sensitivity of bone cells to parathyroid hormone, (2) increased vascularization of the fracture site, (3) increased rates of bone formation by osteoblasts, and (4) for some pulsed EMF signals, decreased intracellular calcium concentrations in chondrocytes, a decrease that promotes replacement of chondrocytes by osteoblasts.

Effects of Electric and Magnetic Fields on Signal Transduction in Bone

Although the effects of exposure to electric and magnetic fields at the tissue level have been clarified somewhat by research over the past three decades, the primary biochemical and biophysical effects at the molecular or ionic level remain obscure. One clear likelihood for the effects of such fields on bone is that the plasma membrane of target cells is likely to be the major site of action, regardless of subsequent cellular mechanisms. The current and voltage involved in these effects are much lower than those that might be required to overcome the resistance of the plasma membrane and induce intracellular effects directly (Adey 1983). Several laboratories showed that exposure to electric and magnetic fields produces modifications in the activities of the plasma membrane of skeletal tissue cells. For example, Luben and colleagues (Luben et al. 1982; Cain et al. 1987; Cain and Luben 1987) demonstrated that exposure of bone and bone cells in vitro to pulsed electric and magnetic fields causes a membrane-mediated desensitization of the osteoblast to parathyroid hormone. Colacicco and Pilla (1983) examined calcium-transport and sodium-transport processes, factors that are likely to be related to osteoblast function, in chick tibia exposed to pulsed electric and magnetic fields. Fitton-Jackson and Bassett (1980) demonstrated positive effects of pulsed magnetic fields on chondrogenesis and osteogenesis. Rodan and colleagues examined the effects of mechanical and electric stimulation on the activity of adenylate cyclase in skeletal tissues (Norton et al. 1977; Rodan et al. 1978). A number of other membrane effects of pulsed electric and magnetic fields were reported in a variety of systems (Schmukler and Pilla 1982; Adey 1983; Borgens 1984). McLeod and colleagues used a number of in vivo and in vitro systems to study the biophysical and cellular biologic properties of bone exposed to electric and magnetic fields. They showed that electric fields induced by devices promoting bone healing are the most likely operative influence on bone-cell function (Rubin et al. 1989) and that those induced fields can prevent bone loss associated with immobility (disuse osteoporosis). Studies with different frequencies of electric fields showed that bone cells are dependent on frequency in responding to electric fields (McLeod and Rubin 1990); the most effective frequencies are in the range of 10 to 30 Hz, closely matching the frequencies most often observed in living animal bones. Field strengths were calculated to be approximately 300 mV/cm at the most effective frequencies. Further studies with isolated osteoblast-like cells (McLeod et al. 1991) suggested that 20-Hz and 60-Hz electric fields at 1-10 mV/cm could cause transient increases in cytosolic calcium-ion concentrations, a finding that corresponds with that of Ozawa et al. (1989), who used a different osteoblast-like cell line to show that calcium ions were involved in activation of DNA synthesis by pulsed electric-field exposure.

McLeod and colleagues investigated bone-cell proliferation as a function of exposure to 30-Hz electric fields at varying plating densities of cells. The cells treated with electric fields responded at medium densities by exhibiting a lowered rate of proliferation coupled with an increased alkaline phosphatase content (McLeod et al. 1993), suggesting that the field promoted differentiation toward a more active matrix-forming osteoblastic phenotype rather than a proliferative stem-cell phenotype. Cell densities above and below the responsive densities showed no effects of the field, suggesting that some cooperative effect might be operating between cells. Such cooperative effects could include cell communication through gap junctions or more complex electric phenomena, such as the "cell-array" model of dielectric impedance proposed by Pilla (1993). A related approach was taken by Fitzsimmons et al. (1989, 1992), who suggested that the effects of exposure to low-frequency, low-strength electric fields on proliferation and differentiation of bone cells in vitro are related to the generation of growth factors or their receptors, especially insulin-like growth factor II. Field-induced changes in the differentiation state of cartilage cells in vitro were shown by Hiraki et al. (1987), whose findings indicated an increased expression of osteoblastic phenotypes in cultured rabbit chondrocytes exposed to a clinically effective bone-healing device. Changes in cell-proliferation rates were also studied by Ozawa et al. (1989), who showed that pulsed electric fields increased DNA synthesis in rapidly growing bone cells but not in bone cells that had already reached a contact-inhibited more-differentiated status. The uptake of calcium ions was correlated with those changes in DNA synthesis, suggesting a membrane-mediated mechanism. These studies indicate that exposure to electromagnetic fields induce changes in the differentiation state of treated osteogenic cultures such that increased matrix synthesis, increased calcification activities, and altered sensitivity to growth factors and systemic regulatory hormones combine to substantially increase formation and decrease resorption of bone. These findings are consistent with the in vivo observations of events during bone healing (Bassett et al. 1982). The basic mechanisms by which such changes in differentiation are brought about have not been thoroughly elucidated, although most researchers suggest that the primary locus of the effects might be at the cell membrane, where responses to most of the hormones and growth factors that regulate bone metabolism are localized (Luben 1991; Pilla 1993). A number of in vitro studies showed that pulsed electric and magnetic fields used in the most widespread clinical fracture-treatment devices (Bassett et al. 1977) produce activation of mouse osteoblasts by means of a strong inhibition of parathyroid hormone (PTH) responsiveness in the cells (Luben et al. 1982), leading to increases in synthesis of collagen (Rosen and Luben 1983), decreases in bone resorption, and accelerated differentiation of osteoblasts from stem cells (Cain and Luben 1987). The effects of electric- and magnetic-field exposure on PTH responses were found to be consistent with decreased coupling of receptors to adenylate cyclase via the stimulatory G protein (Cain et al. 1987). Release of interleukin growth factor II or other growth factors, as suggested by Fitzsimmons and colleagues (1992), could be important in the proliferative responses of less differentiated cells, and the release of transforming growth factor b might be a factor in inducing differentiation (Manolagas and Jilka 1995). However, the above chain of events, although plausible, has not been investigated in detail in any single laboratory or experimental system.

A Hypothetical Scenario for Electric- and Magnetic-Field Effects on Bone

Based on observations of the biochemical effects of exposure to electric and magnetic fields on bone in vitro, a hypothetical model was developed to find ways to induce healing of bone in vivo. One key observation is that osteoblasts exposed to pulsed electric and magnetic fields for as little as 10 min exhibit a persistent desensitization to the effects of PTH on adenylate cyclase (Luben et al. 1982; Cain et al. 1987). Studies using biochemical probes of G-protein coupling (Cain et al. 1987) indicate that the ability of bound hormone-receptor complex to activate G-protein alpha subunits is impaired by treatment of the osteoblast with pulsed fields. The desensitization of the PTH receptor results in an increased rate of synthesis of collagen by the osteoblast (Luben et al. 1982; Rosen and Luben 1983) and a decreased rate of bone resorption by osteoclasts (Cain and Luben 1987). Both effects would tend to increase the amount of bone in a localized area exposed to pulsed fields in vivo, and those effects are in fact observed in healing bone under clinical circumstances (Bassett et al. 1982).

There are some potential clues to the possible molecular mechanism of PTH receptor desensitization by electric and magnetic fields. Desensitization of other G-protein linked receptors is known to be associated with changes in the configuration of the transmembrane domains adjacent to intracellular phosphorylation sites, leading to phosphorylation of the receptor by intracellular enzymes (Sibley et al. 1988). These findings suggest that electric- and magnetic-field treatment might change the conditions at the cell-membrane surface in some as yet unknown way that changes the configuration of key residues of the PTH receptor, leading to desensitization of the receptor and a shift in the balance of osteoblast activities toward increased bone formation. In this regard, PKC is known to be involved in regulation of PTH-receptor desensitization (Ikeda et al. 1991), and the PTH receptor is known to contain phosphorylation sites for PKC but not other known protein kinases (Abou-Samra et al. 1992). Recent studies suggest that a key site of action of electric and magnetic fields might be the PKC enzyme (Uckun et al. 1995), but that finding has not been replicated independently. It should also be noted that studies of magnetic-field effects on ornithine decarboxylase (ODC) (which are well replicated) have all used protocols in which ODC is stimulated by the PKC-dependent phorbol ester signal pathway. Possible increases in cytosolic calcium (Ozawa et al. 1989; McLeod et al. 1991) might also participate in the desensitization of the PTH receptor by PKC.

DISCUSSION

Numerous studies in the laboratory have been initiated to determine the nature of the physical mechanisms involved in electric- and magnetic-field-induced effects and the extent of possible health hazards to living organisms in an environment containing such fields. Biologic responses to exposure to electric and magnetic fields have been shown in many laboratories, and often they appear to be associated with the nervous system. In addition, unconfirmed or controversial data have been reported on observed effects that might be due to field exposure (e.g., changes in brain chemistry and morphology and alterations in reproduction and development). It is not yet known whether confirmed or putative effects are due to a direct interaction of the field with tissue or to an indirect interaction (e.g., a physiologic response due to detection or sensory stimulation by the field).

Whether a biologic effect from exposure to electric or magnetic fields constitutes a health hazard has yet to be answered. Experiments have not confirmed pathologic effects, even after prolonged exposures to high-strength magnetic (10 mT) and high-strength electric (100 kV/m) fields. In the very few tumor-promotion studies that have been reported, results seem to be mixed; most studies show no association between exposure to electric and magnetic fields and increased tumor development.

Although the data are not strong or entirely consistent, some experimental results using animal cancer models suggest a possible association of exposure to electric ar.d magnetic fields and adverse health outcomes. The strongest laboratory evidence for an association between magnetic-field exposure and cancer development is in promotion of mammary carcinogenesis initiated by chemical carcinogens; however, the results have not been consistent. In these experiments, tumors must be initiated with a chemical carcinogen for the magnetic-field exposure to have its apparent effect. Some data also support the possibility that magnetic fields can act as a copromoter; magnetic fields alone, however, have not been shown to be effective in promoting cancer development.

5

Epidemiology

SUMMARY AND CONCLUSIONS

The potential association between childhood leukemia and the presence of power lines reported in epidemiologic studies has raised public concern, particularly among parents. In this chapter, over 15 years of epidemiologic research is reviewed, and the key methods used in epidemiology that affect the interpretation of research are considered.

Based on an analysis of the epidemiologic literature, the committee makes the following general conclusions:

• Wire codes[1] are associated with an approximate 1.5-fold excess of childhood leukemia, which is statistically significant.

Although the literature is not entirely consistent, the combined results from the array of studies that have examined wire codes and related markers of exposure, such as proximity to power lines and calculated magnetic fields from power lines, indicate that an association is present. Biased selection of controls and confounders might have influenced some of the studies, but they are unlikely to account for the overall pattern of association that is identified.

• Average magnetic fields measured in the homes of children have not been found to be associated with an excess of childhood leukemia or other cancers.

---

[1]Used in the context of epidemiology, a wire code is a carefully documented aid to the epidemiologist to classify homes by their presumed correlation with magnetic fields; this use of the term is not similar to that in standard home construction. A detailed description of the term, as used in epidemiology, is given in Appendix B of this report.

• Studies that have examined average magnetic fields measured in homes after a diagnosis has been made have all been severely limited by missing data, and no firm conclusions can be drawn from them. The data that have been generated do not support an association between childhood leukemia and magnetic fields, in contrast to the data generated from wire codes.

• The factors that explain the association between wire codes and childhood leukemia have not been identified.

The original and continued interest in wire codes is their presumed correlation with long-term average magnetic fields in homes. However, epidemiologic studies have generated little evidence that average magnetic fields account for the observed association between wire codes and childhood leukemia. Wire codes are not strong predictors of magnetic-field strengths in homes, although they do distinguish very high fields from outdoor wiring from lesser fields reasonably well. Other explanatory factors, such as neighborhood characteristics, other measurements of exposure to electric and magnetic fields, or air pollution, have received even less support, leaving open the question of what accounts for the observed association.

• Epidemiologic evidence of an association between magnetic fields and childhood cancers (other than leukemia), adult cancers, pregnancy outcome, and neurobehavioral disorders is not, in the aggregate, supported.

A number of studies examined health outcomes other than childhood leukemia, and some of them reported positive associations. However, the number of well-designed studies supportive of such an association is not sufficient to conclude that any of the associations are actually present.

INTERPRETATION OF EPIDEMIOLOGIC EVIDENCE

Epidemiology can be defined as the study of patterns of health and disease in human populations to understand causes and identify methods of prevention. Interpretation of epidemiologic evidence regarding potential causal relations between exposures and health outcomes is a complex process and relies on a wide range of supporting data. No simple checklist can be used to make judgments about the quality of research and the certainty of the results, although a number of considerations might bear favorably or unfavorably on a causal interpretation. As a prelude to the review of epidemiologic studies addressing possible health effects of electric and magnetic fields, some of the key methods used to interpret epidemiologic data are reviewed. For more thorough consideration of these and related principles, see work by Rothman (1986) and Kelsey et al. (1986).

Compared with laboratory approaches to the study of health and disease (e.g., toxicology), observational epidemiology has a number of strengths and weaknesses. The principal deficiency of an observational approach is the inability to assign exposure randomly. This inability introduces the possibility of confounding the effect of the exposure of interest by other disease determinants. Because exposures are observed rather than controlled by the investigator, accurate determination of the exposure is more challenging in observational epidemiologic studies than in experimental studies. However, epidemiology has an advantage in addressing the species of ultimate interest, humans, in their natural environment. In addition, environmental exposure conditions that are difficult to duplicate precisely in the laboratory can be studied directly, at least in general terms, through epidemiology. Epidemiology can include studies of exposures that only produce health effects many years later. The complex array of disease cofactors, genetic heterogeneity, and diversity in human populations is virtually impossible to simulate in the laboratory, yet it is an inherent part of epidemiologic inquiry. Given the strengths and weaknesses of epidemiology relative to other approaches, consistent information from various approaches is desirable to enhance confidence that inferences are valid.

The notion of assigning causality from an observed association is, in a philosophical sense, complex, and the application of epidemiologic data to determination of causality is particularly problematic. Without randomly assigning the potential causes of interest (e.g., magnetic-field exposure) and observing the resulting health event (e.g., a change in cancer incidence), a mistaken inference that a given exposure causes a specific disease can result from a number of potential errors or misinterpretations. Conversely, even when a true causal relationship is present, it will not always be discerned easily. Ultimately, causal inference is enhanced when a number of noncausal explanations have been carefully postulated, tested, and refuted (U.S. Surgeon-General 1964). No universally accepted threshold exists for determining when the process of establishing causality has ended, as indicated by the few remaining skeptics who assert that the causal effect of tobacco smoking on lung cancer is unproved. In fact, rather than asking the broad unanswerable question "When has a causal inference been established?", a somewhat more practical question can be asked: "When is evidence of a causal association sufficient to take a specific action that presumes such a causal relationship?"

In this chapter, published epidemiologic data are reviewed that bear upon the potential association between exposure to low-frequency 60-Hz residential magnetic fields and disease, the potential sources of random and systematic errors common to epidemiologic studies are explored, the possible confounding factors and their potential effects on the findings of the studies are examined, and the effect of these various factors on the conclusions derived from the epidemiologic work is evaluated. The consistency of the results are explored using methods of data pooling, and the criteria for causality will be discussed as they apply to the problem at hand.

Potential Sources of Error in Epidemiologic Studies

Random Error

Random error is perhaps the most easily addressed source of error in scientific studies: results of a given study are subject to variability due solely to random (i.e., statistical) processes, in addition to other sources of error (i.e., bias). An observed risk of 1.0 or 2.0 relating high wire codes to childhood leukemia might be indicative of a "true" risk in the neighborhood of 1.0 or 2.0, in the absence of other sources of error attributable to random processes. The impact of random processes decreases as the number of study subjects increases, resulting in narrower confidence intervals for larger studies. Tests of statistical significance address the probability that, conditional on the observed data, the "true" relative risk[2] is inconsistent with the relative risks under the null hypothesis. Note that even if the null hypothesis is rejected, the statistical test relates to the association between the variables being tested (i.e., high wire codes and leukemia) and not causality. In the light of statistically significant findings, causality in observational epidemiology studies can be inferred only on the basis of design and weight of evidence criteria (see discussion of Hill's criteria below).

Many criticize epidemiology as a nonfalsifiable discipline. Although it is true that epidemiologists cannot prove the absence of an effect, they can identify the smallest detectable effect for a given study, and if this smallest detectable effect is sufficiently small, it is tantamount to the absence of an effect. For example, in a study of high wire codes and leukemia in which an odds ratio of 1.01 is obtained, a possible association might not be ruled out, but with appropriate data, the true relative risk could be stated with a certain probability to lie between 0.95 and 1.05. Thus, one can show that if the effect existed, it would be remarkably small and of little significance to the individual or to the public health.

In regard to observational epidemiology, it should be remembered that the formal methods of statistical significance testing and construction of confidence intervals attempt to address one and only one question: How likely is it that a valid method of randomly assigning exposures to individuals led to results as extreme or more extreme than those that were obtained, assuming that the null hypothesis is correct? Because random assignment is not a feature of epidemiologic studies, interpretation of statistical significance and confidence-interval boundaries have less of a theoretic foundation (Greenland 1994), and therefore less meaning, than they have in results of randomized experiments. In addition, the relative importance of nonrandom error (bias) is generally much greater than random error as a potential source of erroneous results. Therefore, judgments of evidence and causality take into account all available evidence, the level of statistical significance being only one of many considerations.

Information Bias: Misclassification of Disease or Exposure

In assigning exposure and disease status to individuals in epidemiologic studies, error, referred to as information bias or misclassification, can arise. Such error has an effect on the measures of association produced by an epidemiologic

---

[2]The relative risk is defined as the ratio of risk in the exposed population to that in the unexposed.

study. Diseases like leukemia are subject to relatively little misclassification because false negatives are unlikely given the severity of the disease (although misclassification might depend on the stage of the disease at diagnosis), and false positives are unlikely given the medical scrutiny of suspected cases. For prostate cancer, however, false negatives (failure to diagnose disease that is present) are common, as are false positives (misdiagnosing disease when it is not present), and for health events like miscarriage, false positives (e.g., a late menstrual period that is misinterpreted as an early pregnancy loss), and false negatives (e.g., an early pregnancy loss not recognized by the woman) are also quite common.

Exposure misclassification is a pervasive concern in epidemiologic studies on the effects of exposure to electric and magnetic fields. Errors can occur on several levels. In a study of occupational exposures and leukemia, simple errors might occur in assigning the job title. For example, an error in the assignment of "electrical worker" might occur because of erroneous job-title information provided by the coroner or funeral director who fills out a death certificate. When the investigator interprets electrical-worker job titles as "exposed worker" and nonelectrical-worker job titles as "unexposed worker" when examining possible health effects of exposures to electric and magnetic fields, additional error is introduced. If the true historical exposures were known, some of the workers labeled "exposed" based on their job title would not be exposed (e.g., the electrical engineer who almost never works near electric equipment), and some of those labeled "unexposed" would be exposed (e.g., the janitor who routinely uses equipment with electric motors, such as floor polishers or vacuum cleaners).

Any resulting misclassification can produce distortion in the measured association between exposure and disease. If dichotomous exposure comparisons are made, such as for electrical workers versus nonelectrical workers, and if the amount of error in exposure assignment is unrelated to the disease of interest (or vice versa), the direction of the error is predictable: a bias will be evident toward the null value, or an underestimation of any association (Rothman 1986). When exposure is classified into three or more categories, that general principle cannot be assumed to apply, particularly when there is misclassification into nonadjacent categories (Dosemeci et al. 1990). If the errors in exposure assignment described above applied equally to workers who died of leukemia as to workers who died of other causes, the tendency would be to produce measures of association that are closer to the null value (relative risks closer to 1.0) compared with the "true" relative risk. Likewise, because job title is an inherently imperfect marker of exposure and as likely to be similarly imperfect for workers who get leukemia as for other workers, then a bias will be toward the null value when making inferences from job title as a marker of exposure. If, in coding death certificates for leukemia, errors of a similar type and magnitude occur for electrical workers as for nonelectrical workers, the bias will be toward finding no association.

When the error is differential (i.e., the quality of exposure assignment differs for diseased and nondiseased persons, or disease assignment differs for exposed and unexposed persons), the error in the association can go in either direction. The direction and magnitude of error in the association depends on the degree of misclassification of exposure or disease. For example, if funeral directors have heard of the hypothesis linking magnetic-field exposure to leukemia, and they preferentially assign the job of "electrician" to persons who died of leukemia, the bias would be toward a spuriously large association.

Selection Bias

Another potential source of error arises from the constitution of the groups to be compared: those composed of exposed and unexposed subjects in a cohort study or cases and controls in a case-control study. In a cohort study, the primary concern is retaining subjects under observation throughout the study period, because the pattern of losses potentially distorts the comparison of disease rates among exposed versus unexposed subjects. If disease-prone persons were preferentially lost from the exposed group, its disease rate would be biased downward and the relative risk would also be biased downward.

Concern with selection bias is much greater in case-control studies, specifically in regard to control selection. In a case-control study, the goal of control selection is to sample from the study base or the population experience that produced the cases (Rothman 1986). If sampling is done successfully, the results will be identical, on average, to those obtained in a cohort study of the same population.

In studies of residential magnetic-field exposure and childhood cancer, cases have typically constituted a complete roster of all diagnosed children in a specified geographic area and time period. The goal of control selection in such studies is to identify an unbiased sample from the population in that area for the corresponding time period to provide a baseline of exposure prevalence (e.g., prevalence of high-wire-code homes). Any process of control selection that does not yield an accurate indication of the prevalence of high-wire-code homes will yield a biased odds ratio. If high-wire-code homes are underrepresented among controls, then the odds ratio will be biased upward (away from 1.0), and if high-wire-code homes are overrepresented, then the odds ratio will be biased downward. Evaluation of control selection addresses such issues as whether all persons eligible to be cases (i.e., people who, if they became ill, would have been cases) were included in the sampling frame and whether refusal to participate among eligible controls might have altered the prevalence of exposure as compared with controls who were included in the sample.

Confounding and Effect Modification

Confounding, a mixing of effects between the exposure of interest and extraneous risk factors, is not a product of the design or conduct of the study, but results from a natural association among risk factors. For example, assume that children who use electric blankets are more likely to be ill and, because of that pattern of illness, receive medical X-rays more often than children who do not use electric blankets. If medical X-ray exposures caused an increased risk of childhood leukemia and the exposures were not accounted for in the analysis, electric-blanket use would be falsely implicated as being responsible for an increase in risk that actually would be due to X-ray exposure.

The control of confounding is, in principle, easily achieved through statistical methods. For example, measuring X-ray use directly would allow assessment of the association between electric blankets and leukemia among those who did and did not receive X-rays. The results across those strata would be pooled. The potential confounder, X-ray use, could not affect the association of interest within those strata, and therefore the pooled or adjusted result is free of confounding. Obviously, such a solution requires awareness of the potential confounder, accurate assessment of it in the study, and control for it in the analysis. It should be noted that confounding can produce bias in either direction, spuriously increasing or decreasing relative risk, depending on the direction of the association between the exposure, the disease, and the confounder.

A different concept, sometimes confused with confounding, is effect modification in which the association between a given exposure and disease is affected by a third variable. For example, if magnetic fields acted as a late-stage carcinogen or promoter of childhood leukemia and parental tobacco smoking acted as an initiator, parental tobacco smoking can be hypothesized to act as an effect modifier of magnetic-field exposure. The relation between magnetic fields and leukemia would be stronger among those children whose parents smoked than among children whose parents did not smoke. In contrast to confounding, this phenomenon is not a source of distortion or bias, and thus not something to be controlled, but rather it is an observation of interest to be described. Although effect modification is reflective of biologic interdependence, it is commonly treated statistically when the association between two variables (exposure and disease) is determined for subgroups defined by a third variable (effect modifier) and is found to differ.

Criteria for Causality in Epidemiologic Studies

Criteria to assist in the evaluation of whether an association observed in an epidemiologic study is likely to reflect a causal rather than a noncausal association were delineated by Hill (1961) and have been used widely for that purpose, most notably in the U.S. Surgeon-General's report on smoking and health (1964). In his original presentation, Hill carefully stated that his criteria were only general considerations and not, individually or collectively, a checklist or scoring system. As noted above, judgments of causality are not strictly a scientific process but are a subjective interpretation of the accumulation of evidence. At some point, a consensus is reached that sufficient evidence has been accumulated to make some corrective action preferable to no action, indicating an implicit acceptance of a causal association. Nonetheless, the criteria presented by Hill provide useful reminders of issues that are worthy of consideration. Most of the criteria relate to an assessment of the degree to which the data are free from bias of the types described above, as noted by Rothman (1986).

Starting with a reported association between exposure and disease, Hill (1961) suggested several criteria to consider in addressing causality. These criteria are discussed below along with caveats regarding their interpretation:

• **Strength of association:** If a given exposure and disease are strongly associated (i.e., a large relative risk), then unrecognized confounders are less likely to be responsible for the association. For large relative risks, confounders are presumed to be apparent and already identified as important risk factors. This criterion will not be met if a true cause exists that actually has a very small effect. For example, true causes that only affect persons who are more susceptible because of relatively rare genetic or environmental cofactors would appear as weak associations in an epidemiologic study of a general population.

• **Consistency:** If the association is observed in different populations under different circumstances, it is more likely to be a causal relationship and not a product of some methodologic artifact in the study. However, the same error can be made consistently in studies to produce consistent but erroneous results, associations can truly be present under some circumstances but not under others, or inconsistent results can reflect a combination of good and bad studies yielding a mixture of valid and invalid results.

• **Specificity:** A cause should lead to a single effect rather than multiple effects; if multiple diseases are associated with a suspected agent, the associations are more likely to be spurious. Hill acknowledges that this criterion is particularly questionable, and an example of an exposure causing only one disease probably does not exist; examples of exposures with multiple effects are tobacco smoke, ionizing radiation, and asbestos fibers.

• **Temporality:** The exposure must logically precede the disease in time if the association is causal. This criterion is the only one that must be met. In some instances, the possibility of biologic markers being the consequence rather than the cause of the disease should be considered. For example, biologic markers of exposure, such as serum pesticide concentrations, might be disturbed by the occurrence of the disease itself, thus distorting comparisons of cases and controls.

• **Biologic gradient:** A dose-response gradient, in which risk of disease rises with increasing exposure level, is generally more likely to indicate causality than some other pattern of association between exposure and disease. Such an assessment, however, assumes the measurement of a relevant dose indicator. Weiss (1981) discussed in some detail why the presence of such a gradient is supportive of a causal inference, whereas the absence of such a gradient is not sufficient reason for ruling out a causal association. Confounding factors can also follow a dose-response gradient, or the underlying biologic processes might have a threshold or a maximum in their response that obscures the observation of a gradient in risk. Also, the range of exposure under study might be insufficiently broad to cause a dose-response gradient.

• *Plausibility:* Plausibility refers to whether the association is supported by scientific studies or information from disciplines other than epidemiology. The assessment of plausibility is a function of current scientific knowledge, which changes as natural processes are more fully understood. Lack of scientific evidence from other disciplines or conflicting information from other disciplines of course does not confirm that the epidemiologic studies are in error; the nonepidemiologic research, in itself, might be absent or flawed. Nonetheless, the interpretations of the data obtained in epidemiologic studies should be based in part on the agreement or disagreement of findings from other disciplines.

• *Coherence:* A causal interpretation should not be in conflict with current knowledge about the natural history of the disease. This criterion is virtually the same as plausibility, and the same caveats apply.

• *Experimental evidence:* When possible, experimental evidence in the form of randomized trials with prescribed exposures is highly desirable; however, practical considerations can preclude this approach. For example, hazardous exposures that might result in serious adverse health outcomes cannot ethically be tested in this way, although sometimes the removal of an exposure can be randomized to assess possible benefits.

• *Analogy:* If other known and accepted causal agents have been found that are similar to the one under evaluation in their manner of action on the biologic system, then the one under evaluation is more likely to be causal. The ability to identify relevant analogies also depends on the imagination of the investigator, but a documented analogy between a known and a hypothesized causal association is useful in drawing a conclusion of causality.

There are several arguments that should be considered when placing reliance on such criteria. First, these criteria do not provide a substitute for the careful and independent scrutiny of specific studies and their methods, but unfortunately they might provide a seductive shortcut that can make it tempting to do so. For example, absence of a dose-response gradient can reflect a poorly measured dose, saturation of the dose-response curve, or absence of any causal process. Merely to report that no dose-response relation was found as a means of dismissing an association requires much less effort than trying to distinguish among the possible reasons for the absence of a dose-response gradient and adds little to an understanding of the literature or the underlying phenomenon of interest. In spite of carefully stated caveats (Hill 1961; U.S. Surgeon-General 1964), such lists suggesting criteria for causality encourage a checklist approach to interpreting evidence.

Second, most of the criteria indirectly address questions of confounding and bias. It seems preferable to tackle those questions directly by asking whether a given association suffers from distortion due to the study biases.

Third, epidemiologic results should be first evaluated on terms inherent within the discipline, referring to qualities of study design, execution, and analysis. After that, insights and knowledge derived from other scientific disciplines relevant to the association in question can most effectively contribute to judgments about causality.

CANCER EPIDEMIOLOGY—RESIDENTIAL EXPOSURES

Summary of Evidence

The studies that have provided empirical evidence relating residential magnetic-field exposure to cancer are summarized in a series of tables in Appendix A (Tables A5-1, A5-2, and A5-3) that address the study methods. Later in this chapter the methodologic issues are critically evaluated, but this section is intended to provide a summary of the study structure (Table A5-1), of the methods used in control selection in case-control studies (Table A5-2), and of the approaches to exposure assessment (Table A5-3). Although the results are divided into studies of childhood and adult cancers, the summaries of the methods used include both types of studies because the study designs are similar.

At the time these tables were constructed, 12 studies provided relevant data on childhood leukemia and five provided data on adult cancers. Eleven were conducted in the United States or western Europe, and the majority were published between 1986 and 1993 (Table A5-1). All but two of the reports concerned case-control studies, most of which were based on a comprehensive case ascertainment in a geographically defined population. Exposure assessment was based on some form of coding derived from the physical characteristics and distances of nearby power lines and other electric constructions, with varying sophistication in the classification methods, and a number of studies included measurements of magnetic-field strengths in homes (Table A5-3).

Results of the epidemiologic studies are organized into tables that focus on childhood leukemia (Table A5-4), childhood brain tumors (Table A5-5), childhood lymphoma (Table A5-6), other childhood cancers (Table A5-7), childhood cancer in the aggregate (Table A5-8), cohort studies of residential exposure and cancer including all ages (Table A5-9), adult leukemia (Table A5-10), and adult cancers generally (Table A5-11). In each table, the numbers of cases and controls in each group are provided along with the crude and adjusted odds ratios (or other measures of relative risk) with 95% confidence intervals, and the confounders that were considered are noted. The goal in presenting the tables was to provide sufficient information to help readers understand the rationale behind the committee's interpretation and to allow readers to draw their own conclusions.

A decision was made early in the committee's deliberations that the body of studies concerning residential exposure to magnetic fields and occurrence of cancer, particularly childhood leukemia, deserve especially detailed scrutiny. Other exposure sources (e.g., appliances or occupation) and other health outcomes (e.g., reproductive or neurobehavioral efforts) are also considered, but not with the same amount of detail. Residential exposures related to power lines and occurrence of childhood cancer have been and continue to be the principal public concern that drives the broader concern for extremely-low-frequency (ELF) electric- and magnetic-field exposure. As the committee recognized early in its review, an association between proximity to certain types of power lines and childhood leukemia has been replicated with increasingly sophisticated study designs and warrants close examination on that basis alone. Finally, the charge to the committee from the U.S. Department of Energy is, "The committee will concentrate on the electric- and magnetic-field frequencies and exposure modalities found in residential settings." Although the exposures found in residential settings share some features with those in occupational environments and those related to electric appliances, warranting some discussion of nonresidential studies, the most relevant literature is that from exposure in residential settings as outlined in the committee's charge.

Framework for the Interpretation of Evidence Linking Magnetic Fields to Childhood Cancer

In this section, the framework is established for evaluating each of the key linkages pertaining to an association between living in proximity to certain types of electric power lines and childhood cancer; Figure 5-1 is a diagram of the relationships to be discussed; the arrows indicate associations, not causality. If an association between some characteristic of the power lines (as captured by the "wire codes" defined for use in epidemiologic studies) and cancer exists, several factors might be responsible. The association might be explained by magnetic fields produced by power lines, as a number of authors have suggested (Wertheimer and Leeper 1979; Savitz et al. 1988); that factor is designated

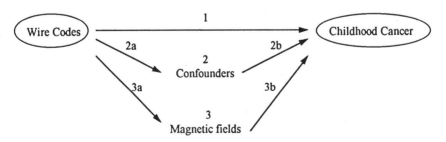

FIGURE 5-1 Conceptual framework for evaluation of evidence on wire codes, magnetic fields, and childhood cancer.

schematically as path 3 in Figure 5-1. Under this scenario, the wire codes are hypothesized to serve as a marker of magnetic-field exposure to the occupants of the home, and it is the magnetic-field exposure that confers an increased risk of cancer. Alternatively, some other agent associated with the wire codes, designated as path 2 in Figure 5-1, might be responsible for the association between wire codes and cancer—thus confounding the magnetic-field and cancer association. Under this scenario, wire codes might reflect the presence of some other agent or attribute (such as socioeconomic status or age of home) associated with the wire codes that is the true determinant of childhood cancer. Path 1 of Figure 5-1 simply refers to whether a link between the wire codes and cancer has been clearly established.

The evaluation of competing hypotheses requires careful consideration of the evidence bearing on each of the paths suggested in Figure 5-1. Besides an evaluation of the evidence supporting or refuting the associations, missing information must be identified that would be helpful in the evaluation. In the next section, the strength of the evidence is considered in the following areas: The evidence linking wire codes to childhood cancer (path 1); the evidence linking wire codes to potential confounders (path 2a); the evidence linking potential confounders to cancer (path 2b); the evidence linking wire codes to magnetic fields (path 3a); and the evidence linking magnetic fields to cancer (path 3b).

Are Wire Codes Associated with Cancer?

Assessing the Association Between Residential Magnetic Fields and Childhood Leukemia Using Techniques of Meta-Analysis to Assess the Role of Random Error

Since the publication of a seminal study by Wertheimer and Leeper (1979), scientists, policy makers, and the public have attempted to make sense of provocative and conflicting studies about the possible association between exposure to electric and magnetic fields and the incidence of disease. As this controversy continues, organizing and reviewing the existing data can provide important insights into the reasons for inconsistent results, gaps in investigative strategies, and limitations in understanding. Toward that end, a meta-analysis is undertaken of the most compelling subset of these data: residential exposure to magnetic fields or their surrogates and the incidence of childhood leukemia. This analysis is an attempt to gain an understanding of the importance of individual studies in prompting the overall conclusions of a possible link between exposure to electric and magnetic fields and cancer and the importance of the constraints needed in any successive study to assure that it would have sufficient statistical power to influence the present conclusions.

Meta-analysis is a statistical method used to provide a single risk estimate that summarizes the results of a set of similar studies (Dickersin and Berlin 1992;

Petitti 1994). In epidemiology, meta-analysis is applied most often to clinical trial data in which the major differences among studies are the differences in specific populations examined rather than in characteristics of the study designs. The validity of broadening the application of this method to environmental epidemiology is controversial because of the differences among studies that might include different methods of exposure assessment, different techniques for identification of controls, and different ways of accounting for such factors as confounders and manner of subject selection. These differences might result in substantial heterogeneity, calling into question the logic of a single summary statistic (see Blair et al. 1995) particularly for case-control studies. However, meta-analysis methods also can be used as an aid to evaluate the strength and consistency of an exposure-disease relationship, to look for design factors that might explain the heterogeneity, to conduct sensitivity and influence analyses, and to evaluate the robustness of the summary of additional studies of similar design.

The characteristics of 19 studies that have examined the possible association between residential exposure to magnetic fields and leukemia are shown in Table A5-1 (Appendix A). Of these, 12 studies have addressed childhood leukemia. Some studies report positive results, and others report no association (see details on the childhood leukemia studies in Table 5-1 below). Scientists disagree about the quality, bias, accuracy, and uncertainties in these studies, resulting in differing interpretations of the likelihood of a possible association overall. Some who examine the evidence find that the positive results are sufficiently compelling to conclude that an overall association exists (e.g., Ahlbom et al. 1993). Others argue that individual study results are artifacts due to systematic or random bias and that proper adjustment has not been made for multiple comparisons. Most conclude that the results, although interesting, do not show a consistent pattern of association (e.g., ORAU 1992; Peach et al. 1992; NRPB 1992). Recognizing the great cost of additional studies, government agencies are grappling with the development of policies in light of uncertainties and controversies.

In this meta-analysis, the results of a number of studies are compared using a variety of different assumptions about the comparability and appropriateness of such combinations of results. The goal is three-fold: (1) to examine quantitatively the consistency of the existing epidemiologic studies; (2) to analyze the influence of any single study on the combined effect measures; and (3) to estimate the sample size or number of studies needed to balance the combined results of previous studies. In short, the purpose of this meta-analysis is to consider the possible role of bias due to random error as an explanation for the observed results in a set of such studies.

Three sets of investigators have previously conducted meta-analyses of childhood cancer and residential exposure to magnetic fields. A report by Great Britain's Advisory Group on Non-ionizing Radiation of the National Radiation Protection Board summarized results of the childhood residential studies, providing pooled odds-ratio estimates for each exposure metric (NRPB 1992). For wire

TABLE 5-1 Childhood Leukemia Case-Control Studies

Study	Table	Exposure Definition	Exposed Cases	Exposed Controls	Unexposed Cases	Unexposed Controls	Expected Exposed Cases	Relative Risk	Standard Error of ln(RR)	p Value	Plot ID
Wertheimer and Leeper 1979; pers. commun.		Wire codes at birth, > end pole	131	124	5	12	51.67	2.54	0.55	0.04	1
		Wire codes at birth, > ordinary low	52	29	84	107	22.77	2.28	0.27	<0.01	2
		Wire codes at birth, > ordinary high	6	5	130	131	4.96	1.21	0.62	0.38	3
		Wire codes at death, > end pole	152	138	3	17	24.35	6.24	0.64	0.01	4
		Wire codes at death, > ordinary low	63	29	92	126	21.17	2.98	0.26	<0.01	5
		Wire codes at death, > ordinary high	12	5	143	150	4.77	2.52	0.54	0.05	6
Fulton et al. 1980	1	Wire code, > very low	131.48	168.75	41.52	56.25	124.56	1.06	0.24	0.41	7
	1	Wire code, > low	86.50	112.50	86.50	112.50	86.50	1.00	0.20	0.50	8
	1	Wire code, > high	41.52	56.25	131.48	168.75	43.83	0.95	0.24	0.59	9
Tomenius 1986	8	≥ 0.3 µT	4	10	239	202	11.83	0.34	0.60	0.96	10
Savitz et al. 1988	7	Wire code, > buried	69	171	28	88	54.41	1.27	0.26	0.18	11

7	Wire code, > low	27	52	70	207	17.58	1.54	0.27	0.06	12
7	Wire code, > high	7	8	90	251	2.87	2.44	0.53	0.05	13
3	≥0.2 µT, low power	5	16	31	191	2.60	1.93	0.55	0.12	14
3	≥0.2 µT, high power	7	29	30	175	4.97	1.41	0.47	0.23	15
VIII	Coleman et al. 1989 <100 m	36	63	48	78	38.77	0.93	0.28	0.61	16
VIII	<0 m	14	15	70	126	8.33	1.68	0.40	0.10	17
VIII	<5 m	3	3	81	138	1.76	1.70	0.83	0.26	18
7	London et al. 1991 Wire code, > underground	200	194	11	11	194.00	1.03	0.44	0.47	19
7	Wire code, > very low	180	167	31	38	136.24	1.32	0.26	0.15	20
7	Wire code, > low	122	92	89	113	72.45	1.68	0.20	<0.01	21
7	Wire code, > high	42	24	169	181	22.41	1.87	0.28	0.01	22
5	24-hr, >67 nT	79	75	85	69	92.39	0.86	0.23	0.75	23
5	24-hr, >118 nT	44	33	120	111	35.68	1.23	0.27	0.21	24
5	24-hr, >267 nT	20	11	144	133	11.91	1.68	0.39	0.09	25
6	spot, >31 nT	73	53	67	56	63.41	1.15	0.26	0.29	26
6	spot, >67 nT	39	25	101	84	30.06	1.30	0.30	0.19	27
6	spot, >124 nT	16	11	124	98	13.92	1.15	0.41	0.37	28

Continues on next page

TABLE 5-1 Continued

Study	Table	Exposure Definition	Exposed Cases	Exposed Controls	Unexposed Cases	Unexposed Controls	Expected Exposed Cases	Relative Risk	Standard Error of ln(RR)	p Value	Plot ID
Olsen et al. 1993	4	>0.1 μT	4	8	829	1658	4.00	1.00	0.61	0.50	29
	4	>0.25 μT	3	4	830	1662	2.00	1.50	0.76	0.30	30
	4	>0.4 μT	3	1	830	1665	0.50	6.02	1.16	0.06	31
Feychting and Ahlbom 1993	5	Calculated, ≥0.1 μT	11	79	27	475	4.49	2.45	0.38	0.01	32
	5	Calculated, ≥0.2 μT	7	46	31	508	2.81	2.49	0.45	<0.01	33
	5	Calculated, ≥0.3 μT	7	32	31	522	1.90	3.68	0.46	<0.00	34
	9	Spot, ≥0.1 μT	5	137	19	207	12.57	0.40	0.51	0.96	35
	9	Spot, ≥0.2 μT	4	70	20	274	5.11	0.78	0.56	0.67	36
	8	Distance, ≤100 m	12	123	26	431	7.42	1.62	0.36	0.09	37
	8	Distance, ≤50 m	6	34	32	520	2.09	2.87	0.48	0.01	38
Fajardo-Gutierrez et al. 1993	2	Distance, <20 m transformer	22	18	59	59	18.00	1.22	0.37	0.29	39
	2	Distance, <20 m distribution line	16	8	65	69	7.54	2.12	0.47	0.05	40
	2	Distance, <200 m substation	5	3	76	74	3.08	1.62	0.75	0.26	41

2		Distance, <200 m transmission line	11	7	70	70	7.00	1.57	0.51	0.19	42
4		Distance, <20 m transformer	9	7	37	42	6.17	1.46	0.55	0.25	43
4		Distance, <20 m distribution line	3	2	43	47	1.83	1.64	0.94	0.30	44
4		Distance, <200 m substation	5	1	41	48	0.85	5.85	1.12	0.06	45
4		Distance, <200 m transmission line	6	3	40	46	2.61	2.30	0.74	0.13	46
1	Petridou et al. 1993	Distance, <50 m	96	132	40	55	96.00	1.00	0.25	0.50	47
1		Distance, <5 m	27	33	109	154	23.36	1.16	0.29	0.31	48
II	Verkasalo et al. 1993	Calculation, >0.01 μT	35	—	—	—	38.03	0.92	0.16	0.70	49
II		Calculation, >0.2 μT	3	—	—	—	1.93	1.55	0.46	0.17	50
III		Cumulative calculation, >0.01 μT	35	—	—	—	38.05	0.92	0.16	0.70	51
III		Cumulative calculation, >0.4 μT	3	—	—	—	2.5	1.20	0.46	0.35	52

codes, excluding the Wertheimer and Leeper (1979) study, the board found a statistically significant increased odds ratio. For data based on the distance from the source of electromagnetic fields and for measured magnetic fields, the pooled odds ratios were found to be increased, but not statistically significant. They concluded that in spite of the increased odds ratios, the small sample sizes (three for each estimate) and methodologic problems in each of the studies precluded drawing definitive conclusions.

Ahlbom et al. (1993) combined the results from three recent studies conducted in the Nordic countries (Olsen et al. 1993; Verkasalo et al. 1993; Feychting and Ahlbom 1993) and argued that because they believed those studies were more similar to one another than to other studies (all used a population registry and estimates of historical exposure), they were appropriate for use in a meta-analysis. By combining the risk ratios of those studies and assigning them weighting factors proportional to the inverse of their variances, Ahlbom and colleagues (1993) found statistically significant increased risk ratios for childhood leukemia.

Washburn et al. (1994) conducted a set of meta-analyses for leukemia, lymphoma, and nervous-system cancers. For the combined results of 13 studies, they found increased risks for all three diseases; those for leukemia and nervous system tumors were statistically significant. Their sensitivity analyses showed that the inclusion or exclusion of data that overlap in the two Swedish studies and the choice of exposure metric had a limited effect on the results.

In contrast to the analysis of Washburn et al. (1994), which provided a single risk estimate for all studies, and the NRPB (1992) and Ahlbom et al. (1993) analyses, which limited their evaluations to three similar studies at a time, the analysis presented here seeks an explanation for the consistency or the heterogeneity of all the studies and estimates the influence of each study on the combined risk estimates. This analysis attempts to bridge the interpretation gap between previous meta-analyses.

Methods

There are three important methodologic components to conducting a meta-analysis: selecting the studies for inclusion, evaluating and weighting the quality of the individual studies, and adopting a method for summarizing results across different studies.

Selecting the Data   To conduct this analysis, 16 studies of residential magnetic-field exposures and childhood cancer were reviewed; the 11 studies used in this analysis are presented in Table 5-1. Of the 16 studies considered, two were excluded from this analysis because the data presentation or the analyses were incomplete (Myers et al. 1990; Lowenthal et al. 1991), and two others were excluded because children were not analyzed independently of older subjects (McDowall 1986; Schreiber et al. 1993). Another study was excluded because it was not published, and the data were inaccessible (Lin and Lu 1989). In the remaining 11 studies, methods vary according to the outcome studied (one mortality study, 10 incidence studies), source of data (hospital records, incidence registry, birth registry, death registry), maximum age of subjects (from 10 to 20), and exposure metric used (wire codes, distance from electric source, magnetic-field spot measurements, and historical reconstruction (calculation) of magnetic fields). Although some would argue that the original study that served as the motivation for the later investigations should be excluded from the meta-analysis (Wertheimer and Leeper 1979, in this case), the committee included this study in its analysis because, as discussed below, one of its primary goals is to assess consistency across studies.

In this investigation, the committee explored the consistency of results across the different types of exposure metrics that were used in the studies by conducting separate meta-analyses for different combinations of the metric and exposure cut points[3] that were used in the studies. These include (1) studies using wire codes and two alternative exposure cut points; (2) studies using only distance from electric source as an exposure metric and applying two exposure cut points; (3) studies using wire codes or distance and considering three alternative exposure cut points; (4) studies using magnetic-field spot measurements for one exposure cut point; (5) studies using calculations of magnetic fields for one exposure cut point; and (6) all studies combined. In the last category, three analyses were conducted: (1) combining results for the exposure category that gave the odds ratios with the smallest probability value (p value) of the hypothesis of no effect; (2) combining results for the exposure category that gave the effect measurement with the largest p value; and (3) using the results for the highest exposure category in each study.

For assessments in which multiple exposure metrics were explicitly stipulated, the committee's preferences for the use of exposure metrics to combine studies were in the following order: wire codes, distance from electric source, calculated magnetic fields, and spot measurements of magnetic fields. This order was chosen because wire codes were used in the initial study that identified an association with childhood leukemia. Distance is one measurement contained within the wire-coding scheme, and calculations would, under some assumptions, represent a better assessment of lifetime exposure than spot measurements. The latter is reasoned because an average can be taken over the daily and seasonal variations (note that these assumptions are untested; see Chapter 2). In contrast, Washburn et al. (1994) used distance in preference to wire codes and wire codes

---

[3] A "cut point" is the value for the exposure variable that, for the binary case, divides the population into "exposed" and "unexposed" groups for the purpose of quantitative analysis. For example, if wire-code rating divided residences into three exposure levels ("low," "medium," and "high"), the cut point might be taken between "medium" and "high" so that those living in "high"-category homes would be counted as "exposed" and others would be "nonexposed."

in preference to spot measurements or calculations. Based on their sensitivity analyses, Washburn and colleagues reported that the cut point chosen when more than one is reported has little impact on the results. In contrast, Wartenberg and Savitz (1993) showed that such choices can affect the results of individual studies substantially.

It should be noted that the data used by Feychting and Ahlbom (1993) included some, but very few, of the same data used by Tomenius (1986). Thus, the results of the two studies should not be considered strictly independent, although they are treated as such in the statistical analyses here.

Evaluating the Quality of Each Study   In the meta-analysis presented here, the decision was made not to evaluate and weight the quality of each study. The intent of these analyses was primarily to evaluate the consistency of the studies and their relative influences on the combined results; it was not clear that an evaluation of the quality of the studies would assist in this assessment. In addition, it was not clear which criteria would be most appropriate to use in evaluating and weighting the quality of each study (Greenland 1994).

Selecting the Statistical Methods   A variety of methods have been used to assess the results of combined studies, to identify heterogeneity, and to conduct influence analyses (Hedges and Olkin 1985; Wolf 1986; Fleiss and Gross 1991; Dickersin and Berlin 1992; Petitti 1994). The simplest method, called *vote count-ing*, relies on tallies of the number of studies with positive results, negative results, and null results. One can calculate expected values and statistical significance and draw a variety of inferences. Many scientists criticize this approach because it has low statistical power and because the summary measurement does not incorporate the observed size of the effect, sample size, or the statistical power (Hedges and Olkin 1985). However, when the null hypothesis is rejected, it is no longer a concern, and inferences are statistically valid. When the null hypothesis is not rejected, it is not known whether the results are due to the absence of an effect or to a limitation of the method. Given the uncertainties associated with this method, results should be used simply for guidance, not for hypothesis testing.

The *combined probability test*, which is nonparametric because it does not rely on an assumption about the probability distribution of the data, combines the logarithms of individual study p values into a chi-squared distributed statistic, P— the degrees of freedom being equal to twice the number of studies combined.

Similarly, one can combine the probability of rejecting the null hypothesis in individual studies to assess the sensitivity of the results to publication bias and determine the number of additional null studies needed to reduce a statistically significant combined effect to nonsignificance. This number is the so-called "fail-safe N."

Statistics that incorporate the individual study effect sizes use either of two statistical models: fixed effects or random effects. The fixed-effects model assumes that a single true underlying effect exists and that the studies included are a sample that allows one to make an inference about the true effect. Intrastudy precision (i.e., an overall treatment effect) is assessed by weighting individual study results by the inverse of the variance. The random-effects model assumes that the true underlying effect varies and is normally distributed. The sample of studies allows one to make an inference about the distribution of true study effects, including interstudy variation (i.e., a sampling effect) and intrastudy precision (DerSimonian and Laird 1986). The choice of model may be based on results of the chi-squared test (or Q test) for heterogeneity. This test assesses constancy of treatment effects. Once heterogeneity is detected, identifying its sources can be one of the most informative aspects of meta-analysis, although some authors simply report the random-effects result (Greenland 1994).

Influence analysis is the recalculation of summary statistics for a set of studies by leaving out one study at a time and doing so for each study. This technique indicates the importance of each individual study in the combined summary statistic to determine whether any of the studies has a disproportionate effect (Olkin 1994).

Using the combined-effect measurement, one can assess how large a study should be to balance the average of reported results, if the variation between study results is due solely to random fluctuations. The size of a single study needed to give a null summary statistic (i.e., an odds ratio of 1.0) can be determined, assuming that the hypothetical study had equal numbers of cases and controls, had an exposure prevalence equal to that observed in reported studies, and had an odds ratio equal to the reciprocal of the reported average. Unlike the fail-safe N, this calculation uses the size of the effect measurement, weights each study result by the inverse of its variance, hypothesizes a study with a protective rather than a null effect, and seeks a null rather than a nonsignificant combined effect.

Results

The results for all the analyses described would constitute over a dozen separate tables. Rather than provide all these data, two tables are presented that describe the data used in the analysis, two tables provide a sample of the results, and a single table summarizes the results of all the selection and exposure definitions in the studies considered. The full set of 16 studies of residential exposure and childhood cancer was considered earlier and is shown in Table A5-1. From these studies, the actual data used in the meta-analyses are shown in Table 5-1. Results of two particular calculations are shown in Tables 5-2 and 5-3. Summaries of all the calculations are provided in Table 5-4, and particularly interesting aspects of the results are described below.

Results of analyses of data from individual studies and selected meta-analyses are shown in Figure 5-2. One striking observation is the preponderance of dots

TABLE 5-2 Wire Codes (Low-Current Configuration)

Study	Exposure Definition	Exposed Cases	Expected Exposed Cases	Drop 1 Combined p's	Individual OR	OR$_{\text{fixed effects}}$	Pr{Q$_{\text{het}}$}	Or$_{\text{random effects}}$
All combined		288	199.31	0.00		1.48 (1.18-1.85)	0.08	1.52 (1.08-2.14)
Wertheimer and Leeper 1979	LCC (birth)	52	22.77	0.01	2.28 (1.34-3.91)	1.35 (1.06-1.73)	0.16	1.36 (0.97-1.91)
Fulton et al. 1980	LCC	87	86.50	0.00	1.00 (0.67-1.49)	1.78 (1.36-2.33)	0.55	1.78 (1.36-2.33)
Savitz et al. 1988	LCC	27	17.58	0.00	1.54 (0.90-2.63)	1.47 (1.15-1.88)	0.04	1.53 (0.97-2.42)
London et al. 1991	LCC	122	72.46	0.00	1.68 (1.14-2.48)	1.39 (1.05-1.82)	0.05	1.48 (0.91-2.42)

OR = odds ratio.

TABLE 5-3 Wire Codes (Low-Current Configuration) and Distance (<100 m)

Study	Exposure Definition	Exposed Cases	Expected Exposed Cases	Drop 1 Combined p's	Individual OR	OR$_{fixed\ effects}$	Pr{Q$_{het}$}	OR$_{random\ effects}$
All combined		448	349.04	0.00		1.36 (1.13-1.63)	0.11	1.38 (1.08-1.76)
Wertheimer and Leeper 1979	LCC (birth)	52	22.77	0.01	2.28 (1.34-3.91)	1.27 (1.05-1.54)	0.26	1.28 (1.02-1.60)
Fulton et al. 1980	LCC	87	86.50	0.00	1.00 (0.67-1.49)	1.47 (1.20-1.80)	0.18	1.47 (1.14-1.90)
Savitz et al. 1988	LCC	27	17.58	0.00	1.54 (0.90-2.63)	1.34 (1.10-1.62)	0.07	1.37 (1.03-1.80)
Coleman et al. 1989	<100 m	36	38.77	0.00	0.93 (0.54-1.60)	1.42 (1.18-1.72)	0.14	1.46 (1.13-1.87)
London et al. 1991	LCC	122	72.46	0.00	1.68 (1.14-2.48)	1.28 (1.04-1.57)	0.11	1.33 (1.01-1.75)
Feychting and Ahlbom 1993	<100 m	12	7.42	0.00	1.62 (0.79-3.30)	1.34 (1.11-1.62)	0.07	1.36 (1.04-1.78)
Fajardo-Gutierrez et al. 1993	<20 m	16	7.54	0.00	2.12 (0.85-5.29)	1.33 (1.11-1.60)	0.09	1.34 (1.04-1.73)
Petridou et al. 1993	<50 m	96	96.00	0.00	1.00 (0.62-1.62)	1.43 (1.17-1.73)	0.12	1.45 (1.12-1.89)

NOTE: Sample size needed to balance observed results = 2,005.

TABLE 5-4 Summary of Alternative Meta-Analyses

Data Set: Cut Point	Number of Studies	Proportion Exposed	Vote Counting Number Positive	Vote Counting Number Statistically Significant	Combined p values Pr{P}	Combined Or_fixed effects All Studies	Combined Or_fixed effects Drop 1 Study	Combined Or_fixed effects Size Needed	Combined Or_fixed effects Pr{Q_Het}	Combined Or_random effects All Studies	Combined Or_random effects Drop 1 Study
Wire codes (LCC)	4	0.40	3 (75%)	2 (50%)	<0.01	1.48 (1.18-1.85)	1.35-1.78	1,294	0.08	1.52 (1.08-2.14)	1.36-1.81
Wire codes (HCC)	4	0.13	3 (75%)	1 (25%)	0.02	1.34 (0.97-1.85)	1.12-1.85	1,319	0.18	1.42 (0.90-2.24)	1.21-1.77
Wire codes (LCC), distance (<100 m)	8	0.39	5 (63%)	2 (25%)	<0.01	1.36 (1.13-1.63)	1.27-1.47	2,005	0.11	1.38 (1.08-1.76)	1.28-1.46
Wire codes (HCC), distance (<50 m)	8	0.19	6 (75%)	2 (25%)	<0.01	1.38 (1.09-1.75)	1.28-1.58	1,838	0.18	1.47 (1.09-1.98)	1.36-1.61
Spot measurement (≥0.2 µT)	4	0.12	2 (50%)	0 (0%)	0.67	0.92 (0.57-1.49)	0.78-1.13	648	0.26	0.89 (0.51-1.57)	0.74-1.08
p value (smallest)	10	0.17	9 (90%)	3 (30%)	<0.01	1.54 (1.24-1.92)	1.48-1.72	2,327	0.01	1.69 (1.16-2.48)	1.54-1.85
p value (largest)	10	0.15	3 (30%)	0 (0%)	0.89	0.90 (0.73-1.12)	0.88-0.94	2,587	<0.01	0.88 (0.08-9.25)	0.82-0.98
Exposure (highest)	10	0.06	8 (80%)	2 (20%)	<0.01	1.37 (1.07-1.75)	1.46-1.72	4,628	0.07	1.45 (1.01-2.08)	1.34-1.60

NOTE: Because the studies used different exposure metrics none of the groupings contained all 11 studies.

141

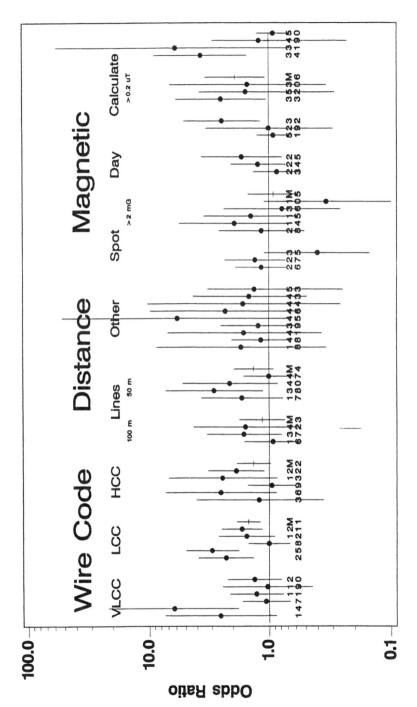

FIGURE 5-2   Odds ratios (dots) and 95% confidence intervals (vertical lines) for each of the dichotomous cut points of each exposure metric of each study as listed in Table 5-1.

(odds ratios) at or above the null-effect line. Only 8 of the 53 odds-ratio dots fall below the null effect; however, it is interesting to note that four of those that do so are for measured magnetic fields. This unweighted vote-counting assessment strongly suggests an association with some feature of the power transmission and distribution system because of a small but consistent positive odds ratio.

Wire Codes   The four wire-code studies were placed in two categories: one using low-current configuration (LCC) as the dichotomous cut point (i.e., ≤ LCC versus > LCC) and the other using high-current configuration (HCC) as the dichotomous cut point (< HCC versus > HCC). The results for the LCC calculations are shown in Table 5-2. Some of the published studies presented more detailed categorization; however, those two categories were the only ones that could be determined for all the studies in which wire codes were assessed. Odds ratios based on the wire codes were fairly similar, although, surprisingly, stronger positive odds ratios were shown on average in the LCC analysis than in the HCC analysis.

Distance Measurements   Five studies reported distance measurements as a metric of exposure. These studies, however, used different distance cut points and measured from different types of electric-power distribution sources. The most commonly reported distance cut points used in these studies were 50 m and 100 m. At each cut point, the proportion of exposed individuals within each of these studies was similar, but the similarity was largely due to inclusion of the study of Petridou et al. (1993) in the 50-m category. Exclusion of this single study brought the proportion exposed in the 50-m category down to 10%, substantially less than in the 100-m category in which 29% were exposed. Stronger positive associations were shown in the high-exposure category (i.e., 50 m) than in the low-exposure category, as would be expected if such an association exists.

Wire Codes and Distance Measurements   Because distance is an important component of the wire-coding scheme, it seemed appropriate to consider both exposure metrics together. To explore this grouping, the LCC data were combined with the 100-m and 50-m cut-point studies, and the HCC data were combined with the 50-m cut point. Where data were available for both distance and wire codes, wire codes were used. Overall, seven studies were available; the results for the combination of LCC wire codes and 100-m distance cut points are shown in Table 5-3. All three combinations gave fairly similar results, showing statistically significantly increased odds ratios, moderate insensitivity to single-study deletions, and large fail-safe N's and sample sizes needed to balance the observed data, indicating substantial robustness of results from additional studies.

Spot Measurements   Spot measurements were reported in four studies. When possible, a similar cut point of 0.2 μT (2.0 mG) was used across all studies for the meta-analyses based on this metric. However, Tomenius (1986) reported the number of subjects above and below 0.3 μT (3.0 mG); therefore, this cut point was used for Tomenius's data. Savitz et al. (1988) reported the number of subjects above and below 0.2 μT with all the appliances turned on (high power) and with all the appliances turned off (low power). For this study, the low-power data were used. London et al. (1991) used exposure quartiles to determine cut points; this resulted in cut points at 0.031 μT, 0.067 μT, and 0.124 μT (0.31 mG, 0.67 mG, and 1.24 mG). For their data, the highest exposure cut point was used. Feychting and Ahlbom (1993) reported data for two exposure cut points—0.1 μT and 0.2 μT. The highest exposure cut point was used for this analysis.

Calculations Three Nordic studies (Olsen et al. 1993; Verkasalo et al. 1993; Feychting and Ahlbom 1993) reported calculations based on historical power use to estimate the magnetic fields in the nearby residences. Note that because proximity to power distribution lines is a key factor in the calculated historical magnetic fields, this method of estimating exposure shares features with wire codes and distance measurements. That is, distance from power lines is an explicit factor in calculating wire code, distance measurements, and historical exposure, but not in spot measurements. The committee's analysis of those studies mirrors that of Ahlbom et al. (1993), which shows an association between exposure and childhood leukemia; cut points of 0.2 μT or 0.25 μT were used.

Combining All Studies Combining all studies using the smallest p value, which biases the result toward the strongest positive association, showed statistically significant heterogeneity and a statistically significant increased odds ratio under the random-effects model. The single largest amount of heterogeneity is attributable to the study of Tomenius (1986). Combining all studies using the largest p value, which biases the result toward the strongest null association, showed homogeneity, and the effect was not statistically significant. The result is far less extreme than that found for the smallest p values. Combining the data using the cut point and metric resulting in the highest exposure showed heterogeneity and a statistically significant increased odds ratio.

Discussion

The purpose of this analysis has been to evaluate the role of random variation in explaining the results observed in the set of epidemiologic studies examining residential magnetic-field exposure and childhood leukemia. When looked at in a variety of analyses, the positive trend in the associations cannot be explained statistically on the basis of random fluctuations, and the results of the combined data show little sensitivity to how they were grouped. Whether the risk association is actually attributable to the magnetic-field exposure or some other factor is not clear. How large the risk might be, if it indeed exists, is also not clear. However, the results of the residential exposure studies to date present a fairly uniform picture supporting an association of childhood leukemia with wire codes, distance from source, and for the three Nordic studies, calculated fields based on historical records of power consumption. Further, as evidenced by the results of the influence analyses, in most instances, deletion of a single study, such as the sentinel study, has little impact on the overall results. It would take a relatively large number of studies with largely negative results to balance this effect to the null. However, the inconsistency of results and the lack of a positive association when spot measurements of magnetic fields are used remain an enigma.

In the graph of the odds ratios (Figure 5-2), the preponderance of positive associations across exposure metrics (except for spot magnetic-field measurements) and exposure cut points speaks to the overall consistency of the data. Although there are issues of independence of results, weighting of multiple results from the same study, and possible biases in each study, the pattern is not random.

In all but 1 of the 12 ways in which the studies were grouped, at least half of them, and often substantially more than half, had increased odds ratios (Table 5-4). If chance alone accounted for the associations seen, one would expect a balance between those showing positive and inverse risks. In 9 of 12 vote-counting analyses, the number of statistically significant odds ratios exceeded the 5% expected by chance. Eight of these nine had at least 25% statistically significant results, and five had over 30% statistically significant results. The analyses that had no statistically significant results were one group of studies using distance from electric source, spot magnetic-field measurements, and one using the largest *p* values. The results of the combined *p*-value test are similar to those of the vote-counting analyses. That is, the same 9 of 12 groups of studies showed statistically significant results for each group as a whole. This suggests that the nine groups with statistically significant results, if not aggregated by a biased method, represent significant associations between the exposure characterizations and the incidence of childhood leukemia. These results, however, were quite sensitive to deletion of individual studies. That can be ascribed to sample size in part, but the data are strongly suggestive but not conclusive of an association.

The results of the combined odds-ratio assessments using the fixed-effects model also show a similar pattern. The risks in 10 of the 12 study groups showed increased odds ratios ranging from 1.14 to 1.90, with the 95% confidence intervals of 7 of the 12 studies excluding the null value of 1.0. These results were not sensitive to the deletion of a single study, as the odds ratio remained larger than 1.0 in each instance. The two groups of studies that did not show an increased risk were those using the spot magnetic-field measurements and those with the largest *p* values. Even so, both were sensitive to deletion of a single study, resulting in an odds ratio greater than 1.0 in some instances.

The random-effects model gave largely similar results. The combination of all studies using the smallest *p* values and the highest exposures showed statistically significant heterogeneity. This result is probably partly due to the different exposure metrics used in studies that likely reflect different aspects of magnetic-field exposure. When separated by exposure metric, none of the combinations shows statistically significant heterogeneity. This finding suggests that statistics summarizing all studies might be misleading. The spot-measurement results are distinct from all other exposure metrics, as would be expected from the discussion above.

One additional way to assess the robustness of results is to determine how large a study must be so that the results of previous studies are fully balanced to yield a combined risk estimate of 1.0. The data for these evaluations show that it would require over 1,000 subjects (500 cases and 500 controls) to refute the analyses of any of the studies (except those that used spot measurements) and over 2,000 for many of the groups of studies evaluated. Some might view this assessment as artificial (i.e., why should the odds ratio be equal and opposite to that observed?), but the number needed is useful in assessing the likely impact of a future study on the combined results. Even a regional study would likely have fewer than 1,000 cases; therefore, another case-control study is unlikely to change the results of a meta-analysis. Unless a new study has a major design innovation or the odds ratio is markedly protective (or quite large), an additional study is unlikely to resolve this controversy.

The largest sample size required for a study that might have the power to refute the existing data is the size in the combined study using the highest exposure cut point in each study. This observation is particularly noteworthy because various researchers have reported anecdotally that the individuals with the highest exposure often seem to drive the study results. Typically, a single or a few subjects have notably higher exposures than others in the study, and an unexpectedly large proportion of these are in case-control studies. The inference from this observation is that if another study were to be conducted, it would be most informative to conduct a study focusing on individuals with high exposure.

Similarly, the fail-safe N's for the results suggest that if the observed excess was due to publication bias, at least a dozen unpublished studies with an inverse association would be required to refute the published data except for those groups with very few studies. In view of the strong interest in this topic among scientists, it seems unlikely that any investigator would have trouble getting even a negative study published. Indeed, 2 of the 12 published studies reported odds ratios less than 1.0, and 8 of 12 did not have p values less than or equal to 0.05 for any exposure cut points, suggesting that negative and nonsignificant results are readily publishable. Nonetheless, the predominance of positive results persists.

Two contradictions remain: (1) Why do spot measurements of magnetic-field strength not show an association with leukemia, even though all the other exposure metrics do? and (2) Why do the data not show a consistent dose-response relationship? With respect to the spot measurements, if the results of the other exposure metrics are not spurious, then at least two potential explanations are possible. First, the surrogate exposure metrics might be markers for the true risk factor and the true risk factor might not be related to magnetic-field strength.

A number of plausible risk factors have been investigated, but none can be used to explain the observed association. Second, the other surrogate exposure metrics might be more biologically relevant measurements of magnetic-field exposure than spot measurements. The surrogate metrics might be more representative of long-term integrated averages of magnetic-field strength or of some other aspect of magnetic-field exposure that is related to the cause of the disease (e.g., peak field strength, field variability, or time above a specific threshold value).

With respect to the inconsistent patterns of dose-response relationship, it might be that the surrogate metrics used are correlated but not perfectly (as certainly must be true) with the true risk factor, whether it be some risk related to the electric-power delivery system or not, and the resulting misclassification gives inconsistent results that on average are positive. Further, wire codes differ among study locations, as evidenced by different magnetic-field strengths reported for the same wire code in different locations (Savitz et al. 1988; London et al. 1991). Finally, the spot-measurement data have various methodologic limitations. For example, in the Savitz et al. (1988) study, the proportion of subjects for whom measurement data were available was only about one third of that for whom the wire-code data were obtained. For the Tomenius (1986) study, measurements were made at the door step rather than inside the house.

The data in hand are not sufficient for the committee to investigate all the alternative exposure metrics or explanations that might lead to understanding the associations observed. The explanations that have been posed generally are post hoc and, therefore, should be used as hypotheses for additional analyses rather than as a priori hypothesis tests. Thus, the finding remains that there are strong and consistent data suggesting a relatively weak increased risk of leukemia for children living in close proximity to power lines. Studies that would best advance understanding of this association would address the inconsistencies in studies and not duplicate studies with simply more precise measurements of exposure and outcome, as suggested by Washburn et al. (1994). The sample-size analyses show that little insight is likely to be provided unless a study of hundreds if not thousands of leukemia cases is undertaken.

Selection Bias and Control Selection in Residential Childhood Cancer Studies

In the absence of any true association between wire codes and cancer, faulty methods of case or control selection in case-control studies can yield spurious associations. If, for some reason, controls in studies of childhood cancer and wire codes were consistently selected in a manner that underrepresented high-wire-code residents, then those studies would consistently yield spuriously increased measurements of association.

In designing an epidemiologic study, the study groups to be compared (either cases and controls in a case-control study or exposed and unexposed in a cohort study) must accurately represent the population from which they are drawn (exposure prevalence in a case-control study or disease risk in a cohort study). The criteria for defining the groups, the methods of evaluating their representativeness, and the assessment of bias in study results depend on the design of the study. In case-control studies, subjects are selected with respect to disease status. If the selection process generates an over- or underrepresentation of exposed cases or exposed controls, a bias will be produced in the measurement of association between disease status and exposure status. In cohort studies, subjects are selected on the basis of exposure status at a time when all subjects are disease free; they are then followed over many years. If the completeness of follow-up differs depending on exposure and disease status, bias might result.

Specific aspects of the residential childhood cancer studies that might have led to bias (Tables 5-5 and 5-6) are considered here. The discussion provides an inventory of possible sources of bias but does not suggest that such bias actually exists, at least not of the magnitude to explain the previously reported associations or lack of associations. The focus is on four main issues: (1) representativeness of cases in case-control study, (2) representativeness of controls in case-control studies, (3) participation (nonresponse) rates, and (4) differential mobility of study subjects.

Representativeness of Cases   Three epidemiologic designs have been used in the studies of residential exposure and childhood cancer. A standard case-control design was used in the first type of study (Wertheimer and Leeper 1979; Fulton et al. 1980; Tomenius 1986; Savitz et al. 1988; Coleman et al. 1989; Myers et al. 1990; London et al. 1991; Olsen et al. 1993). Children with cancer and children without cancer were identified, and their past exposures were estimated. Cases were selected from population-based or hospital-based registries (e.g., birth certificates), by a random-digit-dialing procedure, or from referrals to friends of the cases.

A nested case-control design was used in the second type of study (Feychting and Ahlbom 1993). A cohort of exposed and unexposed individuals was identified. From this cohort, cases and controls were chosen independent of their exposure status and then compared with respect to their exposure status, as in a traditional case-control study.

The third type of study was a historical cohort study (McDowall 1986; Schreiber et al. 1993; Verkasalo et al. 1993). In these studies, investigators identified all persons living near the electric transmission facilities as their cohort and compared their mortality experience to that of the national population.

The case-control studies used registry records to identify all cases diagnosed or dying during a certain time, living in a specified geographic region, and in a specified age range (Table 5-5). Three data sources were used in the studies. Wertheimer and Leeper (1979) and McDowall (1986) used death certificates. One concern with that source is that disease survival would affect identification (e.g., nonfatal leukemia would not be identified). Because diagnosis and treatment

TABLE 5-5 Subject Selection in Residential Childhood Cancer Studies

Design	Study	Source of Subjects		Eligibility			Matching Criteria
		Cases	Controls	Years	Location	Ages	
Case control	Wertheimer and Leeper 1979	Death certificates	Next birth certificate unless a sibling	1950-1973	Colorado birth and Greater Denver resident 1946-1973	<19	File 1: month and county of birth File 2: alphabetic, 5-20 yr range (not sibling)
Case control	Fulton et al. 1980	Rhode Island hospital incidence registry	Birth certificate	1964-1978	Rhode Island resident for 8 yr before diagnosis	<21	Birth year
Case control	Tomenius 1986	Population-based cancer registry	Nearest birth certificate in parish records	1958-1973	Born and diagnosed in Stockholm County	<19	Age, gender, church district of birth, and church district of diagnosis if same as birth for case
Case control	Savitz et al. 1988	Population-based cancer registry and hospital records	Random digit dialing	1976-1983	Denver SMSA	<15	Age ±3 yr, gender, telephone exchange at time of diagnosis of case
Case control	Coleman et al. 1989	Population-based cancer registry (leukemia only)	(1) Solid tumors (not lymphoma); (2) random from Bromley Electoral Roll in 1975	1965-1980	Four London boroughs	(1) All, (2) >17	Age, gender, year of diagnosis, residence
Case control	Myers et al. 1990	Population-based cancer registry and other sources	Nearest birth certificate	1970-1979	Yorkshire health region	<15	Gender, birth in same local area or health district, year of diagnosis

Case control	London et al. 1991	Population-based cancer registry (leukemia)	Friends (first 65) and random digit dialing	1980-1987	Los Angeles County	<11	For most, age ±1-3 yr depending on age, gender, ethnicity
Case control	Olsen et al. 1993	Population-based cancer registry (leukemia, CNS tumor, lymphoma)	Population registry	1968-1986	Denmark	<15	Gender, date of birth ±1 yr
Nested case control	Feychting and Ahlbom 1993	Population-based cancer registry	Population registry	1960-1985	Residence within 300 m of any 220 kV of 400 kV power line in Sweden	<16	For most, in registry during year of diagnosis, birth year, gender, residence in same parish in year of diagnosis or move, near same power line
Historical cohort	Verkasalo et al. 1993	National population registry	National population registry	1970-1989	Residence within 50 m of overhead power lines	<20	5-yr age groups

TABLE 5-6 Number of Subjects in Residential Childhood Cancer Studies

Study	Number of Subjects		Excluded		Moved		Comments
	Cases	Controls	Cases	Controls	Cases	Controls	
Wertheimer and Leeper 1979	344	344	16, no birth address; 72, no birth address		147	128	% HCC of File 1 approximately equal % HCC of File 2
Fulton et al. 1980	119	240	9	15	53		
Tomenius 1986	746 (56 benign)	716	29, bad data; 1, not primary tumor; 43 of 1,172 dwellings	46 of 1,015 dwelling	316		
Savitz et al. 1988	356	278	104 interviews; 228 measurements; 36 wire codes	56 interviews; 71 measurements; 19 wire codes; (78.9% response rate in RDD)	256		
Coleman et al. 1989	811 (84 under age 18	1,614 (141 under age 18), 254	40	18, 223			
Myers et al. 1990	419	656	45	68			
London et al. 1991	331	257	99 interviews; 162 measurements; 112 wire codes	24 interviews; 108 measurements; 50 wire codes (82% response rate for RDD)	57%	66%	4,424 phone numbers resulted in 113 eligible controls

Olsen et al. 1993	1,707	4,788	0	0	1,050	3,125
Feychting and Ahlbom 1993	142	558	1 calculation; 53 measurements	4 calculations; 214 measurements		
Verkasalo et al. 1993	140					

NOTE: RDD = random digit dialing.

can be related to socioeconomic status, which in turn can be related to proximity to power lines, a bias could have resulted. Fulton et al. (1980) used a hospital cancer registry. However, patients at a specific hospital might not accurately represent the total pool of cases in the general population.

The rest of the case-control studies used population-based registries to identify cancer cases (Tomenius 1986; Savitz et al. 1988; Coleman et al. 1989; Myers et al. 1990; London et al. 1991; Olsen et al. 1993; Feychting and Ahlbom 1993). Bias is unlikely to play a large role in these studies provided that ascertainment of cases in the registries is sufficiently high. Finally, two of the cohort studies used population registries to identify cases (Schreiber et al. 1993; Verkasalo et al. 1993). Provided ascertainment is complete, these registries are optimal data sources. Although the case-selection process had minor variations, the methods of most of the studies would be expected to generate reasonably representative study groups.

The age ranges varied among the studies; upper-age bounds for analyses of childhood cancers ranged from 10 to 20. Those ranges are all representative of children with cancer, with perhaps some dilution of cases at older ages. The variation is unlikely to have caused a substantial variation among the studies because of the large overlap of age ranges.

Representativeness of Controls   In the case-control studies, controls were selected in a variety of ways. Some studies used regional birth-certificate files (Wertheimer and Leeper 1979; Fulton et al. 1980; Tomenius 1986; Myers et al. 1990), which limits subjects to those who were both born and diagnosed (or selected) in the same region. In the study of Fulton et al. (1980), however, cases were drawn from a hospital population, and controls were selected from the general population listed in the birth-certificate records, introducing a potential disparity between cases and controls. Further, Wertheimer and Leeper (1980) have argued that because cases were required to live in Rhode Island for 8 years before diagnosis and controls had only to be born in Rhode Island, the control selection might have introduced bias.

Other studies used random digit dialing to identify controls (Savitz et al. 1988; London et al. 1991). Random digit dialing is a method designed to identify a set of controls that comes from a defined geographic region (Waksberg 1978; Robison and Daigle 1984; Ward et al. 1984; Voigt et al. 1992). Although the exact methods vary, in general the procedure is for the investigator to take the case's phone number, discard the last two digits, and replace them with two randomly chosen digits. This number is called; if it is not a residence, another pair of random digits is used; if it is a residence, the resident is asked if a person meeting the matching criteria resides there. If so, that person is recruited as a control. If not, another pair of random digits is used. This process, applied to a childhood study matching gender, age, and ethnicity, typically requires between 25 and 75 phone calls per case to identify an eligible control.

One limitation of random digit dialing is that it samples only homes with telephones. It is important to determine what proportion of residences have telephones and, if possible, to compare those that have telephones with those that do not. Further, even if an individual has a telephone, they might not be easily accessible. Poole and Trichopoulos (1991) argue that people of very low socioeconomic status might be harder to reach by this method and thus might be underrepresented in the sample.

It should also be noted that the initial sampling unit with the random-digit-dialing method is the residence rather than the individual. Each residence is reached via telephone rather than each individual child, as is done in a registry-sampling procedure. If a residence does not have a telephone, none of the children in that residence can be selected. Controls should provide an estimate of the exposure distribution in the population from which the cases were drawn (Rothman 1986) or an estimate of the exposure rate that would have been observed in the cases if no association existed between the exposure under study and disease (Schlesselman 1982). Controls selected by random digit dialing are likely to achieve this goal to a substantial degree, but some potential limitations are apparent in that mode of sampling from the population.

One study combined random digit dialing with use of friends of cases as controls (London et al. 1991). A case was asked to name a close friend who could be recruited for inclusion in the study. Although the rationale of combining that method with random digit dialing was not provided by the study, it was partly due to logistic considerations in adding on to a previously conducted smaller case-control study. One of the problems in using friend controls is that they might be overmatched (Kelsey et al. 1986); that is, they might be similar to the cases solely because they are friends. Precision will therefore be reduced, because the exposures will tend to be very similar and fewer will be discordant pairs. It is also possible that the case exercised some type of selection bias, such as selecting the friend who is most talkative or outgoing.

One study used controls with cancers other than leukemia and lymphoma and controls from the local electoral roll (Coleman et al. 1989). Both of these sources of controls have potential problems. Other cancers might lead to a negative bias (reduction in the possible association) because cancers other than lymphatic and hematopoietic cancers (e.g., brain cancer) might be associated with exposure to magnetic fields. The electoral roll, which was not used for the childhood portion of the study, might not include all persons living in a specific region and might be an ethnically and socioeconomically biased sample from the population.

Finally, the nested case-control study and all cohort studies used a population registry to identify controls (McDowall 1986; Feychting and Ahlbom 1993; Olsen et al. 1993; Schreiber et al. 1993; Verkasalo et al. 1993). Again, if ascertainment of cases is complete and the population registry is complete, then these studies should be free of selection biases. Typically, at least in the United States, investigators do not have access to such convenient data bases for sampling from the general population.

Participation Rates   Another major concern in the case-control studies is the possible bias due to nonparticipation. If nonparticipation rates differ by exposure status only or by disease status only, no bias is produced in the odds ratio (Rothman 1986). However, if participation rates differ by both exposure and disease, a bias might be produced. The greater the magnitude of nonresponse, the more opportunity for differential nonresponse that would distort the measure of association. Although data on nonrespondents are insufficient, by definition, to determine directly whether bias is present, a sensitivity analysis can be conducted to determine the maximum amount of bias that could be present.

The potential for this problem can be examined (Table 5-6). Some of the studies show substantial numbers of exclusions or other losses from the original pool of eligible participants. Without characterizing these individuals, the degree of bias cannot be determined directly. However, a sensitivity analysis often is conducted to determine the size of an effect. In the worst case (e.g., all excluded controls were exposed), the resultant odds ratios would be substantially different from those reported. Lack of information on nonrespondents is not unusual in epidemiologic studies, and it could have a large effect on the results.

Mobility   Differential mobility of cases and controls has been raised as another possible bias in these studies. Although the data needed to evaluate this possibility is generally not published, some of the studies presented information on how many residences were occupied by each study subject (Table 5-6). Jones et al. (1993) argue that the observed associations between wire codes and childhood cancers in one study by Savitz et al. (1988) might have been due to bias induced by differential mobility (controls were required to be residentially stable but cases were not). In a study to investigate that phenomenon, Jones et al. (1993) found 31% more high wire codes in nonstable populations than in stable populations in Columbus, Ohio. Although that is a plausible source of bias, the quantitative impact even in the worst case would have been small.

Conclusions   Issues of selection bias have been raised in various reviews of these residential childhood cancer studies. In one review, the National Radiological Protection Board (NRPB 1992) raised issues of bias in the use of random digit dialing in both the Savitz et al. (1988) and the London et al. (1991) studies and suggested that random digit dialing resulted in an undersampling of controls with low income and in differential mobility between cases and controls in both studies.

In its report, the Oak Ridge Associated Universities (ORAU 1992) reviewed control-selection bias in studies it considered most important (Wertheimer and Leeper 1979; Savitz et al. 1988; London et al. 1991) and then reviewed bias in other studies (Tomenius 1986; Fulton et al. 1980; Coleman et al. 1989; Myers et al. 1990). The ORAU in its report argued that the control-selection procedure used by Wertheimer and Leeper (1979) was not defined with sufficient clarity for critical evaluation, although it acknowledged that the procedure contained no obvious bias. The ORAU report also noted that control-selection bias would be introduced in the study of Savitz et al. (1988) if exposure was related to residential mobility or the chance of being sampled as a control. The question of underrepresentation of those of lower socioeconomic status in random digit dialing was raised with reference to the Savitz et al. (1988) and London et al. (1991) studies. The ORAU report also questioned the representativeness of using friends as controls in the London et al. (1991) study.

For the other studies, the ORAU report identified issues of control-selection bias. For Fulton et al. (1980), it pointed out that cases had to remain near a specific hospital up to the time of diagnosis, but controls did not (residences anywhere in Rhode Island). Coleman et al. (1989) used cancer controls in their study, which might have biased the result downward because children with brain cancer, which is the second most common childhood cancer, were acceptable controls, and brain cancer might be associated with exposure to magnetic fields.

As with any epidemiologic study, the studies of residential magnetic-field exposure and childhood cancer have many possible sources of bias. Each of these possible errors could influence the size of the reported odds ratios, but none is likely to be present or sufficiently large across all the studies to explain the results. Rather, each possible bias might contribute in a small way to the odds ratio in each study, some tending to increase and some tending to decrease the value of the odds ratio determined. Because the study designs and methods are diverse and because no pervasive flaw is found in all of them, the committee believes that any particular selection bias is unlikely to completely explain the reported associations between exposure to magnetic fields, as reflected by the wire codes, and childhood cancer incidence.

Information Bias in Residential Childhood Cancer Studies

Another possible artifactual basis for the reported association between indicators of magnetic-field exposure and cancer is information bias, or misclassification. For example, if homes of children with cancer tended to be erroneously categorized high wire code or homes of control children tended to be erroneously categorized low wire code, then an association would be found in the study even if no association were truly present. This concern is distinct from that of whether the operational definition of exposure is valid (e.g., whether wire codes approximate individual magnetic-field exposure); rather, the concern is whether the wire code per se has been assessed accurately.

Case-control studies that rely on interviews to classify exposure are vulnerable to bias owing to differential recall of cases compared with controls (Rothman 1986). In comparison, exposure measurements used in studies of magnetic fields and cancer have been more objective and, therefore, less vulnerable to such errors. Except for the first study (Wertheimer and Leeper 1979), wire codes were ascertained blindly (i.e., without knowledge of whether the home had been occupied by a case or a control) (Savitz et al. 1988; London et al. 1991; Feychting and Ahlbom 1993). Legitimate concerns were raised in the study by Wertheimer and Leeper (1979) because the investigators did their own wire coding, yet they reported the process to be reliable when a technician reanalyzed the data blinded as to whether the homes were those of cases or controls.

In later studies, a member of the study team made the necessary judgments about such study parameters as the types of electric wiring and distance to electric-power-distribution components without having any other information about the occupant of that residence. Thus, any errors that might be introduced would be similar for cases and controls, and any resulting misclassification would very likely dilute rather than exaggerate the reported associations. The evidence that wire coding can be done quite reliably is good (Savitz et al. 1988; Dovan et al. 1993), so few errors are expected overall, and the opportunity for the errors to be differential is small.

Measurements of magnetic fields are typically not done blindly because contact with the occupants of the home is required. Nonetheless, the protocols for obtaining magnetic-field measurements leave little latitude for the data collector to deviate and intentionally or unintentionally bias the results. Thus, subject to the inherent uncertainties in the instrumentation and the ability to read it accurately, modest errors are expected at most.

Conclusions on an Association Between Electric Wiring Near Residences and Childhood Cancer

Given the preceding discussion, the following question must be considered: How likely is it that some indicators of potential exposure, such as proximity of wiring to residences or current carried in nearby transmission lines, are associated with a higher risk of childhood leukemia or other cancers? Note that the phrasing of this question intentionally ignores, for the moment, what agents, if any, might explain such an association but rather asks whether a study free of all methodologic error would find a statistical association between proximity to certain types of power lines and childhood cancer. The Wertheimer and Leeper (1979) wire code is the most common such measurement of the proximity of power lines, but other measures have also been used (Feychting and Ahlbom 1993; Kaune and Savitz 1994).

Rather than addressing individual studies, the body of research must be considered and a judgment made on whether an error-free design would be likely to yield a positive association. As usual, in epidemiology, one must ask what factors might have accounted for spurious associations in previous studies. A judgment that no association is actually present represents a judgment that some alternative explanation is more plausible. Thus, the assessment must include a careful and explicit consideration of the plausibility of alternative explanations for the evidence obtained to date.

Random Error   The possibility that random error has been misinterpreted as a positive association was explored through a meta-analysis of the epidemiologic data and found to be unlikely. Another mechanism to account for the results would be selective presentation or interpretation of results by investigators, highlighting only those that fit with the hypothesis of an increased risk. It is difficult to place much credence in the hypothesis that the array of reported positive associations are a product of random processes or a product of conscious investigator manipulation. Although the studies are not unanimous in identifying positive associations, most of the methodologically strongest studies do, so the overall weight of evidence supports a modest association.

Selection Bias   There are a number of important concerns about the manner in which controls were selected. These concerns have direct implications for the measurements of association, some being likely to produce bias toward spurious positive associations, such as selection of residentially stable controls who might tend to have lower wire codes (e.g., Savitz et al. 1988), and others being likely to produce bias toward spurious negative associations, such as selection of controls of a period that is earlier for controls than for cases, when high wire codes were more prevalent (e.g., Fulton et al. 1980). Each case-control study is vulnerable to selection bias, although the different approaches in the major studies would require that each study producing positive results would have the postulated bias that artificially increased the relative risk. The shared approach of random digit dialing (Savitz et al. 1988; London et al. 1991) makes the postulation of a common bias more plausible, but two other major studies (Wertheimer and Leeper 1979; Feychting and Ahlbom 1993) did not use that approach. Empirical efforts to characterize the potential bias yielded plausible and testable hypotheses regarding social class, nonresponse, and residential stability, but little direct support that the bias actually occurred.

Information Bias   The assignment of exposure has some element of subjectivity and, therefore, has the potential for bias. In the extreme case, the investigators might assemble all the data and also judge wiring configuration codes (Wertheimer and Leeper 1979) or read the field instrument to record the estimated magnetic-field exposure (Savitz et al. 1988). That procedure was a key concern for Wertheimer and Leeper (1979) in their first study, but in subsequent studies they made serious efforts to collect as much of the exposure data as possible while being unaware of the case-control status of the residences. That procedure does not ensure the absence of errors but makes it highly probable that such errors would be independent of case or control status. Thus, the results would likely be biased toward the null (Rothman 1986).

Summary   Overall, the body of research linking various aspects of wiring near residences to childhood cancer falls short of providing definitive evidence that an association exists, and even if such an association were proved, the causal agent has not been identified. Several specific biases have been postulated, particularly control selection and nonparticipation, and research is insufficient to discount them fully. Nonetheless, the study methods complement one another in that different designs have been used with divergent strengths and weaknesses. Control-selection bias is a major concern in some studies (e.g., Savitz et al. 1988; London et al. 1991) but not in others (e.g., Wertheimer and Leeper 1979; Feychting and Ahlbom 1993). Imprecision due to small numbers of subjects is particularly a problem in some studies (e.g., Feychting and Ahlbom 1993) but not in others (e.g., London et al. 1991). This pattern of results and the committee's analysis of these data suggest that an association is likely to be present and if a flawlessly designed and executed study could be conducted it would identify a positive association between indicators of exposure, such as the proximity of power lines to residences, and childhood cancer.

Is the Association of Cancer with Electric Wiring Near Residences Accounted for by Factors Other Than Magnetic Fields?

Wire Codes and Potential Confounders

Studies have found that wire codes are correlated with factors other than magnetic fields. Various attributes of the neighborhood, such as population density, the home, and its occupants, might well be associated with wire codes.

Socioeconomic Status (SES)   A widely held perception is that homes with higher wire codes are typically less desirable and therefore more likely to have occupants of lower income and social class. However, empirical evidence regarding social class in relation to wire configuration code is sparse. In the study by Savitz et al. (1988), education and income of residents served by underground wiring are clearly distinct (higher SES) from those of residents served by above-ground wiring, but little or no distinction was found among residents served by above-ground wiring.

Distance from wiring related to power distribution depends, in part, on lot size, and high density development is likely to produce higher wire-code categories for pole-mounted distribution wiring. These factors produce a likely connection between income level and socioeconomic status that could confound the effect of wire codes. Multifamily residences often have internal power-distribution wiring that can give rise to magnetic fields that cannot be captured conveniently by wire codes but might be associated with lower socioeconomic status and higher exposure.

Residential Mobility High wire codes in Columbus, Ohio (Jones et al. 1993) have also been associated with homes in which residents are more mobile. In Columbus, the inner city, the oldest region, had the highest proportion of high wire codes, and the suburbs, the newest region, had the lowest proportion of high wire codes. Low electric-energy consumption was also associated with high wire codes (Jones et al. 1993).

Age of Home Older homes tend to have higher wire codes, higher power-line fields, and higher total fields than new homes (EPRI 1993a; Bracken et al. 1992). This information is counter to the argument that average residential exposures have increased over time because of increased use of electricity (Jackson 1992; ORAU 1992). Also, within each wire-code category, the homes with the lowest magnetic fields tend to be the newest homes (less than 30 years old), and except for the very-high-current-configuration (VHCC) category, homes with the highest magnetic fields tend to be the oldest homes (more than 30 years old). Homes with VHCC, regardless of age, are likely to have fields above 0.1 μT (1 mG) (EPRI 1993a). Jones et al. (1993) reported that the inner city of Columbus, Ohio, has the highest proportion of high wire codes, and the suburbs have the highest proportion of low wire codes.

A possible reason for the trend of increasing exposure with increasing age of home could be a greater prevalence of knob and tube wiring in older homes, a practice that can cause higher magnetic fields in homes. Homes newer than 30 years old had no such wiring, but 28.8% of homes over 50 years old and 7.1% of those 30-50 year old did (EPRI 1993a). Homes might also accumulate a larger number of wiring errors as they age, and wiring practices in older homes might have permitted more wiring irregularities. Power lines might be more fully loaded in older neighborhoods. The lines are designed to accommodate the maximum capacity needed, and that might be reached more often in older neighborhoods than newer neighborhoods. Finally, the population movement out of central cities, with small yards and greater housing density, to suburban areas with greater average distance from power lines to residences, might result in reduced wire codes and reduced in-home exposures.

Because older homes tend to fall into the high-wire-code category more often than new homes, confounding might be introduced if a real or imposed disparity exists between the ages of homes of cases and controls. Wertheimer and Leeper (1980) noted that the Fulton et al. (1980) study design, which included dates of birth and diagnosis, generated more recent average occupancy dates for cases than for controls.[4] More recently constructed and occupied homes will

---

[4] Homes at the time of diagnosis were provided for cases, but only homes at the time of birth were available for controls.

have, on average, low wire codes; therefore, the odds ratios in that study would be expected to be biased downward.

Cancer and Potential Confounders

In epidemiologic studies of the association between exposure to power-frequency magnetic fields and cancer, the possibility of confounding by risk factors associated with magnetic-field-exposure indicators must be considered. To control for such confounding, the confounder must be identified, measured as one of the variables in the study, and accounted for in the statistical analysis.

Candidate confounders should be a known or suspected risk factor for childhood leukemia, but the causes of childhood leukemia are generally not known. Although a number of risk factors have been proposed, no consensus has been reached on which ones are sufficiently plausible to be worthy of consideration.

Several demographic markers of risk have been established. Acute lymphocytic leukemia is characterized by a marked age dependence: the incidence is low in the first year of life, increases to a maximum at age 2-3, and decreases down to near baseline by age 7-8 (see Figure 5-3) (Rubin 1983). This marked age dependence has been interpreted as indicative of an infectious origin to leukemia (MacMahon 1992), either due to a specific but unknown infectious agent or the combined effect of all infections since birth (Morris 1991). Studies

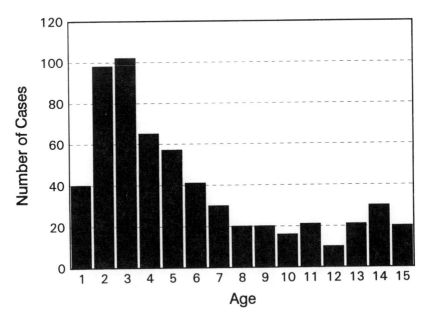

FIGURE 5-3   Age at incidence of childhood leukemia.

in Connecticut and Massachusetts showed that the incidence of acute lymphocytic leukemia varies markedly by year and location (Figure 5-4). An association of childhood leukemia with birth order was noted by MacMahon and Newell (1962), in which the firstborn has about double the risk of a fifthborn child. Studies in Massachusetts also suggested that the incidence of acute lymphocytic leukemia in children differs according to the mother's age at the birth of the child, so the excess risk is about 25% for a child born to a mother older than 35 years of age compared with a mother less than 20 years of age. In addition, several studies suggest increasing risk with increased socioeconomic status (Robison et al. 1991). All the observed variations in the incidence rates make it difficult to interpret small risk ratios in epidemiologic studies.

A number of exposures incurred in utero and postnatally have been implicated but not well established as causal factors, including maternal marijuana use (Robison et al. 1991), maternal tobacco use (John et al. 1991), and consumption of nitrites in processed meats (Peters et al. 1994). Among environmental agents, exposure to ionizing radiation is known to produce excess risk of leukemia in

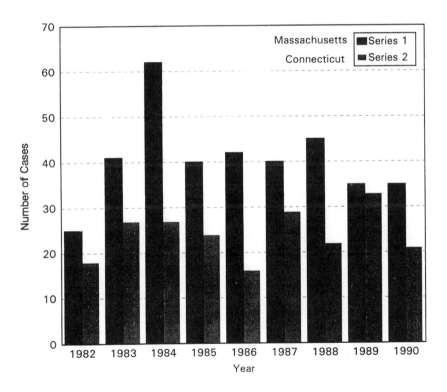

FIGURE 5-4   Incidence of acute lymphocytic leukemia in children in Massachusetts and Connecticut.

adults as well as children (Robison et al. 1991), but no other environmental determinants have been identified with certainty. Suspected causes include a wide range of occupational exposures to the father (Savitz and Chen 1990), including ionizing radiation, heavy metals, and solvents. Benzene, a known cause of leukemia in adults (usually acute myeloid leukemia), has not been directly examined as a potential cause of childhood leukemia (usually acute lymphocytic leukemia); the indirect evidence is based on an association of traffic density (an indicator of air quality) with childhood leukemia (Savitz and Feingold 1989). Traffic density at the location of a residence would have an effect on local air quality; such lowered air quality could be a risk factor for childhood leukemia and could be related to wire codes.

Evidence of Confounding in Previous Studies

If confounding is suspected or anticipated and if the confounding variable is measured in some way in the study, it is possible to control for confounding. If income or socioeconomic status is suspected to be a confounder in a study of the association of wire codes and childhood leukemia, the study subjects can be stratified by income, and the association between wire code and risk of childhood leukemia can then be evaluated separately for each income group. If several potential confounders need to be evaluated, the more usual method of controlling for confounders is to apply regression procedures.

In general, attempts to adjust for those confounders measured in studies of magnetic-field exposure and childhood leukemia appear to have produced evidence against the presence of substantial confounding. Savitz et al. (1988) carried out stratified analyses and found that differences in the crude and adjusted odds ratios were not large enough to include them in the published results. The factors they evaluated for confounding included income and education. London et al. (1991) found that adjustment for potential confounders reduced but did not eliminate the association they found between wire code and childhood leukemia. Feychting and Ahlbom (1993) adjusted their odds ratios of leukemia risk for various demographic variables, socioeconomic status, and concentrations of nitrogen dioxide in the outdoor atmosphere and found that the adjusted results were virtually identical to the unadjusted results.

Conclusions About Confounding

In observational studies that lack randomization, confounding by some as yet unknown risk factor can never be fully disproved as a spurious source of results. Imaginative critics can suggest possible but currently unestablished causes of childhood leukemia and other cancers that are closely associated with wire codes. Those causes that can be postulated and deserve serious attention should be incorporated into future studies to evaluate their impact on the wire-code and cancer association. This search is severely hampered by the scarcity of established or even strongly suspected causes of childhood leukemia. This ignorance should not be misinterpreted as evidence against confounding, which can only be produced by identification, measurement, and adjustment for such causes.

Among the sociodemographic and environmental factors considered previously, none has been identified as a strong confounder on empirical or theoretic grounds. A leading contender is socioeconomic status (SES) or some correlated characteristic, yet neither the association between SES and childhood cancer nor the association between SES and wire code is very strong. Air pollution related to traffic density remains a candidate (Savitz and Feingold 1989), yet empirical evaluation has not shown this to be a source of confounding in one study (Savitz et al. 1988), and at present, little evidence exists to establish air pollution as a risk factor for childhood cancer. Nevertheless, a more accurate indicator of air pollution from motor-vehicle traffic might yield a different conclusion.

At present, confounding remains a possible explanation for the wire code and cancer association. However, past efforts to identify such confounders have failed, and few strong candidates can be postulated at present. Either the evidence of an association of possible confounders with cancer is weak, as in the case of age of home, or evidence against an association is fairly good, as in the case of parental smoking.

Is the Association with Cancer Accounted for by Magnetic Fields?

Review of Exposure-Assessment Methods in Residential Studies

A variety of approaches were used in past epidemiologic studies to assess potential magnetic-field exposure; the approaches were largely based on the description of nearby power lines or on spot measurements of fields in and around the home. To interpret past studies, it is useful to identify what causative factors might be represented by the exposure indicators used in the studies. To the extent that associations are found between possible indicators of exposure and cancer, the challenge is to determine the actual exposure responsible for the association. Most epidemiologic studies examine several exposure metrics, and some studies have been reanalyzed with several metrics, so some inferences concerning the appropriateness of the different indicators might be possible. All indicators used, including wire codes and field measurements, are surrogates for some unknown "causal exposure" (if one is indeed present) and must be considered with regard to their effectiveness as substitute measurements for that unknown exposure. In all instances, the exposures of interest took place in the past, so the relationship between contemporary and historical indicators of exposures is of interest. Thus, all other things being equal, the more applicable an exposure indicator is to historical periods, the more useful it is.

To evaluate the hypothesis that the causal exposure is some aspect of power-frequency magnetic fields, several steps should be taken in analyzing the exposure-assessment methods. First, the exposure indicators can be examined to determine what aspects of the magnetic field they might represent. A comprehensive analysis of that sort is impossible because of the virtually infinite number of field descriptors that could be considered. For power-frequency magnetic fields alone, descriptors include annual average field, annual peak field, time above some field-strength value, cumulative dose, and field variability. Largely because of unavailable instrumentation, most researchers to date have considered only the average magnetic field over some period as the exposure of interest, although some work has been done with other field metrics, and other research is under way to examine additional electric- and magnetic-field descriptors. As noted by Bracken et al. (1993), "Without a clear understanding of the mechanism of interaction relevant to health consequences, selection of an exposure metric for assessment purposes is arbitrary." Some research has shown that many of the possible descriptors are correlated with each other, suggesting that whatever descriptor is used, it may be considered a surrogate for the "true metric" (Theriault et al. 1994). Many descriptors may be correlated, but some will capture a unique aspect of the field that might not be present in other descriptors. So a particular metric (such as mean or cumulative exposure) might not be a viable surrogate for the true metric or causal exposure.

After identifying the exposure of ultimate interest, research needs to examine the accuracy and specificity with which the available exposure indicator approximates the candidate exposure. Accuracy and the potential for bias can be influenced by whether the methods of calculation or scoring are objective, capable of reflecting the measurements of ultimate interest, and consistent across studies, and whether the exposure assessment was done blindly. For example, a method of calculation that arithmetically adds the contributions of the several transmission wires (Fulton et al. 1980) is likely to overestimate the field to an extent that depends strongly on transmission-line configuration. Using different methods to estimate transmission-line load for different time periods (Verkasalo et al. 1993) might result in inconsistencies. If total accumulated exposure is of interest, a deficiency common to all the studies is the failure of the exposure metrics to incorporate away-from-home exposures, although some studies suggest that away-from-home exposure might not contribute much to total exposure (Kaune et al. 1994).

To the extent that the exposure indicator imperfectly reflects the relevant exposure and assuming that the imperfection is similar for cancer cases and controls, the measurement of any real association obtained from a dichotomous exposure comparison will be diluted (Rothman 1986). For example, including in the exposed group all dwellings within 150 m of substations, transformers, and electric railroads and subways (Tomenius 1986) or including twisted first-span secondaries (Wertheimer and Leeper 1979) would likely cause the exposed group to include dwellings without increased magnetic fields. On the other hand, most dwellings within 40 m of 200-kV transmission lines, or near open first-span secondaries, would be likely to have increased magnetic fields. Because power-line fields vary greatly as a function of time, spot measurements, or short-term averages, are also imprecise measurements of the long-term average field. When the quality of these exposure indicators can be ranked, there is an opportunity to examine whether better indicators are more closely associated with cancer risk than poorer indicators, as would be expected if better indicators are subject to less misclassification and a true causal association is present. Table 5-7 summarizes the exposure metrics used in each of the childhood-cancer studies with regard to accuracy, blindedness, and specificity. Because wire coding and contemporary spot measurements are used so often in the studies, the logic and evidence of adequacy for these two metrics warrants a detailed examination. This examination is given in Appendix B.

Evidence Linking Magnetic Fields to Cancer

Epidemiologic research on indicators of magnetic-field exposure other than wire codes is needed to help assess how likely it is that associations between wire codes and cancer are due to the fields rather than to some other correlate of wire codes. Several areas of evidence are relevant, most notably the studies of measured fields and cancer. Studies that address divergent magnetic-field sources, such as appliances or workplace exposures, provide useful information in that confounders are likely to be different, if not absent, for those field sources. Convergent information would suggest that the common component of magnetic fields might be the underlying cause.

Measured Magnetic Fields and Childhood Cancer   The key issue that puzzles many who conduct and follow epidemiologic research is the so-called "wire-code paradox": If wire codes operate through magnetic fields, why are measured magnetic fields less strongly associated with cancer risk than the wire codes themselves? In other words, if wire codes are functioning as an indirect indicator of exposure, then why is the more direct indicator (measured magnetic fields) not more strongly associated with cancer risk? To examine this issue, both the underlying premises and empirical evidence bearing on the argument need to be evaluated. The premise is that in studies of residential exposure and childhood cancer, present-day magnetic fields (those being the only fields that can be measured) are superior to wire codes as a surrogate for the historical exposures of interest. As discussed in Appendix B, the evidence regarding which is the better indicator of long-term past exposure is inconclusive.

Measurements have the potential to integrate a wide range of sources, including grounding currents, outdoor power lines, and indoor wiring, many of which would remain stable over extended periods. To the extent that those measurements capture multiple stable sources, they should be superior to wire configuration

TABLE 5-7 Evaluation of How Well the Exposure Assessment Represents Average In-home Magnetic-Field Exposure

Study	Exposure Assessment	Blind	Accuracy	Specificity
Wertheimer and Leeper 1979	Wire code: LCC and HCC; power lines within 40 m	No	Fair, except wire size ambiguous	Good, problem with twisted secondary
Fulton et al. 1980	Quartile relative exposure; power lines within 50 m	Unknown	Very crude model	Crude model
Tomenius 1986	Visible electric structures within 150 m	Unknown	Fair	Many sources are insignificant
	200-kV power lines less than 150 m	Unknown	Fair	Fields beyond 50 m insignificant
	Front-door spot measurements of magnetic field; 0.3-μT cut point	Yes	Good	Field varies over time
McDowall 1986	Substations and power lines within 15, 35, or 50 m	Yes	Fair	Poor, most were substations which have low fields
Savitz et al. 1988; Severson et al. 1988	Wire code: 5-level and 2-level	Yes	Fair, wire size ambiguous	Good, problem with twisted secondary
	Spot measurements of magnetic field: 0.2, or 0.3 μT; cut points of 0.065, 0.1, and 0.25 μT	No	Good	Field varies over time
Myers et al. 1990; Youngson et al. 1991	Calculated annual maximum magnetic field (poor estimate of average field); cut points of 0.01, 0.03, and 0.1 μT	Yes	Poor, used system maximum for some, line maximum for others	Poor, only five homes had fields over 0.1 μT
	Distance from power line; cut points of 25, 50, 75, and 100 m.	Yes	Fair	Probably small beyond 50-m fields
London et al. 1991	Wire code: 5-level	Yes	Fair, wire size ambiguous	Good, problem with twisted secondary

Study	Exposure assessment			
Coleman et al. 1989	Substations and power lines; cut points of 25, 50, and 100 m	Yes	Fair	Poor, most were substations
	Magnetic-field measurements in the bedroom and other locations; various cut points	Yes	Excellent, some 24-hr measurements	Good
Feychting and Ahlbom 1993	Calculated annual average magnetic field; cut points of 0.1, 0.2, and 0.3 μT	Yes	Excellent, transmission line models good	Excellent
	Spot measurements of magnetic field; cut points of 0.1, 0.2, and 0.3 μT	Yes	Good	Field changes over time
	Distance from 220 and 400 kV lines; cut points of 50 and 100 m	Yes	Fair	Good
Olsen et al. 1993	Calculated average annual magnetic field; cut points of 0.1, 0.25, and 0.4 μT	Yes	Fair, loads were estimated	Good
Verkasalo et al., 1993	Calculated average annual magnetic field; cut points of 0.05, 0.1, and 0.2 μT	Unknown	Fair, three methods of estimating load	Good
	Cumulative exposure of 0.05, 0.1, 0.2, 0.4, 1.0, and 2.0 μT-yr	Unknown	Fair, based on calculated field	Good
Schreiber et al., 1993	Distance from 150-kV line or substation; cut point of 100 m	Unknown	Fair	Poor, exposed group over inclusive

codes as an indicator because wire codes address only one source. Conducting measurements under conditions of low power use in the home is intended to eliminate historically unstable sources, such as appliance placement and use, ostensibly providing a measurement that reflects outdoor power lines and grounding currents in plumbing. Nonetheless, measurements represent arbitrary samples in both time and space. Wire codes consider structural characteristics of the power-distribution system that are quite stable, but at best wire codes constitute an indirect indicator of only one source of exposure. It is not known whether wire codes might be a better indicator than spot measurements of the average magnetic field over the space and time of interest in relationship to the development of childhood leukemia.

Perhaps the most tenuous part of the paradox is the assumption that epidemiologic studies have obtained valid estimates of the association between measured fields and childhood cancer. The studies that have examined that association either found a weaker association than the wire-code and cancer association or have found an absence of association.

There can be no doubt that the use of measured magnetic fields rather than wire codes has not yielded stronger associations of magnetic-field exposure with childhood cancer as one might expect if the measurements were notably superior to wire codes as indicators of the magnetic-field exposure and if the magnetic-field exposures were the cause of the disease. Isolated from wire codes, however, the data on measurements and cancer are surprisingly weak and inconclusive. The basis for asserting that measured magnetic fields are not strongly associated with childhood cancer comes from three studies (Savitz et al. 1988; London et al. 1991; Feychting and Ahlbom 1993). These three case-control studies were summarized in a previous section but need to be reexamined to evaluate their effectiveness in measuring the association between measured magnetic fields and cancer.

Savitz et al. (1988) were able to obtain measurements in only 36% of case homes, in contrast to 75% of control homes. The low coverage of cases was due to the large number who had moved out of the homes they occupied before diagnosis. Given this grossly incomplete coverage of the cases, it is difficult to draw any meaningful conclusions about measured fields per se or the contrast of results from magnetic-field measurements and wire codes (obtained in 90% of case and 93% of control homes). The direction of bias resulting from such incomplete coverage is not easily predicted. A modest gradient in risk was found with increased measured low-power magnetic fields, resulting in an odds ratio of 1.49 with a 95% confidence interval (CI) of 0.62-3.60, contrasting measured fields of greater than 0.25 μT with fields of less than 0.065 μT. For leukemia alone, a cut point of 0.20 μT yielded an odds ratio of 1.93 (95% CI = 0.67-5.56). These results are not definitive due to incomplete coverage, nor are they incompatible with the results from wire codes.

London et al. (1991) were not notably more successful in obtaining measurements of the magnetic fields on a high proportion of their cases and controls. Starting with 331 eligible cases, 50% had 24-hr measurements and 42% had spot measurements in the bedroom in at least one residence. Among 257 eligible controls, 56% had 24-hr measurements and 42% had spot measurements in the bedroom. Regardless of the pattern of results, this magnitude of nonresponse makes the results questionable. Mean 24-hr measurements were identical in case and control homes, with slight increases in the proportion of time above 0.25 μT among cases in the spot measurements. Both 24-hr and spot measurements showed very weak associations with leukemia risk, and the highest odds ratios tended to appear in the highest exposure intervals, but no dose-response gradients were found, and no relative-risk estimates were above 1.5. One unexplained result of their study is the grossly different magnitudes of magnetic field found using 24-hr versus spot measurements; spot measurements were approximately half the strength of 24-hr measurements. One would expect that those deficiencies that might exist in the spot measurements would be ameliorated somewhat by integrating for 24 hr. The failure of 24-hr magnetic-field measurements to produce clearer associations with cancer than the spot measurements constitutes evidence against an association between magnetic fields and cancer.

A significant anomaly in comparing the studies in Denver and Los Angeles is the notably consistent magnitude of relative risks for childhood leukemia based on wire codes in spite of what would appear to be different field strengths that correspond to those codes. For ordinary and very high current configurations, the mean values were 0.122 and 0.212 μT in Denver, 0.072 and 0.115 μT in the Los Angeles 24-hr measurements, and 0.033 and 0.061 μT in the Los Angeles spot measurements. If the measurements are correct, the fact that the two studies yielded very similar relative-risk estimates for the wire code and cancer association is evidence that some aspect of wire code other than the magnetic field as determined by average or spot measurements is operative.

Feychting and Ahlbom (1993) examined present-day spot measurements in a subset of the homes that their cases and controls had occupied before diagnosis, up to 25 years in the past. Among 142 cases, 63% had measured fields, and among 558 controls, 62% had measured fields. These figures are only slightly better than those for the studies in the United States. The results relating measured fields to cancer are highly unstable due to small numbers, and they show no association for total cancers, a modest inverse gradient for leukemia, and a modest positive association for brain cancer. Even without knowledge of the absolute validity of measured magnetic fields as an indicator of historical exposure, Feychting and Ahlbom (1993) had the ability to construct an exposure index that was better than spot measurements. Their key contribution to exposure assessment was the development of a predictive model to estimate historical field strengths by using power-line load data from the relevant time in the past. Absence of such a model assumes historical loads were identical to present-day loads, an assumption known to be erroneous. In this study, the improved estimate of magnetic fields using wiring plus load information yielded notably stronger associations with childhood cancer, arguing in favor of magnetic fields as the causal agent.

Cumulatively, these three studies provide a very weak test of the hypothesis that measured residential magnetic fields are associated with childhood cancer. In each study, coverage of wire codes (or calculated fields in Feychting and Ahlbom's (1993) study) was more complete, and results were more strongly positive. Nonetheless, the contrast between the measured fields and wire-code results is clouded by the differences in completeness of response. In absolute terms, very little information in these studies indicates whether measured magnetic fields constitute a risk factor for childhood cancer. This situation calls for refined study of the issue.

Wire-Code Categories in Relation to Magnetic Fields and Cancer   In examining the pattern of cancer risk in relation to the Wertheimer and Leeper (1979) wire codes, consideration should be given to whether the uneven relationship in field strength (see Appendix B) is similarly uneven with respect to cancer incidence. As noted above, several studies showed that the Wertheimer and Leeper codes are predictive of higher-strength fields in very-high-current-configuration (VHCC) and ordinary-high-current-configuration (OHCC) homes, but there is little or no difference in measured magnetic fields below those categories. The pattern of risk related to wire code in the two studies that considered multiple categories (Savitz et al. 1988; London et al. 1991) is consistent with the pattern of magnetic fields: little or no fluctuation in the odds ratios was observed across categories of buried lines, very-low-current-configuration (VLCC) lines, and ordinary-low-current-configuration (OLCC) lines, and a rising risk was observed for OHCC and VHCC. Although not entirely consistent or clear, the tendency for excess risk to be concentrated in the same categories as the higher-strength magnetic fields appears to support a link between magnetic-field exposure and cancer.

Several refinements have been made to the Wertheimer and Leeper wire code to make it a more accurate indicator of magnetic-field exposure. It is logical to ask whether such refinements are accompanied by corresponding increases in the strength of association with cancer. Under the assumption that misclassification of magnetic-field exposure is reduced by the refinements and the hypothesis that magnetic fields are the mediating agent, increased odds ratios would be expected.

Leeper et al. (1991) reported that spun (wrapped) secondaries are not good predictors of increased field strengths but that open first-span secondaries are. Using this information for the Savitz et al (1988) Denver study increases the odds ratio from 1.5 to 1.8, a modest increase because spun secondaries were rare in the area (Leeper et al. 1991). A three-level wire code was used in the study of Severson et al. (1988) of adult leukemia, and no association with the disease was found.

A second refinement in the wire codes was made by Kaune and Savitz (1994) that resulted in a smaller number of categories, a larger proportion in the highest interval, and a similar predictiveness for magnetic fields. The risk estimates produced by using this modified (three-category) code are larger than those based on the dichotomized results (HCC and LCC) and more precise than those from the five-level wire code used in the 1988 study (Savitz and Kaune 1993). These increases in relative risks combined with improvements in accuracy of the modified wire codes as an indicator of residential magnetic-field strength lend some support to the hypothesis that magnetic fields might be associated with cancer.

Magnetic-Field Exposures from Appliances and Cancer   Some appliances, such as electric blankets, constitute another exposure source that is entirely distinct from outside power lines or grounding currents. Little research has been done to address potential adverse health effects related to these devices; weakly positive results for different appliances have been reported in two studies (Savitz et al. 1990; London et al. 1991). At present, the state of knowledge on this topic can only be described as uncertain. More definitive research on this potential association would be useful given the clear documentation that exposures are incurred from at least some appliances, given the amenability of appliance use to historical recall without extensive effort on the part of the investigators, and given the independent test it provides of an effect of the magnetic fields per se.

Magnetic-Field Exposures in Schools and Other Settings   Although exposures to magnetic fields outside the home, particularly in schools, have received substantial public attention, no data are available whatsoever on an association between exposures in other settings and childhood cancer. Any setting in which substantial time is spent, most notably, day-care and school, would make an important contribution to total exposure. In contrast to the potential socioeconomic confounders associated with home selection and thus with home wire codes, exposure in school is likely to be largely independent of such potential biases. Logistic challenges would be present, particularly in identifying exactly which classroom was occupied, because the fields are not likely to be homogeneous throughout a school. In addition, daytime and nighttime exposures could have different biologic effects. Nonetheless, this research gap could be addressed as another source of information to help evaluate whether magnetic fields cause cancer.

Occupational Exposures to Magnetic Fields and Adult Cancer   The research exploring a link between electrical occupations and cancer, particularly leukemia and brain cancer, is extensive. (This research is discussed in detail later in the section Cancer Epidemiology—Occupational Exposures.) This research suggests a relatively small increase in risk for those workers in the aggregate, and fails to show that the most highly exposed workers are at greatest risk. The relevance of this literature to an evaluation of residential exposures and childhood exposures is limited for several reasons: (1) the age groups differ and presumably the time course of cancer development differs in adults and children; (2) workplace exposures occur during the day and residential exposures during the day and night; (3) the pattern of exposure might differ, workplaces having irregular patterns of high and low exposure compared with a more steady exposure in homes; and (4) the histologic types of cancer, particularly leukemia, differ in children and adults, further challenging the ability to extrapolate. Nonetheless, the suggestion of similar cancers associated with similar agents should not be ignored. The sources of bias are largely distinct for the studies of residential exposure and childhood cancer versus occupational exposure and adult cancer, so that if both research avenues have been misleading, they have been so in different ways. If occupational studies are pursued to clarify the issue and if they provide more conclusive evidence that magnetic fields can cause brain cancer and leukemia in adults, they will add more substantial indirect support to the proposition that magnetic fields can cause cancer in children. Conversely, if additional study should find no evidence for an association of cancer in adults occupationally exposed to higher than average magnetic fields, it would tend to support a proposition that there is no association between exposure to residential magnetic fields and childhood leukemia.

Integrated Personal Exposures to Magnetic Fields   Improved efforts at comprehensively assessing magnetic-field exposure in children are being developed. Time-integrated exposure to magnetic fields for periods of a week or two can be fully addressed, but applying this knowledge to epidemiologic studies has not been straightforward. Residential average fields are known to be important contributors to total exposure (Kaune and Zaffanella 1994), but appliances and sources outside the home can also make substantial contributions to exposure in some instances. Explicit efforts to integrate magnetic fields across multiple sources in a manner that can be extrapolated to the historical time periods of interest is essential to judge whether magnetic fields per se account for previously observed associations. If those refinements yield stronger associations with cancer, then the likelihood that magnetic fields account for the wire-code associations would be markedly strengthened. On the other hand, if validated indicators of exposure that are superior to residential wire codes do not yield stronger associations with childhood cancer, then the likelihood that magnetic fields per se cause cancer would be diminished. Improvements in assessments of magnetic-field exposure of children is under way, but close collaboration of engineers, biostatisticians, and epidemiologists is needed to ensure that the refinements are technically valid and applicable in field settings.

Evaluation of Epidemiologic Evidence

If a true association is present between power lines near residences and childhood cancer, how likely is it that such an association reflects a causal link between magnetic fields and cancer?

Putting aside those issues raised above that diminish certainty of a link even existing, we must contend with interpretation of such a link, should it exist. A positive association must be subject to one of two categories of explanation: either the wire codes are related to childhood cancer through magnetic-field exposures or they are serving as an indicator of some other agent or process. To make this evaluation, both the evidence that the association results from magnetic fields and the evidence implicating something else must be considered. Strong evidence supportive of some other explanation for an association would diminish the plausibility of magnetic fields accounting for the association.

Quality of Magnetic-Field Indicators and Strength of Association

In evaluating the studies addressing residential magnetic-field exposure and childhood cancer, a number of indicators of exposure have been applied in and across studies. Features of the power lines have been more closely associated with childhood-cancer risk than present-day magnetic-field measurements, but uncertainty about which indicator is the more useful one makes such a pattern largely uninformative. The one study that included 24-hr measurements as well as spot measurements found little difference, in spite of the advantages of a longer measurement period. Recently, an improved wire code (one that explains more of the magnetic-field variance) was shown to yield stronger associations with cancer risk (Savitz and Kaune 1993), but it has not been tested elsewhere. Although a gradient of risk corresponding to a gradient in the quality of measurement could be an informative criterion for judging the likelihood that wiring associations reflect magnetic-field effects, at present very little information supports or refutes that contention.

Dose-Response Gradients

Evaluation of the entire spectrum of wire codes and cancer risk rather than dichotomized exposure classification consistently shows a pattern in which the increased risk is most pronounced or even restricted to the highest wire-code levels (Savitz et al. 1988; London et al. 1991; Feychting and Ahlbom 1993). The absence of a monotonic relation across the full range of wire codes might be argued to be counter to a dose-response gradient, but measurements suggest that the only clear distinctions of exposure are at the high end of the wire-code spectrum. Thus, the pattern of risk in relation to wire codes corresponds rather closely to the likely gradient in magnetic-field exposure. When measurement data (Wartenberg and Savitz 1993) or estimated field strengths (Feychting and Ahlbom 1993) have been examined in detail, higher cut-point scores tend to show the largest relative risks. Overall, the data from published studies support an argument for an increased risk with higher exposure level; however, the anomaly between measured magnetic fields and wire codes in different cities severely weakens this interpretation.

Confounding

There is little evidence that known or suspected childhood-cancer risk factors introduced confounding into earlier studies. Efforts to adjust for those risk factors have had little impact on the results, which is not surprising in light of their having no strong association with childhood cancer or with wire codes. Approaching the question of confounders from both ends—risk factors for childhood cancer or correlates with wire codes—few contenders, other than magnetic fields, explain the association. Childhood-cancer correlates that remain under contention are viral exposure or some other phenomenon associated with housing density, which, in turn, is associated with wire codes. There is, for example, epidemiologic evidence from British experience during wartime evacuation of children to rural areas that the development of childhood leukemia has an infective component (Kinlen and John 1994). The British experience was that a higher incidence of childhood leukemia was associated with children living in densely populated cities relative to those living in the countryside. In addition, introducing children from the city to the country tended to lead to an increase in leukemia in the region where they were relocated.

We might also ask what exposures, other than magnetic-field exposures, might be indicated by the wire code that are the true causes of childhood cancer? An intriguing suggestion that currents on plumbing lines result in exposure to copper ions was offered (Kavet 1991) but was not supported. Application of herbicides near power-line poles or leakage of polychlorinated biphenyls (PCB) from transformers near homes in the high-exposure group might be a candidate for an increased risk. Again, a chain of tenuous assumptions is required: high wire code corresponds to overall closer proximity to poles and transformers, herbicides are commonly sprayed and transformers commonly leak, children spend time in the vicinity of the herbicides and PCB, and herbicides and PCB cause cancer in children. Each link in these various assumptions has questionable credibility, and cumulatively the proposition has so little support that it is of limited interest. Air pollutants associated with living on busy streets (where major power lines are often placed) constitute a more credible candidate, yet the studies that have evaluated that exposure have not found it to explain the association (Wertheimer and Leeper 1979; Savitz et al. 1988; London et al. 1991), and there is very little independent evidence that air pollutants cause childhood cancer.

Whatever other exposures or characteristics might be associated with wire codes, average residential magnetic-field strengths are predicted, albeit imperfectly, by wire codes. Putting aside questions of the biologic consequences of such exposure, children who live in homes with the highest wire codes have an average ambient exposure in their homes that is higher than children who live in homes with the lowest wire codes. Although other factors are also likely to be associated with wire code to some extent (yard size, housing density, and family income), none constitutes exposures in the biologic sense, and they have only indirect value as indicators of such exposures.

Clearly, more work is needed in assessing the implications of wire code on magnetic-field exposures and on other exposures.

Consistency with Secular Trend Data

The argument has been made that magnetic fields could not be a causative factor in childhood cancer because substantial increases in residential consumption of electricity (assumed to be linked to increases in personal magnetic-field exposures) have occurred over a period of many years, but there have not been corresponding increases in cancer rates during the same period (Jackson 1992; ORAU 1992). More specifically, per capita residential electricity consumption rose by a factor of 20 from 1940 to 1990, whereas mortality from all cancers (except respiratory cancer) did not increase. Instead, cancer mortality slowly declined during this period, falling from 1.12 deaths per 1,000 persons per year in 1940 to 0.93 deaths per 1,000 persons per year in 1987 (Jackson 1992). On the surface, this argument would seem quite persuasive; however, it is weakened considerably because of the use of cancer mortality data. Variations in cancer mortality over long periods of time can be attributed to a number of determinants, such as improvements in treatment and reliability of diagnosis, which make it difficult to draw inferences about the relationship of this variation (or lack thereof) to any one determinant. It might be more meaningful to examine the relationship between cancer incidence data and increases in the use of electricity over time. When that is done for the period in which reliable incidence data are available (1969-1986), the incidence is essentially unchanged, particularly for childhood leukemia, whereas the rate of residential electricity usage increases by a factor of 2 (Jackson 1992). Thus, it would seem quite clear that increased electricity consumption, as well as all other potential determinants in the environment, has not produced an increase in the incidence of childhood leukemia. Only the strength of the link between magnetic-field exposure and electricity consumption needs to be tested to complete the argument that magnetic-field exposures are not related to childhood leukemia incidence. However, major difficulties are found in assessing the strength of this link. First, it is extremely difficult to assess the actual changes in residential magnetic-field exposures to individuals that have occurred during 1969-1986. Measurements do not exist. With the increased use of electricity came many improvements in the technology for the distribution and use of electricity that might, by themselves (ignoring for a moment the trends in electricity usage), decrease individual exposure. The more extensive use over the past 20-30 years of three-phase transmission and distribution lines, double-circuit lines, and underground lines substantially reduces electric and magnetic fields in comparison to earlier use of single-phase and single-circuit overhead lines. In addition, the more spacious layout of suburban (in comparison to urban) house lots has moved residents farther from power lines. The change to Romex cable for wiring houses (instead of knob and tube wiring) also substantially reduces exposure of residents. The practice of making things smaller and more energy efficient (e.g., home radios and televisions) further reduces exposure. Even the recent proliferation in the use of remote control devices tends to reduce individual exposures by making it unnecessary for individuals to move near the appliances.

On the other hand, power lines, houses, and appliances are more numerous today. The residential use of electricity has certainly increased over the years, as documented above. Because of the developments in technology, it would not be surprising if there were periods (perhaps 5, 10, or 20 years long) when exposures tended to increase and other periods when average exposures actually decreased.

In summary, large increases in the residential use of electricity clearly have not been accompanied by comparable increases in the incidence of childhood leukemia. The apparent persuasiveness of the argument based on the observation that magnetic-field exposures are not related to the incidence of childhood leukemia is diminished, however, by the large and indeterminate uncertainties in the implicit assumption that population exposures to magnetic fields increase proportionately to residential energy use.

CANCER EPIDEMIOLOGY—APPLIANCE EXPOSURES

The determination of an effective exposure of a population to power-frequency electric and magnetic fields is an extremely difficult task because such fields are ubiquitous and exhibit considerable variation with time and place. At the same time, there are no clear or agreed upon indications of which metric or field characteristics might be most closely associated with biologic effects.

A number of epidemiologic studies of the possible association between magnetic-field exposures and various disease outcomes have explored exposures ascribed to the use of various electric appliances. If such associations are to be explored successfully, it is desirable to consider electric appliances that are capable of producing substantial exposures, that are not in universal use, and can be assessed for exposure on the basis of subject recall of use. Mader and Peralta (1992) evaluated the reconstruction of such exposures due to appliances and compared those exposures with residential exposures due to other sources. They concluded that the time-weighted contribution to whole-body exposure made by normal assortments of household electric appliances is small compared with the usual background exposures, but that appliances might be the dominant source of exposure to the body's extremities. They estimated the time-weighted whole-body exposure to be of the order of 0.016 μT. That estimate is consistent with the estimate of 0.02 μT reported in the recent 1,000-homes study (EPRI 1993a). Mader and Peralta (1992) also pointed out that extremity exposure due to appliances can be estimated as 0.97 μT. That estimate can be compared with an average of the order of 0.1 μT whole-body (and extremity) exposure due to all other sources (including power lines and grounding systems).

Electric blankets can make substantial contributions to whole-body exposure to magnetic fields. Electric-heated water beds have also been included in some studies, but some doubt is cast on the significance of their exposures by studies of Delpizzo (1990) showing that mattress heating pads contribute insignificantly to the overall exposure. Preston-Martin et al. (1988) evaluated the effect of electric-blanket use on the incidence of myelogenous leukemia in adults in a case-control study. In 224 pairs of cases and controls, the regular use of electric blankets produced odds ratios of 0.9 (95% confidence interval (CI) of 0.5-1.6) for acute myelogenous leukemia and 0.8 (CI = 0.4-1.6) for chronic myelogenous leukemia. Cases did not differ from controls in duration of use, period of use, or socioeconomic status. These authors estimated that use of electric blankets increased the exposure of a subject to magnetic fields by no less than 31% and no more than 35% of the total estimated exposure without the electric blanket. They concluded that no major leukemogenic risk is associated with electric-blanket use in the population studied.

Verreault et al. (1990) conducted a case-control study of testicular cancer in which the cases and controls were asked about their use of electric blankets. The investigators did not report any attempt to estimate actual exposure other than through recall of frequency and length of use during the 10-year period before diagnosis. The number of cases and controls reporting electric-blanket use was almost identical, as was the duration of use. The age-adjusted odds ratio of the use of electric blankets was reported as 1.0 (CI = 0.7-1.4); the authors concluded that increased exposure to electric and magnetic fields from electric-blanket use contributes little, if at all, to the risk of testicular cancer in adult white men.

Savitz et al. (1990) evaluated the association of electric-appliance use with childhood cancer in a case-control study in Denver. Exposure estimates were based on parental recall of the use of a number of appliances by the mother during pregnancy and of exposures of the child after birth. Associations were tested for all childhood cancer, leukemia, and brain cancer. For some of the appliances (e.g., electric blankets), recall of settings that might influence exposures was sought. After testing for a number of appliances, prenatal exposures to electric blankets were reported to be associated with increased risk of brain cancer (odds ratio (OR) = 1.8, CI = 0.9-4.0). Adjustment for confounding by income level resulted in an odds ratio for electric-blanket use of 1.3 (CI = 0.7-2.2) for total childhood cancer, 1.7 (CI = 0.8-3.6) for leukemia, and 2.5 (CI = 1.1-5.5) for brain cancer. The authors concluded that this study "provides some indication that pre- and post-natal exposure to electric blankets may increase the risk of developing childhood cancer." The size of the study population, however, produced too low a power to detect anything but strong associations. In addition, the nature of the exposure assessment is insufficiently precise, which could introduce substantial but presumably random misclassification. That plus the possibility of recall bias could easily lead to inaccurate estimates of the association.

The study of childhood leukemia by London et al. (1991) had larger numbers of subjects with leukemia than did the study by Savitz et al. (1988). The London et al. study also evaluated the effects of electric-appliance use. Self-reported use of electric appliances did not produce strong associations with precise estimates. Only 2 out of 20 tested associations were statistically significant (use of electric hair dryers on the child produced an odds ratio of 2.82 (CI = 1.42-6.32), and use of a black and white television produced an odds ratio of 1.49 (CI = 1.01-2.23)). In this study, as in the Savitz et al. study, recall of use of appliances was relied on for exposure assessment, a method that inevitably introduces misclassification of actual exposures. In addition, when only a small fraction of the study population uses a given appliance, the precision of risk estimates suffers.

Vena et al. (1991) studied the relationship between electric-blanket use and risk of postmenopausal breast cancer in 382 cases and 439 community controls in western New York State. Exposure to extremely-low-frequency magnetic fields, through a suppression of normal nocturnal rise in melatonin release, was hypothesized to produce an increased risk of breast cancer (Stevens 1987a,b). Vena et al. (1991) found point estimates of a slightly increased risk (OR = 1.31, CI = 0.88-1.95) in persons who reported continuous use of electric blankets in season for 10 years or more. Subjects reporting less use had lower point estimates, and none of the findings reached statistical significance. The hypothesis being tested might respond only to fields imposed on the pineal region, and electric blankets would probably deliver a lower dose to that region than to the rest of the body. The study had a minimum detectable odds ratio of 2.1, and the power to detect an odds ratio of 1.5 was only 0.28. A relatively high nonresponse rate of 44% among cases and 54% among controls makes interpretation of the study more difficult also. The authors concluded, "Generally the findings of this study do not support the hypothesis that electric-blanket use is associated with an increased risk of breast cancer."

Studies in which particular electric appliances were evaluated in terms of their association with adverse health outcomes are no more persuasive than other forms of assessments of exposure to electric and magnetic fields. In part, this conclusion might be due to inadequate statistical power in the studies that were published or to the rarity of the outcomes reported. The most important general conclusion that can be drawn is that assessment of the effect of use of different appliances associated with increased exposure to magnetic fields does not provide insight that can be applied across different studies.

CANCER EPIDEMIOLOGY—OCCUPATIONAL EXPOSURES

Extensive literature has been accrued on the pattern of cancer among workers in electrical occupations, primarily men, who are presumed to encounter increased exposures to electric and magnetic fields in their jobs. Initial suggestions by Wertheimer and Leeper (1979) and Milham (1982) have been followed by dozens of surveys of workers in electrical occupations. The focus has been on leukemia and brain cancer, but studies of lymphoma, malignant melanoma, and breast cancer (among both men and women) have also been conducted. Detailed reviews have been published (Theriault 1990; Savitz and Ahlbom 1994) and will not be repeated here. Following a brief summary of the earlier literature, research of current interest—the impact of refined exposure measurements on study results and breast cancer as an outcome of particular interest—will be reviewed.

Across a wide range of geographic settings (mostly in North America and Europe) and diverse study designs (proportionate incidence or mortality, case-control, or cohort), workers engaged in electrical occupations have often been found to have slightly increased risks of leukemia and brain cancer (Savitz and Ahlbom 1994). The general structure of these studies is to identify one or more groups of electrical occupations, such as linemen, electricians, electric-equipment assemblers, and electrical engineers, and compare their cancer incidence or mortality rates to that of nonelectrical occupations. The guiding assumption in these studies is that the job title provides an indication of encountering increased exposure to electric and magnetic fields on a regular basis during the workday. A related assumption is that characteristics of the job other than exposure to electric and magnetic fields that are related to cancer risk can be identified and controlled as needed in the analysis.

Averaged across these studies, a modest increase in leukemia, particularly acute myeloid leukemia, and brain cancer is found. Although formal meta-analyses have not been reported, relative risks on the order of 1.2-1.5 would be expected (Savitz and Calle 1987; Theriault 1990). It is reasonable to ask whether, among electrical occupations, those who are more likely to have increased exposure to electric and magnetic fields (e.g., linemen) have greater increases in cancer risk than those less likely to have increased exposure (e.g., electrical engineers), but the diversity of workers represented by these broad administrative categories that include clerical workers, technical staff, professionals, and management offers little promise of refinement. Nonetheless, on the basis of recently published data on typical exposures across a range of electrical and nonelectrical occupations (London et al. 1994), no obvious association was found between increased exposure to magnetic fields and cancer among electrical workers. In addition to having a greater potential for increased exposure to electric and magnetic fields, some electrical workers might encounter hazardous chemicals, such as solvents, in their jobs and might be prone to higher rates of leukemia or brain cancer for reasons other than workplace exposures. The obvious test of the hypothesis that exposures to electric and magnetic fields account for the reported increases in risk is to refine the measurement of these fields and assess whether the associations become more pronounced.

Over the past several years, a series of publications have examined more refined measurements of exposures to electric and magnetic fields in relation to cancer. Matanoski et al. (1993) studied leukemia other than chronic lymphocytic leukemia among telephone workers in relation to magnetic-field exposures estimated through job titles and a series of measurements. They found little support for increased risk due to increased average fields, but increasing field levels at peak exposure were associated with increased leukemia risk.

Floderus et al. (1993) conducted a community-based study of leukemia and brain cancer in Sweden. They evaluated exposure by taking workplace measurements in the employment locations of cases and controls to classify exposure more accurately. On the basis of a quantitative index of magnetic-field exposure, the most highly exposed workers were estimated to have a 3-fold increased risk of chronic lymphocytic leukemia and a 1.6-fold increased risk of total leukemia. Brain-tumor risk was increased by a factor of 1.5 in the highest category.

Three studies of electric-utility workers have yielded inconsistent results. A study at Southern California Edison (Sahl et al. 1993) yielded no associations between exposure and leukemia, lymphoma, or brain cancer; all relative risks were less than 1.4. In contrast, a large, well-designed study of utility workers in Canada and France provided evidence for a 2- to 3-fold increased risk of acute myeloid leukemia among men with increased magnetic-field exposure (Theriault et al. 1994). Brain cancer showed much more modest increases (relative risks of 1.5-2.8) with increased magnetic-field exposure. The most recent study (Savitz and Loomis 1995) examined cancer mortality in a large cohort of U.S. electrical workers. Leukemia mortality was not found to be associated with indices of magnetic-field exposure, whereas brain-cancer mortality was associated. Brain-cancer mortality generally was found to increase in relation to accumulative exposure, reaching a relative risk of 2.3-2.5 in the most highly exposed workers. All three studies found no evidence of confounding by the presence of workplace chemicals. A smaller study of Norwegian hydroelectric-power-company workers did not find increased risk associated with estimated magnetic-field exposure, but did report an increased rate of melanoma (Tynes et al. 1994a).

Recently, two studies examined cancer among electric-railway workers in Sweden (Floderus et al. 1994) and Norway (Tynes et al. 1994b). Select groups of workers in this industry are believed to have chronic exposure to fields at 16.66 Hz. Tynes et al. (1994b) found no indication of increased leukemia or brain-cancer incidence for any historical period, whereas Floderus et al. (1994)

found men who were employed in exposed occupations during the 1960s (but not the 1970s) to have increased leukemia. They suggested that exposure was decreased due to changing work practices in the later time period.

Methodologically, these more recent studies are clear improvements on the earlier studies that relied on job titles alone. The investigators developed rather elaborate approaches for classifying exposures more accurately and taking potential confounders into account. In spite of those refinements, the patterns of association have not become more consistent and pronounced, nor have they gone away. The relative risks in the upper categories of 2-3 reported in the high-quality studies of Floderus et al. (1993) and Theriault et al. (1994) cannot be ignored. However, the inconsistency in which cancer types show increased risks, the presence of contradictory studies (e.g., Sahl et al. 1993), and the irregular dose-response gradients make the interpretation problematic. Overall, the most recent studies have increased rather than diminished the likelihood of an association between occupational exposure to electric and magnetic fields and cancer, but they have failed to establish an association with a high degree of certainty.

Another avenue of research to be noted is the concern with occupational exposure to electric and magnetic fields and breast cancer. A series of three studies reported an association between electrical occupations and male breast cancer (Tynes and Andersen 1990; Matanoski et al. 1991; Demers et al. 1991), which were similar in character to the initial studies of leukemia and brain cancer. More recently, a report of no association was published (Rosenbaum et al. 1994). Female breast cancer in relation to electrical occupations was evaluated by Loomis et al. (1994) among a large number of decedents in the United States. A modest increase in risk was found for women in electrical occupations, particularly telephone workers, encouraging further evaluation of a potential link between exposure and this common cancer.

REPRODUCTION AND DEVELOPMENT

Epidemiologic studies of potential adverse reproductive effects of exposure to electric and magnetic fields are limited in quantity and, to some extent, in quality. There are a multiplicity of exposure sources of potential interest (including residential exposures from power lines, occupational exposures from video-display terminals (VDTs), and other exposures from electric appliances, such as electric blankets) as well as a diversity of reproductive health end points (including infertility, miscarriage, congenital defects, growth retardation, and preterm delivery). Most of those areas have been addressed in fewer than three studies, the exception being VDTs in relation to spontaneous abortions. Although grouping data for exposures from all types of sources with the relevant outcomes is tempting for review purposes, comparing studies of similar types of exposures and outcomes without the assumptions required for aggregation is more constructive. The absence of efforts to replicate these studies is the predominant source of uncertainty in this literature. Some excellent reviews of the topic are available (Hatch 1992; Shaw and Croen 1993; Delpizzo 1994).

Video-Display Terminals

The epidemiologic literature on VDTs was most recently covered in a review by Delpizzo (1994). In the VDT literature, which is large and of reasonably high quality, the evidence is clear that VDT use per se is not associated with increased risk of adverse reproductive outcomes, such as spontaneous abortion, congenital defects, or intrauterine growth retardation. However, the use of VDTs is not synonymous with exposure to extremely-low-frequency electric and magnetic fields. In fact, VDT use is a complex mixture of some increase in exposure primarily to very-low-frequency fields (3-30 kHz), potential psychological stress associated with repetitive tasks, physical inactivity associated with a sedentary job, and a modest increase in exposure to extremely-low-frequency fields (30-300 Hz). Depending on the particular machine, the location of the operator relative to the VDT in use, and the number and location of VDTs surrounding the operator, field exposures are typically in the range of 0.1 to 0.3 μT (Delpizzo 1993). Thus, the sizable literature on VDTs and reproductive outcomes has little value in addressing questions concerning prolonged exposure to increased power-frequency electric and magnetic fields. Only one study (Lindbohm et al. 1992) carefully related the VDT use to electric- and magnetic-field exposure (described below in Workplaces).

Residences

In contrast to the large number of cancer studies, only two published studies to date address a possible link between residential sources of exposure to electric and magnetic fields and adverse reproductive outcomes. Juutilainen et al. (1993) evaluated the residential magnetic-field exposures of 89 women who had suffered early pregnancy loss (as diagnosed by assays of human chorionic gonadotropin in urine) compared with the exposures of 102 women who had not. Magnetic fields were measured at several locations in the home to classify subjects. After adjustment for cigarette smoking, risk of early pregnancy loss was increased among women who resided in residences with measured fields above around 0.25 μT, particularly above 0.63 μT. On the basis of 6-8 cases of exposure in the highest exposure strata and 2-3 controls, the odds ratios were found to be substantially increased (in the range of 3 to 5) but highly imprecise. As the authors recognized, these results are imprecise and subject to uncertainty related to possible confounding or response bias. Robert (1993) recently evaluated the risk of birth defects in relation to residence in municipalities with and without high-tension power lines. An inverse association was found; the communities without power lines were at higher risk. However, because the estimation of exposure was based on the condition that a power line was anywhere in the community, as opposed to the Wertheimer and Leeper wire codes in which power lines within 150 feet of the home are considered, the assignment at the community level is unlikely to reflect any information about individual exposure. Thus, the study does not contribute to the question of reproductive health effects of exposure to electric and magnetic fields.

Electric Appliances

Wertheimer and Leeper (1986, 1989) evaluated fetal loss in relation to electric-blanket use in Colorado and ceiling cable-radiant heat in Oregon. Those seasonal field sources were considered particularly useful for study, given that risk could be compared among users of those devices across seasons. Data from the Colorado study were based on birth announcements and a telephone survey, and the data from Oregon were derived from birth certificates. The methodologic details, particularly of the Colorado study (Wertheimer and Leeper 1986), are difficult to interpret, but the authors' conclusion from both studies was that spontaneous abortion risk was greatest in seasons in which uses of the heating devices was increasing. The unconventional design of the study and the pattern in which risk was not highest when exposure was highest diminish the credibility of the overall results, although no clear bias was evident that would have produced the reported pattern. It should also be noted that the Colorado study reported that birth weights were lower among those who used electric blankets, whereas gestational duration was not shortened; this finding was interpreted by the authors as an indication of fetal growth retardation.

Another major study of home electric-appliance use addressed congenital defects, specifically oral clefts and neural tube defects, in New York State (Dlugosz et al. 1992). In this well-designed study, the New York State Congenital Malformations Registry served to identify 121 cases with isolated cleft palate, 197 cases with cleft lip with or without cleft palate, and 224 cases with anencephalus or spina bifida. Controls were selected from birth-certificate files. Mailed questionnaires elicited information on electric-blanket and heated-water-bed use along with an array of potential confounding factors. Relative-risk estimates suggested no increase in risk, the odds ratios being 0.8, 0.7, and 0.9 for cleft palate, cleft lip, and neural tube defects, respectively, for electric-blanket use. Uncertainty arises from the reliance on self-reported electric-appliance use several years in the past and the potential bias from nonresponse. For the specific question of electric-blanket and heated-water-bed use in relation to the specific congenital defects studied, the data provided some assurance of no association.

The most recent and detailed study of reproductive health consequences of magnetic fields focused on electric-blanket use in relation to fetal growth (Bracken et al. 1995). Women were interviewed and enrolled in this prospective study by 16 weeks of gestation, and subsets of women were assigned to variably detailed magnetic-field-assessment protocols. Multiple exposure sources were considered, including ambient residential fields, electrically heated beds, and wire codes. Among the 2,709 women enrolled, approximately 4% delivered low-birth-weight infants and approximately 7% delivered infants below the tenth percentile of weight for gestational age (labeled as small-for-gestational age). Electrically heated bed use on a daily basis was associated with relative risks on the order of 1.1-1.3 for the two outcomes; the association was not stronger with longer use. No clear associations were seen with the other sources of magnetic fields that were considered, leading the authors to conclude that "risk of low birth weight and intrauterine growth retardation is not increased after electrically heated bed use during pregnancy." For the field types and outcomes examined, the data suggest little or no association.

Workplaces

The earliest study of exposure to electric and magnetic fields and reproductive health concerned males exposed at high-voltage substations (Nordstrom et al. 1983). Assessment of exposure was based on working in a high- as opposed to low-voltage switchyard. A larger proportion of high-voltage switchyard workers reported having had children with malformations (8%) as compared with other workers (1-3%). The usual proportion is typically about 5%, depending on how narrowly or broadly malformation is defined. Given the poor quality of reporting of malformations, particularly by fathers, and the minimal evidence of electric- and magnetic-field-exposure gradients, this study adds little information.

Magnetic-field exposure from VDTs was examined in relation to spontaneous abortion by Lindbohm et al. (1992). Spontaneous abortion cases (191 cases) and live-birth controls (394 controls) were identified among Finnish clerical workers. Exposure to extremely-low-frequency magnetic-fields from VDTs was identified by combining self-reports with measurements of specific VDT units used by the women in the study. Work with VDTs in general showed no association with spontaneous abortion (OR = 1.1), whereas among women who worked with VDTs emitting the highest magnetic-field strength, the odds ratio rose to 3.4. Combining the number of hours of use with the estimated field strength produced by the unit yielded increased risks in relation to exposure to extremely-low-frequency and very-low-frequency magnetic fields. Risk increased steadily from women who used low-field-strength units for brief periods to women who used higher-field-strength units for longer periods; the odds ratios were 1.7 (95% CI = 0.8-3.6) for medium and 3.8 (95% CI = 1.6-8.8) for high relative to low-estimated extremely-low-frequency exposures.

Methodologic Issues

The methodologic issues for reproductive studies are largely the same as those for childhood cancer studies. The greatest limitation for reproductive studies and cancer studies concerns exposure assessment. The evolution of cancer epidemiology studies has progressed far more, however, than reproductive studies because refinements are being made successively in the studies. At least one major reproductive study is in progress at the California Department of Health Services directed by Dr. Shanna Swan. The study in progress and recently completed ones use the most sophisticated exposure assessment methods available to investigate spontaneous abortion, late fetal loss, fetal-growth retardation, and preterm delivery. Given the relatively short time course of pregnancy and the relatively high frequency of some adverse outcomes, the opportunity to monitor exposure prospectively or at least closer in time to the development of reproductive effects is much greater than that for cancer. The newer studies include home measurements as well as reports of appliance use and measurements. Given how limited previous studies of reproductive effects have been, these new results could completely change the picture in ways that cannot be predicted.

Confounding is a somewhat greater concern for reproductive health outcomes than for childhood cancer, largely because so much more is known about risk factors. Across many reproductive outcomes (with the exception of many congenital malformations), there are strong associations with social class, mother's age, tobacco use, and, to a lesser extent, alcohol and illicit drug use. The more sophisticated studies take such factors into account, and failure to do so could easily lead to confounding of such sources of electric- and magnetic-field exposures as electric blankets or residence in high-exposure homes.

Selection bias is a concern here as well, but the source population can be defined unambiguously with greater ease from birth records or prenatal care records. Potential for recall bias and response bias are also relevant to interpreting reproductive studies, particularly because such exposures as electric blankets are increasingly perceived by the public as potentially harmful. These perceptions could affect reporting by women who have had an adverse outcome (recall bias) or affect their willingness to participate in the study at all (response bias).

LEARNING AND BEHAVIOR

The scientific literature on the association between exposure to power-frequency electric and magnetic fields and behavior includes a series of studies that relate exposure to a wide range of outcomes, including suicides, depressive symptoms, headaches, and neuropsychologic performance. In general, the studies of behavioral outcomes used potentially biased designs and obtained results that are inconsistent and of poor quality. Few studies used a validated measurement instrument to assess subjective symptoms, opportunity for misclassification of exposure and outcome was ample, most did not adjust adequately for confounding (especially demographics), and few used exposure measurements with adequate temporal and spatial resolution. Nonetheless, the consistent lack of association seen in this set of studies is notable. The committee reviewed the details of these studies, as has Paneth (1993). Note that this section does not consider studies of acute effects (e.g., see Stollery 1986; Gamberale et al. 1989; Cook et al. 1992) nor reports of hypersensitivity to electric or magnetic fields because they are beyond the committee's charge.

Suicide

Suicide was the first outcome evaluated in modern studies of exposure to electric and magnetic fields and behavioral response. It is listed on death certificates and medical records and thus particularly amenable to evaluation. One set of studies conducted in England compared exposures to high-voltage electric-power transmission lines (equal to or greater than 32 kV) among 598 suicides and 598 randomly selected electoral-register controls (Reichmanis et al. 1979; Perry et al. 1981). The first study used magnetic-field exposures calculated from transmission-line configurations to compare subjects, and the second used measured magnetic fields as well as a variety of other metrics.

In the study by Reichmanis et al. (1979), exposures to electric and magnetic fields were calculated at all residences of subjects. Values for case and control residences then were compared among three categories of increasing exposure. The overall results showed statistically significant differences among the suicide cases and the controls, although no exposure-effect trend was observed. Next, cases and controls were ranked individually by exposure, and a paired comparison was conducted. The control exposure values were statistically significantly higher than the cases' exposure, suggesting that suicide subjects were more likely to live in lower-level electric- and magnetic-field environments than controls. The authors were equivocal in their interpretation of those results, but noted that the data are consistent with the notion that exposure to electric and magnetic fields does not induce suicide.

In the study by Perry et al. (1981), the same data were compared on the basis of additional exposure metrics: (1) type of residence, (2) distance from school, major road, church, or open water, and (3) doorstep measured magnetic field. No significant differences were found for metrics 1 or 2. However, a statistically significant number of cases compared with controls were found to be at or above the median doorstep-exposure value of 0.04 μT (0.4 mG) (52% vs. 43%), and, overall, case homes had statistically significantly higher measured magnetic-field exposures than control homes. As noted by the authors, the median magnetic field was measured at 0.04 μT, whereas that calculated in the study of Reichmanis et al. (1979) was 0.005 μT, suggesting that transmission lines might not have been the main source of exposure. That discrepancy also provides a possible explanation for the disparate results of the two studies.

Commenting on those studies, Bonnell et al. (1983) pointed out that suicide rates vary several-fold by gender, age, socioeconomic status, and urban or rural character, but none of these potential confounding variables was adjusted for. In addition, measurements were made only once, measurements were taken on the doorstep rather than in the living space, measurement takers were not blinded to the house status, and measurements were taken many years after the occurrence of the event. Further, Bonnell et al. (1983) questioned the potential biases in the control-selection procedure (e.g., mobility). Smith (1983) noted an additional potential problem of confounding in that the correlation between the number of suicides and the average exposure in the homes of the suicide cases was negative, a condition that might indicate an association between lower socioeconomic status of suicide cases and the tendency of persons of lower socioeconomic status to use less electricity because of its cost.

Suicide cases were also assessed as part of a cohort mortality follow-up study conducted by McDowall (1986). McDowall reported no evidence of an increased rate of suicide among those living within 50 m of a substation or 30 m of an overhead line.

Baris and Armstrong (1990) reviewed British occupational mortality data on the proportion of deaths from suicide among electrical workers. They found that the category composed of all electrical occupations did not show excess suicides, but the categories of radio, radar, and television technicians showed excess proportion of death from suicide in 1970-1972 and 1979-1983. The categories of telegraph and radio operators showed excess suicides in 1970-1972 but a deficit proportion of suicides in 1979-1983. Baris and Armstrong (1990) noted the imperfections in their exposure data and, because only age was adjusted for, the possibility of uncontrolled confounding. Overall, they concluded that their results were negative. Nondifferential misclassification of both exposure and outcome might have introduced bias.

Depression

The first reported studies of the association between residential proximity to power lines and depression were by Dowson et al. (1988) and Perry and colleagues (Perry and Pearl 1988; Perry et al. 1989). Dowson et al. (1988) conducted a study in England among people living near 132-kV overhead power lines. They compared that population with a population living away from overhead lines and closely matched in house type. The study was designed to assess recurrent diseases, health decline during the previous year, and work-time lost due to illness; adjustments were made for age, sex, social status, and duration of residence. The residents near power lines were younger, they reported more days off from work, had more headaches or migraines, and suffered more depression. Adjusting for sex and years of residence did not alter the results.

Perry and Pearl (1988) studied hospital admissions among residents of multi-story buildings and found no statistically significant differences in overall incidence of heart disease, drug overdoses, and psychiatric problems among those living near and those living away from electric cables, but a statistically significant increase in depression and personality defects was found among those living near the cables. Gender and age were similarly distributed in each case and control group and could not have introduced confounding. Banks (1988) criticized the exposure characterization and argued that the proximity method used was likely to have misclassified many subjects.

Perry et al. (1989), responding to critics, conducted a larger case-control study of discharge diagnosis or cause of death of patients in local hospitals with spot measurements of magnetic fields. They again found a statistically significant association between depressive illness and high magnetic-field exposures. The myocardial infarction results were null, in agreement with the earlier study, which showed a statistically nonsignificant increase in myocardial infarction. Issues of possible confounding and selection bias still were not addressed adequately.

Two recent studies incorporated a validated measurement of depression (e.g., the CES-D scale) in their study designs. Poole and Kavet (1993) conducted a telephone interview survey of 382 persons to assess the prevalence of depressive symptoms and headache in relation to visual proximity of residence to overhead power lines. They found a statistically significant positive association for depressive symptoms but not for headaches or migraines. Adjustment for demographic factors did not account for the observed effects. As noted by the authors, the assessment of exposure was crude. The overall participation rate was 69%.

McMahan et al. (1994) studied depression in women living adjacent to and one block away from overhead transmission lines in Orange County, California. Field measurements of magnetic fields were used to confirm exposure classifications. Confounders considered were age, income, education, and length of residence. Personal interviews of 152 women (61%) were conducted on the subject of depression. Questions were also asked about general health, life events, family history, health habits, occupation, and home life. Depression (CES-D score above median) was positively associated with shorter tenure at residence, less education, younger age, nonwhite ethnicity, and higher income and negatively associated with living near a power transmission line, although none of the reported associations was statistically significant.

Two other studies investigated the association between occupational exposures to electric fields and depressive symptoms. Broadbent et al. (1985) interviewed 390 electric-power transmission and distribution workers. Electric fields were measured for 287 subjects. No significant associations were found between electric-field exposure and headaches, depression, or related conditions. Savitz et al. (1994) analyzed data from the Vietnam Experience Study. Using the Diagnostic Inventory Survey and the Minnesota Multiphasic Personality Inventory, they compared results for 183 electrical workers and 3,861 nonelectrical workers. Electrical workers did not show increased depression or depressive symptoms overall, although electricians were approximately twice as likely to be depressed, but this association was not statistically significant.

Headaches

Headache frequency was reported mainly in conjunction with other symptoms. Dowson et al. (1988) reported more headaches and migraines among people living near 132-kV overhead power lines than among those in a comparison population living away from overhead lines. Poole and Kavet (1993) did not find an association between reported headache frequency and living within sight of an overhead line. Broadbent et al. (1985) found no association between headache frequency and electric-field exposure.

In a study designed to follow up results from Dowson et al. (1988), Haysom et al. (1990) used a standardized questionnaire to investigate the incidence of self-reported headaches and migraines. Subjects lived on large estates adjacent to overhead power lines in Southampton, England. Large estates were used in the study to control for age and social status and to allow for a wide range of exposures by including houses close to (less than 100 m) and far from (greater than 100 m) the power lines. The study group comprised 592 adults classified as exposed and 592 classified as unexposed. Of those subjects determined to be eligible, response rates were similar among exposed (63.5%) and unexposed (66.2%). A lower rate of headache was reported by subjects living 100 m or more from the power lines, although a chi-squared analysis showed the results were not statistically significant. The highest rate of headache was reported by the group living 50-100 m from the power lines. Reported migraines showed a similar pattern and were statistically significant, both for indices of severity and frequency. Finally, the incidence of headaches was more pronounced near a 400-kV power line than a 132-kV power line, although this difference also was not statistically significant. No explanation or postulated physiologic mechanism is given for the findings.

Neuropsychologic Performance

Two studies of occupational exposures to electric and magnetic fields and neuropsychologic effects were conducted. Knave et al. (1979) conducted a matched (age, location, and length of employment) cross-sectional study of exposed and unexposed workers. Exposed workers performed better on memory tests, one-hand manual dexterity tests, reaction-time tests, and perceptual speed tests. However, these subjects also were more educated, and the authors speculated that the educational difference could be responsible for the performance difference.

Baroncelli et al. (1986) conducted a cross-sectional study of railroad workers. They compared results of a variety of laboratory and performance tests among those exposed to electric and magnetic fields. Reaction times and anxiety tests showed no statistically significant differences among four exposure categories.

Summary

The results of studies of neurobehavioral responses to exposure to electric and magnetic fields are inconsistent and of mixed quality. The exposure measurements used—job titles, calculated fields, spot measurements and visual proximity—all have known limitations and are likely to result in substantial misclassification. A range of symptoms for clinically relevant outcomes were reported in the studies, and only some of the studies used standardized instruments to assess their occurrence. Hospital and medical records are likely to result in incomplete ascertainment and are likely biased by educational level and socioeconomic status of the subject.

6

Risk Assessment

SUMMARY AND CONCLUSIONS

This chapter presents the committee's assessment of the risks to human health from exposure to electric and magnetic fields. Assessment of the risk from exposure to a chemical or physical agent usually begins from a supposition—born of observation—that some hazard exists; for example, that cigarette smoking is linked to cancer and lung disease or that animal tests with chemicals of concern show excess cancer. If data are sufficient, the assessor may extrapolate these data and arrive at a quantitative estimate of the magnitude of the risk; for example, that one in seven lifetime smokers of a pack of cigarettes a day will contract lung cancer or that a specific dose-response relationship exists in at least one animal species and can be used by analogy to estimate human-health risk. That process is called risk assessment. The task of risk assessment is made more accurate and compelling if the epidemiologic and laboratory data are consistent in positively identifying the suspected agent of disease as the causal factor; a good estimate of the exposure is possible; a dose-response relationship is evident in the data (that is, if one smokes more cigarettes, one is more likely to get lung cancer); and a biologically plausible explanation exists for the relationship between the action of the presumed causal agent and the observed effect.

In previous chapters, the committee reviewed the data concerning the possible relationship between exposure to electric and magnetic fields and biologic effects in both in vitro and in vivo biologic systems and in epidemiologic studies. The conclusions reached regarding these data are summarized at the beginning of Chapters 3, 4, and 5. The set of studies that investigated the relationship between childhood leukemia and residential proximity to particular configurations of external electric wires has captured the interest and concern of the public, scientists, and government officials. The committee carefully considered the reports of weak positive associations and found them consistent and not explainable by other known factors. This chapter considers the use of those data for estimating the nationwide childhood cancer risk from residential exposure to magnetic fields.

The conclusions in this chapter about the possible risks of exposure to electric and magnetic fields relate mainly to cancer, as is typical in risk assessments. At the end of the chapter, the committee comments on what the data might say about any possible association between exposure to magnetic fields and other health-effect end points, such as adverse effects on reproduction or on learning ability.

RISK ASSESSMENT

Risk assessment is a method designed to evaluate human-health risk (NRC 1983; Wartenberg and Chess 1992). Together with information about costs, benefits, and sociopolitical concerns, risk assessments are used to formulate policies to reduce risks, a process called risk management. The basic tenet of risk assessment is that data on health effects detected in small populations exposed to high concentrations of suspected hazardous agents, usually chemicals, can be extrapolated to predict health effects in large populations exposed to lower concentrations of the same agent. Thus, if workers in a factory are exposed to a high concentration of the solvent toluene and no one develops cancer after follow-up testing for a sufficient number of years, toluene can be inferred to be less carcinogenic than another substance known to produce cancer in workers. Furthermore, if community residents are exposed to toluene at a low concentration, it can be inferred that they will not incur a substantial increased risk of cancer either.

In concept, risk assessment is meant to be an objective approach to risk evaluation for making informed public policy. Although components of the method allow for some flexibility, permitting risk assessors to accommodate a range of circumstances, many federal agencies have explicit guidelines for carrying out risk assessment. Thus, some federal officials believe that little room is left for subjective judgments if agency guidelines are followed. In practice, however, many staunch supporters of risk assessment acknowledge the imprecision of the technique, recognizing that it entails a considerable measure of subjective judgment.

Risk assessment is a four-stage process (NRC 1983). The goals of the first stage, known as hazard identification, are to catalogue situations in which an agent can pose a risk to human health and to predict all possible adverse health effects. It is meant to address a hazardous agent regardless of the amount present and to characterize all possible health end points. At a later stage, specific agents or health effects can be excluded if they are deemed to be unimportant.

The second stage of risk assessment is dose-response assessment—determining how much exposure to a given hazardous agent is harmful to public health. For carcinogens, the results of this stage of risk assessment typically are reported as the "cancer potency." That is, risk assessors fit a mathematical equation to the experimental data to describe how the risk of disease increases with the amount of an agent to which a person is exposed and report a standard index of that rate of increase.

The third stage of risk assessment is exposure assessment: investigators estimate the amount of a given agent a typical person is likely to encounter. These data can be obtained in many ways, including direct measurement, historical records of releases, and extrapolation of similar processes occurring in other locations and adjusted for the specifics of the current study. The final result is a combination of where a person spends time and what activities a person engages in coupled with the source, concentration, and movement of the studied agent.

In the fourth stage of risk assessment, risk characterization, information from the three other stages is combined into a single overall estimate of risk. For the agent identified in the first stage, the assessor determines in stage two if a dose-response relationship exists, and in stage three, quantifies the exposure. That exposure is then multiplied by the potency estimated in stage two to derive a predicted risk of cancer. In this final stage, a quantitative risk assessment is made; that is, a numerical estimate of the magnitude of human risk.

The assessments from all these stages are combined with other insights, such as the presence or absence of a biologic explanation for the relationship between exposure and effect, to reach a judgment about the overall concern warranted by exposure to an agent. The strength of individual studies and of groups of studies is weighted by assessors when making risk-assessment judgments. More weight might be ascribed to studies that have been replicated in different laboratories and to studies that are well designed and carefully performed.

In the case of exposure to magnetic fields, the most compelling data available are derived from human studies rather than animal experiments. The most reliable positive laboratory data come from studies of mammary-cell tumors in test animals (Beniashvili et al. 1991; Mevissen et al. 1993; Löscher et al. 1993). However, tumors occurred only when the animals were treated with a chemical carcinogen or ionizing radiation before exposure to the magnetic field, and even then, the tests did not show a classic exposure-effect relationship. Given that exposure to electric and magnetic fields is not genotoxic and that this type of exposure is qualitatively and mechanistically different from both chemical exposure and ionizing-radiation exposure, the committee felt it was premature to use those data further.

Using human-response data from epidemiologic studies of exposure to low field strengths presents a somewhat different situation for evaluation and extrapolation than that typically encountered in risk assessment. In terms of more accurate and precise estimates, the human data obviate the need to extrapolate from animals and, if the studies identified earlier in this report are used, extrapolation from high experimental doses to low environmental exposures is unnecessary because typical environmental exposures were used in those studies. In terms of sources of error that are not taken into account, the committee has added considerations of variation from person to person, which is controlled for genetically in animal studies; variation of dose and inaccurate measurement of dose, which are prescribed and carefully assessed in animal studies; and failure to adjust for each subject's characteristics (e.g., ethnicity or socioeconomic status) or exposure to other agents (e.g., ionizing radiation) that are known to modify the effect of exposure or directly cause the disease under study. Adjusting for these effects is not a part of traditional risk assessment, although failure to do so might result in inaccurate estimates of risk.

After reviewing the available data, the committee does not believe it is appropriate to perform a complete assessment of the risks of exposure to power-frequency electric and magnetic fields through the four formal stages described above. The committee believes that the data are too uncertain to result in a meaningful analysis. Some members of the committee considered the data so inconclusive as to preclude any attempt at risk assessment. Others believe that, even though a quantitative prediction of risk is technically possible, the interpretation of the resulting risk number would be problematic and likely lead to misinterpretations. However, many persons are concerned about the possible risks of exposure to residential electric and magnetic fields, and in light of its examination of the entire body of evidence, the committee recognizes the need to contribute to a better understanding of the risks by assessing them in a more limited context. That is attempted below. These observations might help put the risk into a context that can be considered for possible personal actions or government policies. The formal framework of risk assessment is used to the extent possible.

Hazard Identification

The conclusions presented in previous chapters concerning the possibility of risk from exposure to electric and magnetic fields are the following:

• There is no evidence of effects of electric- and magnetic-field exposure on bacterial DNA. Negative results from such tests are generally accepted to imply that DNA was not damaged by the agent to which the bacterial DNA was exposed.

Cancer is generally assumed to be initiated because DNA has been damaged; therefore, findings of effects, or no effects, in this type of test might be important (subject to verification by tests in other biologic systems) in identifying an agent as hazardous.

• There is no consistent or convincing evidence of effects of electric- and magnetic-field exposure on cell systems cultured in the laboratory that might imply a human-health effect at typical environmental exposure levels. Although exposure to electric and magnetic fields at 50-60 Hz has been shown to induce changes in cultured cells, the exposure conditions exceed ordinary human residential exposure by factors of 1,000 to 100,000 times. In studies at exposure levels similar in magnitude to residential exposure, no effects have been reported and replicated.

Results from tests performed using mammalian cells cultured in the laboratory have broader implications than studies using bacteria. The cell systems are closer in their complexity to those of humans, and measurements of effects on organelles, such as chromosomes, that are present in humans can be performed.

• There is no consistent or convincing evidence of effects of power-frequency electric- and magnetic-field exposure on whole animals that might imply a human-health effect at typical environmental exposure levels. Laboratory animals have shown neurobehavioral and neuroendocrine changes in response to electric and magnetic fields. However, these responses are not convincing evidence of adverse effects for humans, and they occur at exposure levels well above the human experience. Results obtained on whole animals, when the animals tested respond in a way analogous to the way humans might respond, are important in determining risk for human beings.

• There is a moderately consistent, statistically significant association between wire codes, an indirect measurement of electric- and magnetic-field exposure, and childhood leukemia. Average magnetic fields measured in the home after diagnosis of disease have not been found to be associated with excess childhood cancers.

Studies have not identified the factors that explain the association between wire codes and childhood leukemia. Wire codes are not strong predictors of exact magnetic-field strengths in the home, though they do distinguish high versus low field strengths reasonably well.

• There is a moderately consistent, statistically significant association between indirect measurements of occupational exposure to magnetic fields and both leukemia and brain cancer. The associations have been primarily with job titles that are expected to provide a greater than average exposure to magnetic fields; in some more recent studies, measurements have also been made of contemporary magnetic fields. Although the magnitude of the risks vary from study to study, most studies show increases in those two cancers.

Evidence from epidemiologic studies is the most important class of data when performing a risk assessment. The studies involve humans, and the results of the studies can be applied directly. In other types of studies, even when the results are positive, the findings are indirectly related to humans and must be extrapolated from the tested biologic system to humans.

Dose-Response Assessment

In general, hazardous substances exert their effects in proportion to the amount of exposure received—for example, two packs of cigarettes per day should be more hazardous than one pack per day.

In reviewing the data, the committee perceived no clear indication that exposures to electric and magnetic fields of different magnitudes could be related to variations in responses in systems tested in laboratory studies. The epidemiologic data, however, are open to varying interpretations. Some study results have been interpreted as showing a weak dose response; that is, some analyses indicate that those groups of subjects with higher rates of disease are more highly exposed. Nonetheless, the committee concludes that electric- and magnetic-field-exposure data overall do not provide sufficiently convincing evidence of a dose-response relationship to use these data to develop a mathematical model.

This finding is important because a demonstration of a dose-response relationship can serve as a strong confirmation that any effect that might have occurred is real and not an artifact of the experiment.

The epidemiologic data that support a dose-response relationship are sparse, and the nature of the studies and the data published preclude rigorous analysis.

Exposure Assessment

It is essential when assessing risk to ascertain whether humans or test species have been exposed to the putative causative agent and to quantify the extent of their exposure. There is no doubt that humans are exposed to electric and magnetic fields in their daily lives. In fact, exposure is so universal and unavoidable that even a very small proven adverse effect of exposure to electric and magnetic fields would need to be considered from a public-health perspective: a very small adverse effect on virtually the entire population would mean that many people are affected.

Extensive data are available about human exposure to 60-Hz magnetic fields and, to a lesser extent, to electric fields in various environments, including the residential environment. Surrogates for magnetic-field exposures, such as wire codes, have been used in some epidemiologic studies. Typical values for U.S. populations, such as the percentage of homes in various exposure categories, are presented in Chapter 2.

It is critical to the understanding of our risk assessment to recognize that the epidemiologic studies showing an association between wire codes and childhood leukemia do not establish an association between directly measured electric and magnetic fields and disease, because wire codes have not been validated as an appropriate indirect measure of the fields.

Risk Characterization

The conclusions reached in previous chapters of this report and the summary risk assessments made above indicate that the data on the effects of exposure to electric and magnetic fields on biologic systems are either negative or so uncertain that making such an estimate would be injudicious and misleading, though a theoretic risk assessment could be performed.

Biologic Mechanism of Action

As mentioned above, scientists understand cancer to be associated with damage to DNA. Thus, if an agent is found to damage DNA, the carcinogenic potential of that agent has been established as biologically plausible. That is an example of a general principle—in attempting to assess the risks of an agent, one should examine whether some biologically plausible means exists for that agent to cause disease.

No biologically plausible explanation for a putative relationship between exposure to electric and magnetic fields from power lines and an adverse effect in biologic systems has been proposed, tested, and shown to be valid.

The occurrence of a copromotional effect of exposure to very-low-frequency electric and magnetic fields, which has been reported in cell systems and animals, deserves consideration. Cancer is thought to occur after an initiator (a chemical or physical agent) has damaged DNA and started a process leading to the disease. A copromoter might not have the ability to initiate the cancer process, but the risk of cancer can increase when the test biologic system is subjected to a copromoter after the system has been exposed to an initiator.

Overall Conclusions for Risk Assessment

The body of evidence, in the committee's judgment, has not demonstrated that exposure to power-frequency electric and magnetic fields is a human-health hazard. However, some epidemiologic data support an association between surrogate measurements of magnetic fields and an increased risk of childhood leukemia. Further research for understanding the various ways of measuring exposure and their possible association with adverse health outcomes in model and human systems will be needed to resolve the uncertainty.

The committee's overall conclusion is based on the weight of the evidence after review and analysis of biologic data at the molecular, cellular, and whole-animal level that are considered relevant in assessing the possibility that exposure to electric and magnetic fields in the environment causes cancer in humans. First, in vitro studies were found to observe biologic effects at field strengths that are 1,000 or more times greater than would be experienced in residential situations. Even then, the results are inconsistent and often not reproducible. The demonstration of in vitro effects does not necessarily imply potential adverse human-health risk. Animals see because a neural effect occurs; light enters the eye, strikes the retina, and the brain forms messages on the basis of information it receives. Yet, the effect of vision is not a deleterious human-health effect, even though an effect, indeed, has been demonstrated. For example, although tests have shown that many animals can detect electric fields at 1 to 5 kV/m, there has been no evidence that these effects imply a health hazard.

Exposure to electric and magnetic fields is not genotoxic at any level of exposure. After considering the effects of exposure to electric and magnetic fields on laboratory animals, the committee concludes that the effects of such fields have no consistent pattern as a direct carcinogen. Evidence, not yet replicated, does show that exposure to electric and magnetic fields combined with exposure to high concentrations of known carcinogens increases the number of rat tumors and accelerates their appearance. Such evidence does not identify electric and magnetic fields as a probable carcinogen.

Finally, after analyzing the epidemiologic data, the committee concludes that the evidence of an association between exposure to electric and magnetic fields and cancer is not convincing, although residential wire codes have been associated with cancer. In addition to these data, no plausible biophysical mechanism has been identified that would suggest that the action of electric and magnetic fields is carcinogenic.

The data at different biologic complexities, taken in total, do not provide convincing evidence that electric and magnetic fields experienced in residential environments are carcinogenic. No tests or studies can prove that an agent is not carcinogenic at some dose level, in combination with some other biologic agent, or for some sensitive populations of humans. All that can be stated is that, under the exact experimental conditions of an extremely large number of studies, exposure to power-frequency electric and magnetic fields at environmental strengths does not produce patterns of data similar to those found for other agents that have been shown to be carcinogens. Possibly, such fields can be hypothesized to act as a nongenotoxic cocarcinogen or to act through hormonal pathways to suppress protective molecules, such as melatonin. Although such hypotheses must be tested carefully when scientific justification exists to do so, the overall conclusion that must be drawn from all the data in the studies of cells, lower animals, and humans is that the data are negative or inconclusive. Electric and magnetic fields are neither genotoxic in cells, nor a direct carcinogen in animals, nor associated conclusively with cancer in exposed humans.

The association between residential proximity to high-wire-code configurations and increased rates of childhood leukemia remains unexplained, as do the associations between occupational exposures and leukemia and brain cancer. Positive human epidemiologic data are the strongest evidence in evaluating any human-health risk. The associations for childhood leukemia have been shown to be statistically reliable and robust findings that must be considered carefully in drawing conclusions about overall risk. Uncertainty is introduced because the associations found with wire codes are not found with measured fields, and that raises serious questions about the interpretation of the positive findings and their use in quantitative modeling. The human epidemiologic studies stand alone as suggesting possible adverse health effects, and the results themselves indicate small risks (e.g., relative risk of 1.5) relative to other adverse exposures that epidemiologists consider.

Other Possible Human-Health Effects

The committee was asked to examine not only data for cancer but also data from studies investigating a possible association between exposure to electric and magnetic fields and health effects related to reproduction and development and neurologic effects expressed as learning or behavioral disorders. Although fewer studies of these effects have been conducted, the committee concludes that no studies to date have shown an association between exposure to residential electric and magnetic fields and an adverse human-health effect.

The following conclusions are based on studies with animals:

• There is no convincing evidence of an association between exposure to power-frequency electric and magnetic fields and reproductive or developmental effects.

• There is no convincing evidence of an adverse neurobehavioral effect in association with exposure to residential electric and magnetic fields.

The following conclusions are based on studies in humans:

• There is no indication of an adverse human-health impact from exposure to power-frequency electric and magnetic fields, although there is some evidence for electric- and magnetic-field-induced neuroendocrine changes. The data from epidemiologic studies on exposure to electric and magnetic fields and adverse pregnancy outcomes do not support the existence of an association. Epidemiologic studies have not been performed with the specific aim of determining the existence of neurobiologic effects, and no statements can be made regarding the occurrence of this human-health effect.

• There is no consistent or convincing evidence of adverse effects of power-frequency electric and magnetic fields on reproduction or development in human studies. However, the number and quality of the epidemiologic studies is limited, making any inferences tentative.

7

Research Needs and Research Agenda

Power-frequency electric and magnetic fields of the strength found in residences have not been shown to constitute a threat to public health that would warrant an adjustment in national research policy. However, within the range of funding that might be available for issues relating to the biologic effects of electric and magnetic fields, certain avenues of research could be pursued to resolve uncertainties that remain in the epidemiologic and laboratory findings.

EPIDEMIOLOGIC STUDIES

Epidemiologists have found an apparent weak association between childhood leukemia and wire codes, an index describing the neighborhood electric-power distribution system. Wire codes were used because they are considered an index of long-term exposure to electric and magnetic fields. Yet, subsequent attempts to replace estimates based on wire codes with measurements of electric- and magnetic-field exposure made after cancer was diagnosed have resulted in a weakening, not a strengthening, of the association. This difference in results could be resolved by further studies aimed at tracing the association, if it persists, either to some specific property of the ambient fields or to some completely different source of the disease.

From the beginning, epidemiologic evidence has been at the forefront of the concerns with possible health effects of magnetic fields. In the committee's review of the epidemiologic data, the research was found to show patterns indicative of associations but also to show some notable inconsistencies. Additional epidemiologic studies might resolve these concerns.

Various researchers have reported anecdotally that the most highly exposed individuals often seem to drive the results of an epidemiologic study. That is, a relatively small number of subjects typically have notably higher exposures than others in the study, and these higher exposures are associated with an increased risk of cancer. The obvious inference from that finding is that conducting another study focusing on the individuals with higher exposures would be most informative. If the risk factors for exposure to electric and magnetic fields operate the same as most chemical risk factors, then individuals with the highest exposures would have the highest risks, and studies of them would have the greatest statistical power and would best address issues of dose-response relationships. Therefore, it might be particularly fruitful to consider the study of children in very-high-current-configuration homes.

Two contradictions remain: Why is no association with childhood leukemia found with spot measurements of magnetic-field strength after cancer diagnosis, even though the association is found with all the other exposure metrics, and why is no consistent dose-response relationship found in these data? With respect to the spot measurements of magnetic fields, if the results of the other exposure metrics are not spurious, two obvious potential explanations exist. First, the surrogate exposure metrics (wire codes, distances from power lines, and calculated fields) might be indicators for the true risk factor and the true risk factor is not related to magnetic-field strength. A number of such plausible risk factors have been investigated, but none has explained the observed association. Second, other surrogate magnetic-field exposure metrics might be more biologically relevant measurements of magnetic-field exposure than spot measurements. These other surrogate metrics might be more representative of long-term integrated averages of magnetic-field strength or of some other aspect of magnetic-field exposure that is related to the cause of the disease (e.g., peak field strength, field variability, frequency of transients above a given field strength, or time above a specific threshold value). What would best advance understanding of this issue are studies that address the inconsistencies, not just more studies similar to those that have been done. The committee's sample-size analyses show that a study would be unlikely to change markedly the existing pattern of results unless hundreds if not thousands of leukemia cases were included in the study.

Another research approach would be to consider different exposure regimes, an approach that might enable investigators to untangle the conflicting results. Therefore, a study of children with high exposure from sources other than outdoor distribution lines might be fruitful. People with homes near high-voltage transmission lines have high exposures and are different from most of the populations studied so far in the United States. Although such studies have been conducted in the Nordic countries, there have been very few cases because of the small populations involved. Much more high-frequency variation (i.e., more on and off switching), reflecting the usage patterns of the source, is likely to be shown in distribution-line sources than in high-voltage lines, which impart much more consistent fields.

The areas of possible research parallel the issues discussed in this report. Those areas can be organized on the basis of the arrows shown in Figure 5-1 (see Chapter 5). Each of the items noted in Figure 5-1 is subject to uncertainty and could be clarified either by building upon completed studies or by developing entirely new studies of the issues. Countering the claim that epidemiologic studies have gone as far as they can in addressing the potential role of exposure to electric and magnetic fields in cancer etiology, it is quite likely that, at least in the near term, only further epidemiologic research can more strongly implicate or exonerate magnetic fields. That is not to argue for simply conducting more studies to reach consensus, but rather to design studies, some of a purely methodologic nature, that can address the specific gaps in our understanding.

Wire Codes and Childhood Cancer

Although the committee has concluded that an association probably exists between living in homes with high wire codes and childhood cancer, the epidemiologic evidence is not entirely persuasive for two reasons: potential control-selection bias and imprecision due to the modest number of homes within the high-wire-code categories.

Potential for selection bias could be mitigated by conducting studies in settings in which the roster of eligible controls is more readily defined (e.g., identifying and selecting controls at the approximate time that cases are identified) or in settings where a complete roster of the population is available as a sampling frame for controls. An alternative is to scrutinize the control selection procedures used in completed studies with additional analyses and possibly to acquire new data to evaluate the potential for such bias. In broader terms, the generic problem of identifying suitable control groups for case-control studies in the United States could be addressed, and the hypothesized deficiencies associated with random digit dialing need to be identified.

The limited precision of the wire-code categories of greatest interest, including the very-high-current configuration defined by Wertheimer and Leeper (1979), suggests identifying locations in which these categories are more prevalent than found in previous studies. Perhaps a systematic effort to estimate the prevalence of high-wire-code homes in different geographic locations could guide investigators in selecting a more optimal location.

Wire Codes and Confounders

The correlates of wire codes, such as age of home and sociodemographic characteristics of the home occupants, are not well understood. An effort to characterize the neighborhood and personal correlates of wire codes would help in postulating and testing potential confounders that could account for the link between wire codes and childhood cancer. Completed studies could be examined more systematically than they have been if descriptions were given of the geographic areas and personal attributes associated with different wiring configurations.

Confounders and Childhood Cancer

As part of a broader research agenda, more knowledge of the causes of childhood cancer would be of great benefit in evaluating the role of exposure to electric and magnetic fields. Delineation of risk factors allows control for them as potential confounders. Well-defined risk factors also would provide clues regarding mechanisms for possible biologic activity of magnetic fields (e.g., the critical time periods for cancer etiology), which could then be tested. Obviously, the desire for a better understanding of the causes of childhood cancer extends far beyond the scope of the health effects of electric and magnetic fields.

Wire Codes and Magnetic Fields

In addition to the more engineering-based questions regarding sources of magnetic fields in the environment and the role of power lines as a source (see following section on exposure and physical interactions), the empirical relationships between wire codes and human exposure are a concern. Clearer understanding of the relationship of exterior power lines to patterns of magnetic fields in homes, time spent in homes by occupants, and additional sources of magnetic fields, such as appliances, could be gained; more broadly, an understanding of the sources of human exposure and the role played by wire codes and specific aspects of electric power lines could be gained. Determinants of human exposure could be understood more broadly to assess whether wire codes, though weakly predictive of average magnetic fields in homes, are more strongly predictive of some other parameter of exposure.

Magnetic Fields and Childhood Cancer

The relationship between magnetic fields and childhood cancer is the single issue of greatest uncertainty and importance in regard to future research. Two distinct research avenues for extending knowledge can be defined.

Improved Studies of Measured Residential Magnetic Fields and Childhood Cancer

As noted previously, incomplete coverage of homes in past studies calls the largely negative results into question. Combining more advanced methods of characterizing home exposure (integration over time and consideration of grounding currents) with complete coverage of cases and controls could yield important insights into the possibility that magnetic fields account for the wire-code and cancer association.

Studies of Magnetic Fields from Sources Other Than Power Lines

Evidence that sources of fields other than those from power lines are related to childhood cancer would provide convergent evidence and strongly bolster the hypothesis that it is actually the magnetic fields associated with wire codes that influence childhood cancer risk. Other sources of exposure to electric and magnetic fields in the home, such as appliances, or sources outside the home, such as those encountered in schools, could have different potential confounders and other methodologic issues, and observed associations would be unlikely to suffer from the same validity concerns as the associations of power lines and cancer in previous studies.

LABORATORY STUDIES

Engineering Studies

Both epidemiologic and biologic studies rely on an accurate determination of the exposure variables. Characterization of the environmental fields provides a basis to test associations of epidemiologic variables with the measured field parameters and provides information on relevant features of the fields for use in biologic studies. Additional refinements in engineering techniques and in field characterization could be considered an integral part of any research agenda.

Instrumentation and Transient Currents

Instrumentation is needed that can measure more rapid changes in magnetic-field strength (changes on the order of 0.1 sec). Currently, instrumentation can capture a sample of the magnetic field every few seconds, but in biologic experiments, cells can respond to more rapid changes. For example, the ear can distinguish events that are separated by more than about 0.1 sec; it fuses events that occur more rapidly. The ability to measure magnetic-field change on the order of 0.1 sec could be used to test some of the more recent hypotheses of magnetic-field interactions. In addition, before epidemiologic or laboratory studies can be designed to test hypotheses involving transient currents, a standard way to produce and measure transient currents must be developed.

Biophysical Modeling

Realistic models are needed to evaluate induced electric currents and fields at the cellular and subcellular levels. An intermediate step could involve the development of better models of the body, such as the incorporation of anisotropies; the complexity of today's models is generally limited to inhomogeneities in tissues.

Wire Codes

Studies are needed to better understand the correlation between wire codes and magnetic fields for different utility-service areas; this work is needed before multi-city epidemiologic studies (using wire codes) can be undertaken and to better interpret the studies already completed. Such data might also be useful for extending meta-analyses to more meaningful exposure estimates.

Grounding System Currents

Grounding system current is one of the three major sources of residential magnetic fields. It seems reasonable to assume that ground currents will be greater in areas with higher wire codes or near substations (the ultimate destinations of ground return currents). How grounding system currents vary as a function of distance from substations and how these might correlate with wire-code categories are questions that could be studied.

Contemporary Versus Historical Exposures

Studies of the correlation between personal-exposure measurements (or time-weighted-average measurements) taken days, months, and years apart might contribute to an understanding of the relationship between contemporary magnetic-field strengths and magnetic fields measured at the time of disease etiology.

Wire Codes Versus Contemporary and Historical Magnetic-Field Measurements

The stability of wire codes (they change little with time over reasonably long periods) suggests that wire codes should correlate as well with contemporary spot (or time-weighted-average) measurements as they would with historical (prediagnosis) measurements. However, several trends complicate that hypothesis, and studies could examine the relationships. For example, many regions tend to increase the current demand over time, causing power lines to become more heavily loaded. Eventually, the power lines in a particular area might have to be upgraded to handle the load; thus, the wire configuration is changed, and the cycle of increased loading begins again. If the wire configurations remain unchanged over the course of a particular epidemiologic study, contemporary magnetic-field measurements might be somewhat higher than historical measurements, and the absolute magnetic-field cut points between wire-code categories would be somewhat higher than they would have been at the time of disease etiology. This trend causes no problem for the long-term validity of wire codes if the relative discriminating ability of wire codes is consistent, a point which could be examined.

There is a tendency, seen in the 1,000-homes study (EPRI 1993a) and elsewhere, for older homes to have higher wire codes, higher power-line fields and higher total fields. Possible reasons for such a trend could be that most homes built over 50 years ago and 7.1% of homes built 30-50 years ago have knob and tube wiring (which can cause higher in-home fields); none of the homes built less than 30 years ago had knob and tube wiring (EPRI 1993a). In addition, homes seem to accumulate a larger number of wiring errors as they age; wiring practices in older homes might have permitted more wiring irregularities; or power lines might be more fully loaded in older neighborhoods.

Assuming that the trend is correct, it could be interpreted to mean that in the historical period, when homes were newer, a greater spread in the magnetic fields was represented by the wiring codes. Therefore, wire codes might have somewhat greater discriminating power when applied to a historical period rather than a current period, a point which should be examined.

Biologic Studies

A serious barrier to acceptance of a possible weak connection between human health and exposure to extremely-low-frequency electric and magnetic fields in residences is the absence of a plausible physical mechanism to account for such a connection. Because any biologic effect of exposure to electric or magnetic fields at residential field strengths is, at best, at the margins of detectability, laboratory experiments to identify such a mechanism at those field strengths appear not to be feasible. However, at higher magnetic-field strengths, there appear to be genuine biologic effects, which are accessible to experiment. Mechanistic studies at high field strengths, accompanied by attention to dose-response relationships, might illuminate the situation at lower field strengths.

Laboratory experiments must be designed with appropriate positive and negative controls, conducted under blind or double-blind conditions as appropriate, and described in sufficient detail to permit independent replication. The committee found that a large fraction of published reports failed to meet those criteria.

The areas of biologic research that seem most likely to be productive include the following:

• ***Bone healing and other therapeutic applications.*** The clinical application of electric- and magnetic-field exposure at relatively high field strengths for bone healing is of limited but real benefit. Additional research could clarify the molecular mechanisms involved. Better understanding of these mechanisms might allow design of electric and magnetic-field treatments for other conditions, such as osteoporosis, and lend insight into understanding possible interactive mechanisms that might result in adverse health effects of such exposure.

• ***Characterization of the dose-response relationship for in vitro effects.*** Reproducible effects have been documented in cultured cells for electric- and magnetic-field-induced changes in calcium flux and gene expression, but only at very high magnetic-field flux densities or high electric-field strengths. When a robust effect can be observed in such studies, special effort should be made to define the change in effect as a function of the strength of the applied fields. The shape of the relationship between the field strength and the biologic effect must be established precisely to permit extrapolation and use in predicting effects at lower strengths. Exposure-response studies could also be extended to characterize the effects of the frequency of the applied fields and the effects of transient currents on the biologic response.

• ***Signal-transduction events.*** Further replication and validation studies could be carried out to investigate the apparent effects of magnetic fields on signal-transduction events, such as Ca^{2+} flux, protein-kinase cascades, and membrane-receptor activities. These pathways are important in both normal and neoplastic cell proliferation and differentiation, and possible effects of magnetic fields on these pathways might be related to the observed copromotion activity of exposure to magnetic fields in animal studies.

• ***Gene expression.*** Previously reported effects of magnetic fields on gene expression (e.g., changes in differentiation markers of bone cells and changes in signal-transduction effects) could be investigated further. Studies of putative direct effects of magnetic fields on transcriptional events could have low priority.

• ***Biophysical mechanisms.*** Research could be directed at plausible biophysical mechanisms to explain the observed in vitro and in vivo effects at relatively high magnetic-field strengths (e.g., 1-10 G). The possible role of transient currents on electric- and magnetic-field-induced effects also could be examined.

• ***Cocarcinogenesis.*** There are unreplicated data in animal studies that reported increased tumor incidence when magnetic fields were applied in combination with chemical carcinogens. Those data require replication, and if replication reinforces the reported positive results, these observations should be pursued in detailed experiments. Such experiments should focus on the dose-response relationships of magnetic-field exposures, the interacting exposures, and the temporal relationships between the different exposure conditions. Positive results could be tested in different cell or animal systems to determine whether the response is peculiar to specific biologic systems.

• *Magnetic-field-exposed initiated animals.* Early studies that suggested an influence of magnetic-field exposure on mammary-cancer development could be extended in initiated animals. Studies should include a rigorous investigation of mammary-tumor development in exposed animals. Additionally, the possible changes observed in relevant hormonal factors in magnetic-field-exposed animals should be investigated to examine potential mechanisms related to mammary cancer.

RESEARCH STRATEGY

The characteristics of electric and magnetic fields generated by the production, transmission, and use of electric energy and the possible effects of these fields on biologic systems have been the subject of extensive research for the past two decades. Due to the uncertainties and inconsistencies in the results of much of this work, Congress passed legislation in 1992 to fund an enhanced program of scientific research in this area. The Energy Policy Act of 1992 established a 5-year program of enhanced study of the characteristics of environmental exposure to power-frequency electric and magnetic fields and enhanced study of the in vitro and in vivo biologic responses to exposure to low-strength 60-Hz magnetic fields. This program is now in its third year of research and is focusing on the replication of experiments considered important to understanding the mechanism by which such fields might interact with the living system. This program is an important part of the research strategy for resolving the issues related to the possible biologic effects of magnetic-field exposure.

The work supported by the Energy Policy Act of 1992 is not anticipated to answer all the questions regarding the possible health effects of exposure to electric and magnetic fields by the program's end in 1997. Following the enhanced study supported by the Energy Policy Act of 1992, research in engineering, dosimetry, biology, and epidemiology could be initiated based on the scientific merit of the proposed work and could follow leads to plausible mechanisms that have been uncovered in previous studies. Continued research is important, however, because the possibility that some characteristic of the electric or magnetic field is biologically active at environmental strengths cannot be totally discounted. If ongoing or future research should uncover evidence of potential mechanisms that could lead to such a result, research should be continued to follow those leads and address that possibility.

Appendix A

Tables

TABLE A3-1 In Vitro Assays of Electric- and Magnetic-Field Exposure and Genotoxicity

Study	Cell Type	Exposure Characteristics	Electric-Field Strength of Culture Media	End Points Evaluated	Outcome
Hungate et al. 1979	Salmonella TA100 or TA98 exposed 20 hr in liquid nutrient broth suspension	200-800 kV/m electric field in air	Cannot be determined from report	Mutation	1.5-3-fold increase in mutation frequency in TA100 at 800,000 V/m
Moore 1979	Salmonella TA98 and TA100 tester strains exposed during growth in nutrient broth for 5-24 hr	0.3-Hz triangular magnetic field at 0.015 and 0.03 T	Induced electric field cannot reliably be estimated from report	Reverent assay	No significant effects observed
Wolff et al. 1980	CHO cells exposed 4 hr (SCEs) or 13 hr (chromosomal aberration)	NMR gradient field; 1.82 pulses/sec, 4.6 T/sec; coexposed 0.352-T static magnetic field and 5-mW/cm^2 magnetic field at 15 MHz	Cannot be determined from report	Chromosomal aberrations and SCEs	No significant effects observed
Wolff et al. 1980	CHO cells exposed 4 hr (SCEs) or 13 hr, 40 min (chromosomal aberration)	0.35 T plus coexposure to RF field at 15 MHz, 5 mW/cm^2 and time-varying magnetic-field changes at 4.6 T/sec and 1.82 T/sec	0	Chromosomal aberrations and SCEs	No significant effects observed
Cooke and Morris 1981	Human lymphocytes exposed 1 hr	0.5-1.0 T	0	Chromosomal aberrations and SCEs	No significant effects observed

Reference	Exposure conditions	Field	Calculated field	Assay	Results
Thomas and Morris 1981	*E. coli* AB1157 exposed 5 hr (agar plates)	1.0 R ± coexposure to RF field at 1 mW/cm² and gradient magnetic field at 1-12 T/sec	Not calculated	Revertant assay	No significant effects observed
Thomas and Morris 1981	*recA, uvrA,* or *recA uvrA E. coli*: exposed 5 hr in Petri dishes	1.0 T	0	Survival (*recA, uvrA,* or *recA uvrA E. coli* mutants) compared with wild type	No significant effects observed
Thomas and Morris 1981	*E. coli recA, uvrA,* and *recA uvrA* mutants or *E. coli* AB1157 exposed 40 min or 5 hr on agar Petri plates	Gradient magnetic field at 1-12 T/sec; coexposure to 0.094-T static magnetic field and 1-mW/cm² RF field	2-30 mV/m calculated from exposure apparatus by McCann et al. (1993)	Revertant assay	No significant effects observed
Mileva 1982	Human peripheral lymphocytes: exposed 15-360 min	0.3 T	0	Chromosomal aberrations	No significant effects observed
Nordenson et al. 1984	Human peripheral lymphocytes exposed 3 hr to phytohemagglutinin stimulation	50-Hz sinusoidal field applied through agarose bridges	14 V/m (10 A/m²) calculated from exposure apparatus by McCann et al. (1993)	Chromosomal aberrations	No significant effects observed
Nordenson et al. 1984	Human peripheral lymphocytes in whole blood exposed 1 min before phytohemagglutinin stimulation	10 spark discharge pulses, 2 msec wide	250-350 kV/m	Chromosomal aberrations	At the highest dose, a significant increase in chromosomal breaks

Continues on next page

TABLE A3-1 Continued

Study	Cell Type	Exposure Characteristics	Electric-Field Strength of Culture Media	End Points Evaluated	Outcome
d'Ambrosia et al. 1985	Bovine lymphocytes in liquid culture medium exposed 72 hr by applying external electrodes to side walls of culture flasks	50-Hz sinusoidal field with 11% THD applied through capacitive coupling	0.016 V/m (0.024 A/m²)	Chromosomal aberrations	A significant increase (~3-fold) in chromosomal aberrations for three experiments
Cohen 1986; Cohen et al. 1986a,b	Peripheral blood lymphocytes from normal individuals (Cohen et al. 1986b) and individuals with chromosomal instability syndromes exposed 69-hr culture period	60-Hz sinusoidal field applied through agarose bridges and coexposure to 60-Hz sinusoidal magnetic field at 10-200 μT (38-75 mT/sec)	0.24 V/m (0.2 A/m²) (no reliable estimate available from published report)	Chromosomal aberrations and SCEs	No significant effects observed
Cohen et al. 1986a,b	Peripheral blood lymphocytes normal (Cohen et al. 1986b) and with chromosomal instability exposed 69 hr in culture	60-Hz sinusoidal field, circularly polarized, at 10-200 μT (38-75 mT/sec)	0.7-13 mV/m calculated from exposure apparatus by McCann et al. (1993); coexposure to 60-Hz sinusoidal electric field, 0.24 V/m (0.3 A/m²) (McCann et al. 1993)	Chromosomal aberrations and SCEs	No significant effects observed

Reference	Test system	Field conditions	Field strength	Assay	Result
Juutilainen and Liimatainen 1986	*Salmonella* TA100 and TA98 exposed in top agar or liquid nutrient broth culture for 48 or 6.5 hr, respectively	100-Hz sinusoidal field at 0.13, 1.3, 13, and 130 μT	0.2, 2.0, 20, and 200 μV/m (Petri dishes); and 1.5, 15, 150, and 1,500 μV/m (flasks) calculated from exposure apparatus by McCann et al. (1993)	Revertant assay	No significant effects observed
Livingston et al. 1986, 1991	Human lymphocytes or CHO cells exposed 24-96 hr or 72 hr, respectively	60-Hz sinusoidal field applied through agarose bridges	0.024-24 V/m (no reliable estimate available from published report) (0.03-30 A/m^2)	Chromosomal aberrations	No significant effects observed
Livingston et al. 1986, 1991	Peripheral blood lymphocytes or CHO cells exposed 24-96 hr or 72 hr, respectively	60-Hz sinusoidal field, circularly polarized, at 0.22 mT (0.082 T/sec)	0.7-13 mV/m calculated from exposure apparatus by McCann et al. 1993; coexposure to 60-Hz sinusoidal electric field at 0.024-24 V/m	SCEs and micronuclei	No significant effects observed
Whitson et al. 1986	Normal human fibroblasts previously or post irradiated with UV light (254 nm) exposed up to 48 hr	60-Hz applied through capacitative coupling; field in air outside media 10 kV/m	0.4 mV/m	DNA single-strand breaks assayed via 5-bromodeoxyuridine photolysis; pyrimidine dimers assayed using hydrolysis then two-dimensional paper chromatography, or	No significant effects observed

Continues on next page

TABLE A3-1 Continued

Study	Cell Type	Exposure Characteristics	Electric-Field Strength of Culture Media	End Points Evaluated	Outcome
				by treating cells with a UV-specific endonuclease followed by a fragment sizing analysis on sucrose gradients	
Takahashi et al. 1987	Chinese hamster V79 cells exposed 24 hr	100-Hz saw-toothed field at 0.180-2.500 mT (7.2-100 T/sec)	0.02-0.33 V/m calculated from exposure apparatus by McCann et al. (1993)	SCEs	No significant effects observed
d'Ambrosia et al. 1988-1989	Bovine lymphocyte cultures exposed 3 or 45 hr	50-Hz sinusoidal field applied through agarose bridges	0.77-7.7 V/m (1-10 A/m²)	Chromosomal aberrations	Significant increases in chromatid breaks at high exposure level reported after 45-hr exposure and in total aberrations in one of two cultures tested after 3-hr exposure
Reese et al. 1988	CHO cells exposed 1 hr	60-Hz sinusoidal field applied through agarose bridges; coexposure to 60-Hz sinusoidal field at 0-2 mT	1-38 V/m	DNA single-strand breaks	No significant effects reported

Reference	Exposure	Electric field	Endpoint	Results
Reese et al. 1988	CHO cells exposed 1 hr	60-Hz sinusoidal field at 2 mT (0.75 T/sec); coexposed to 60-Hz sinusoidal electric field at 0-38 V/m	8 mV/m calculated by McCann et al. (1993)	DNA repair measured by alkaline elution	No significant effects observed
Bersani et al. 1989	Human peripheral lymphocytes or two human cell lines were exposed 48 hr	50-Hz saw-toothed field at 2.5 mT peak strength (1 T/sec; induced pulse 2-msec wide at 2 mV/m)	2 mV/m	DNA single-strand breaks	No significant effects observed
Cossarizza et al. 1989; Bersani et al. 1989	Human lymphocytes exposed 6 hr after some cultures irradiated with 100-Gy ^{60}Co	50-Hz saw-toothed field at 2.5 mT peak strength (1 T/sec; induced pulse 2-msec wide at 2 mV/m)	2 mV/m	Unscheduled DNA synthesis	No significant effects observed
Peteiro-Cartelle and Cabezas-Cerrato 1989	Human lymphocytes exposed 3 hr or simultaneously cultured and exposed 72-96 hr	0.045-0.125 T	0	Chromosomal aberrations and SCEs	No significant effects observed
Rosenthal and Obe 1989	Human peripheral lymphocytes cultured 72 hr in magnetic field	50-Hz sinusoidal field at 0.1-7.5 mT (0.031-2.4 T/sec)	0.1-8 mV/m calculated from exposure apparatus by McCann et al. (1993)	SCEs	No significant effects observed
Rosenthal and Obe 1989	Human peripheral lymphocytes pretreated with NMU, DEB, or trenimon and cultured up to 72 hr in presence of magnetic field	50-Hz sinusoidal field at 0.5-2 mT (0.16-0.63 T/sec) with coexposure to NMU or trenimon	0.61-2 mV/m	SCEs	Statistically significant ($p < 0.05$) increase in SCEs only in cells treated with NMU or trenimon

Continues on next page

TABLE A3-1 Continued

Study	Cell Type	Exposure Characteristics	Electric-Field Strength of Culture Media	End Points Evaluated	Outcome
Takatsuji et al. 1989	Human peripheral lymphocytes exposed <30 min	1.1 T + coexposure to protons and alpha particles	0	Chromosomal aberrations	Proton coexposure significant dose-response effect; frequency of dicentrics increased for both coexposures
Frazier et al. 1990	Human peripheral lymphocytes previously exposed to γ-irradiation (5 Gy) exposed 0-30 min during repair	60-Hz sinusoidal fields applied through agarose bridges; coexposure to γ radiation, 60-Hz sinusoidal magnetic field at 0-0.001 T	1-20 V/m	DNA single-strand breaks	No significant effects reported
Garcia-Sagredo et al. 1990	Peripheral blood lymphocytes or CHO cells exposed 24-96 hr or 72 hr, respectively	4.4-kHz saw-toothed pulses of 5 msec width, 14 pulses per sec at 1-4 mT peak strength (50-200 T/sec)	0.07-0.27 V/m calculated from exposure apparatus by McCann et al. (1993)	SCEs	No significant effects observed
Balcer-Kubiczek and Harrison 1991	C3H/10T½ cells exposed 24 hr; post-exposure of some cells with TPA, either preceded or followed by X-rays given at 0.5, 1, or 1.5 Gy	2.45-GHz microwaves pulse modulated at 120 Hz with electric fields at 18, 56, or 120 V/m and magnetic fields at 0.09, 0.27, or 0.56 µT	Not calculated	Cell survival and neoplastic transformation	EMF alone demonstrated no effect; transformation due to EMF plus TPA highly significant; neoplastic transformation dependent on level of

					EMF exposure and additive of X-rays given as a cocarcinogen
García-Sagredo and Monteagudo 1991	Human peripheral lymphocytes cultivated in vitro 72 hr and exposed over the last 24 hr to magnetic fields	Quasi-rectangular pulses lasting 26 μsec, frequency 4.4 kHz, in trains of 5 msec at 14-Hz repetition rate with peak strength at 1, 2, and 4 mT	Not calculated	Chromosomal aberrations	Significant effect observed at 4 mT; no significant effects observed at 1 and 2 mT
Khalil and Qassem 1991	Human lymphocytes grown 24, 48, or 72 hr in presence of the magnetic field	50-Hz pulsed field at 1 mT (0.72 T/sec)	0.043 V/m	Chromosomal aberrations	Significant decreases in mitotic index; increases in chromosomal aberrations for all exposure periods; slight increase in SCEs ($p < 0.05$) only for 72 hr
Novelli et al. 1991	*Saccharomyces cerevisiae* cultures exposed up to 24 hr and then examined by pulsed-field gel electrophoresis (PFGE)	50-Hz electric- and magnetic-field exposure consisting of 4 units: 1. uniform magnetic field; 2. uniform electric field; 3. orthogonal uniform electric and magnetic field; and 4. no field control with electric field from 0.1-20 kV/m and magnetic field from 0.2-200 μT	Not calculated	DNA double-strand breaks	No significant effects observed

Continues on next page

TABLE A3-1 Continued

Study	Cell Type	Exposure Characteristics	Electric-Field Strength of Culture Media	End Points Evaluated	Outcome
Scarfi et al. 1991	Human lymphocytes grown 72 hr in the magnetic field	50-Hz saw-toothed field at 0.025 T (some cell cultures coexposed to mitomycin C)	0.005 V/m	Micronuclei	No significant effects observed
Fiorani et al. 1992	Cultured K562 human tumor cells exposed 1, 4, 6, 12, or 24 hr	50-Hz electric field at 0.2-20 kV/m and magnetic field at 0.2-200 μT	Not calculated	DNA single-strand breaks and cell growth	No significant effects observed
Chahal et al. 1993	E. coli K-12 strain AB1157, and its derivatives TK702 umuC (deficient in error prone repair) and TK501 umuC uvrB (lacking both error prone and excision repair) exposed 1 or 16 hr	1-Hz electric field at 3 kV/m for 1 hr or 1 kV/m for 16 hr alone or in combination with UV and/or mitomycin C	Not calculated	Mutations	No significant effects observed
Fiorio et al. 1993	Chinese hamster V79 cells exposed 10 days	50-Hz sinusoidal magnetic field at 200 μT	Not calculated	Chromosomal aberrations, SCEs, and cell survival	No significant increase in chromosomal aberrations or SCEs; cell viability decreased by 50% after 10 days with

Reference	System/Exposure	Field	Dose	Endpoint	Results	Comments
Scarfi et al. 1993	Human peripheral lymphocytes exposed 72 hr and assayed using the cytokinesis-block micronucleus assay	50-Hz ac sinusoidal electric field at 0.5, 2, 5, and 10 kV/m	Not calculated	Micronuclei	No significant effects observed	only 100 plated; however, no reduction in viability with 2×10^5 seeded cells
Zwingelberg et al. 1993	Cultured rat peripheral lymphocytes exposed 7-28 days, 24 hr/day	Homogenous 50-Hz, magnetic field at 30 mT	Not calculated	SCEs and chromosomal aberrations	No significant effects observed	
Fairbairn and O'Neill 1994	HL-60 cells, Raji cells, HeLa cells, and human peripheral lymphocytes exposed 2-30 min	50-Hz magnetic field with peak amplitude at 5 mT and pulse duration of 3 msec	Not calculated	DNA single-strand breaks	No significant effect observed	
Libertin et al. 1994	HeLa cells transfected with a CAT construct transcriptionally driven by HIV-LTR promoter exposed 24 or 48 hr	ac field: 10 Hz-1.6 kHz, 0.07-35 µT; dc field: 170 µT	Not calculated	HIV-LTR expression	No significant effects observed	
Nordenson et al. 1994	Human amniotic cells exposed 72 hr continuously and intermittently (15 sec on, 15 sec off; 2 sec on, 20 sec off)	50-Hz magnetic field at 30 µT (rms) and 300 µT	Not calculated	Chromosomal aberrations	A significant increase observed in intermittently exposed cells; no significant increase seen in continuously exposed cells	

TABLE A3-2 Peer-Reviewed Reports on Power-Frequency Electric- and Magnetic-Field Exposure and Calcium, October 1990-1994

Study	Cell Type	Exposure Characteristics	Electric-Field Strength of Culture Media	End Points Evaluated	Outcome
Moses and Martin 1992	Early chicken embryos	NA	NA	Levels of 5-nucleotidase, acetylcholinesterase (NT), and alkaline phosphatase	Nine exposed and 13 controls with morphologic anomalies; in normal embryos NT activity decreased; in abnormal embryos, all three were decreased
Karabakhtsian et al. 1994	HL-60 cells	NA	NA	c-fos and c-myc	Experiments demonstrated that calcium is necessary in the cell response to electric and magnetic fields
Carson et al. 1990	HL-60 cells	Radiofrequency EMF, static magnetic field, and time-varying magnetic field supplied by a magnetic resonance imaging unit for 23 min	NA	Calcium-sensitive fluorescent indicator indo-1	Cells treated with all three fields in combination exhibited increase in calcium; cells exposed only to the time-varying magnetic field also had calcium higher than controls
Walleczek and Budinger 1992	Thymic lymphocytes (rat)	3-Hz pulsed magnetic field (PMF) with peak flux densities at 1.6, 6.5, or 28 mT	Induced electric fields at 0.04, 0.16, or 0.69 mV/cm	Concanavalin-A (Con-A)-induced calcium-ion signaling	Exposure of Con-A responsive cells to the 1.6-, 6.5-, and 28-mT fields resulted in 29.8, 45.7, and 95.6% inhibition of $^{45}Ca^{2+}$ uptake, respectively; decreases induced by the 6.5- and 28-mT fields were statistically significant; PMFs having flux densities nearly 10^4 times greater than those found in the average human environment were shown to stimulate or inhibit Ca^{2+} signal transduction

Yost and Liburdy 1992	Thymic lymphocytes	16-Hz, 42.1-μT with collinear static magnetic field at 23.4 μT (ac/dc field ratio 1.8)	Dosimetry carefully controlled	Mitogen-activated (Con-A) calcium transport using $^{45}Ca^{2+}$	Increase in Ca^{2+} observed 100 sec after mitogen stimulation
Smith et al. 1993	Seeds of *Raphanus sativus*	60-Hz magnetic field tuned to the ion cyclotron resonance frequencies for calcium and potassium	24 hr/day for 21 days	Germination rate, growth, size	Seeds exposed to calcium-tuned fields slow to germinate but grew more rapidly and finally larger than the controls; potassium-tuned fields produced rapid germination but inhibited all but root growth, which was larger than controls
Schwartz and Mealing 1993	Atrial strips of frog heart	Exposed for 32 min to continuous-wave (CV) or amplitude-modulated (AM) magnetic field at 1 GHz; modulation at 0.5 Hz in synchrony with the preparation or at 16 Hz	Specific absorption rate (SAR) ranging from 3.2 μW/kg to 1.6 W/kg	Calcium efflux using $^{45}Ca^{2+}$	No effects observed
Schwartz et al. 1990	Isolated frog hearts	30 min in Crawford cell; 240 MHz, either CV or AM at 0.5 or 16 Hz	Calculated SAR between 0.15 and 3.0 mW/kg	Calcium efflux using $^{45}Ca^{2+}$	No effect at 0.5 Hz; movement of calcium ions observed at 16 Hz; 18% change at 0.3 mW/kg and 21% at 0.15 mW/kg
Dutta et al. 1992a	Neuroblastoma cells NG108	147 MHz AM at 16 Hz for 30 min	NA	Acetylcholine esterase activity	Monitoring AChE activity in power density and time windows confirms earlier work on neuroblastoma cells in culture where calcium efflux was monitored

Continues on next page

TABLE A3-2 Continued

Study	Cell Type	Exposure Characteristics	Electric-Field Strength of Culture Media	End Points Evaluated	Outcome
Dutta et al. 1992b	Neuroblastoma cells NG108 and IMR-32	147 MHz AM at 16 Hz for 30 min	SAR at 0.001–0.5 mW/g	Enolase activity	SAR at 0.05 mW/g had significant enhancement in AChE, calcium, and enolase; all three depressed at 0.01 mW/g
Parkinson and Sulik 1992	Diatom, *Amphora coffeiformis*	Collinear dc and ac magnetic fields tuned to the cyclotron resonant condition for calcium at 16, 30, and 60 Hz	NA	Fractional motility	No effect; unable to reproduce earlier work by Smith et al. (1987)
Reese et al. 1991	Diatom, *Amphora coffeiformis*	16-Hz magnetic field at 20.9 µT parallel to the horizontal component of the dc field	NA	Fractional motility	Observed field-associated increase in diatom motion at 0.25 mM calcium (similar to that reported by Smith et al. 1987); however, percentage of moving cells not sufficiently reproducible to allow examination for frequency dependence
Klavinsh et al. 1991	Chick small intestine	Steady magnetic field at 1.14 mT and pulsed magnetic field at 80 Hz	NA	$^{45}Ca^{2+}$ uptake	Steady and pulsed magnetic fields enhanced uptake of $^{45}Ca^{2+}$ by 40–50%
McLeod and Rubin 1992	Turkey ulnae (wings)	Pulsed electric field with flux ramp of 550 T/sec and 380 msec duration, repeated at a rate of 75 Hz; 15-, 75-, and 150-Hz sinusoidal fields	No more than 0.01 mV/cm at the tissue	Bone area	Osteogenic stimulation −3%, +5%, and +20% for 150-, 75-, and 15-Hz sinusoidal electric fields, respectively

Note: NA, not available.

TABLE A4-1 Electric- and Magnetic-Field Exposure and Carcinogenesis

Study	Species (Sex)	Number per Exposed Group	Exposure Characteristics	Exposure Duration	End Point Evaluated	Outcome	Comments
Spontaneous Tumor Development							
Beniashvili et al. 1991	Rats (female)	25-50	20 μT, 50 Hz	0.5-3 hr/day for up to 2 yr	Mammary tumors	Increase of tumor incidence at 3 hr/day exposure	Minimal exposure system information
Rannug et al. 1993a	Mice (female)	36	50 or 500 μT, 50 Hz	19-21 hr/day for 103 wk	All tumors	Reduction in survival time and trend to increase in leukemia at 500 μT exposure	
Svedenstal and Holmberg 1993	Mice (female)	63	15 μT, 20 kHz	Life long	Lymphomas	No effects of field exposure	X-ray, 4 × 0.31 Gy used as an initiator
Implantation of Tumor Cells							
Thomson et al. 1988	Mice (female)	20	1.4, 200, or 500 μT, 60 Hz	6 hr/day for 5 days/wk until death (~2 wk)	P388 leukemia cells	No effect on survival time	Severe time limitation with model dynamics
Promotion of Tumors							
Rannug et al. 1993b,c	Rats (male)	9-10	50 or 500 μT, 50 Hz	19-21 hr/day for 12 wk	Liver tumor	No tumor promotion	

Continues on next page

TABLE A4-1 Continued

Promotion of Tumors

Study	Species (Sex)	Number per Exposed Group	Exposure Characteristics	Exposure Duration	End Point Evaluated	Outcome	Comments
McLean et al. 1991	Mice (female)	32	2 mT, 60 Hz	6 hr/day, 5 days/wk for 21 days	Skin tumors promoted by DMBA/TPA	Trend to more rapid development of tumors	Copromotion with TPA suggested
Stuchly et al. 1992	Mice (female)	48	2 mT, 60 Hz	6 hr/day, 5 days/wk for 21 days	Skin tumors promoted by DMBA/TPA	Decrease in tumor latency; increase in tumor incidence	
Rannug et al. 1993a	Mice (female)	36	50 or 500 μT, 50 Hz	19-21 hr/day for 12 wk	Skin tumors promoted by DMBA/TPA	No effect on tumor latency or tumor incidence	
Beniashvili et al. 1991	Rats (female)	50	20 μT, 50 Hz	0.5-3 hr/day for up to 160 hr	Mammary tumors promoted by NMU	Increase in tumor number; decrease in tumor latency at 3 hr/day	Minimal information on system and experimental design
Mevissen et al. 1993	Rats (female)	15-18	30 mT, 50 Hz	24 hr/day for 3 mon	Mammary tumors promoted by DMBA	Increase in tumor number per animal	Not reproduced upon repeat
Löscher et al. 1994	Rats (female)	36	0.3-1 μT, 50 Hz	24 hr/day for 3 mon	Mammary tumors promoted by DMBA	Trend to decrease in tumor latency; decrease in nocturnal melatonin secretion	

Löscher et al. 1993	Rats (female)	99	100 μT, 50 Hz	24 hr/day for 3 mon	Mammary tumors promoted by DMBA	Decrease in tumor latency; strong increase in tumor incidence	Strong experimental protocol and methods
Baum et al. 1995	Rats (female)	99	100 μT, 50 Hz	24 hr/day for 3 mon	Mammary tumors promoted by DMBA	Increase in incidence; increase in median tumor size	Strong experimental protocol and methods

TABLE A4-2 Electric- and Magnetic-Field Exposure and Reproductive and Developmental Effects

Study	Species	Developmental Stage	Number of Controls and Exposure Characteristics	Number of Exposed and Exposure Characteristics	End Points Evaluated	Outcome	Comments
Andrienko 1977	Rats	Parental male and female before and during gestation	270 animals	270 animals; 5 kV/m, 50 Hz, 1.5-4.5 mon including gestation	Reproductive processes and in utero development	Decrease in weight of newborns and survival to 21 days	No apparent relationship to exposure; statistical design and lack of description of experimental design do not meet scientific criteria for evaluation
Algers and Hultgren 1987	Cows (Swedish red and white)	4-mon gestation	58 animals; same environment below 100 V/m and 70 nT	58 animals; 4 (1.4-8.4) kV/m, 50 Hz, 2 (0.4-4.7) μT; exposure under 400-kV power line	Fertility, estrous cycle, progesterone levels, intensity of estrous, viability of offspring, malformations	No changes detected	Blinding in study not indicated
Berman et al. 1990	Chickens (white leghorn)	Embryo during first 48 hr of development	100 animals each in six laboratories; no field applied	100 animals each in six laboratories; unipolar pulses, 100 Hz, 9.5 msec duration, 1 μT peak, 2 nsec rise time	Fertility, developmental stage, morphology	Two of six laboratories detected a decrease in percent of normal embryos as a function of fertile eggs and live embryos; effect	Field uniformity 5% electric field, dc magnetic field, 50 and 60 Hz evaluated, also vibrations; all laboratories agreed on viability, stage, and somite

Reference	Species	Age	Control/Sham	Endpoint	Exposure conditions	Results	Comments
						significant when results of all laboratories pooled	development; disagreed on malformations; four laboratories had no increase in malformations, one had a 4-fold increase, and one had a 2-fold increase; evaluations blinded
Blackman et al. 1988b	Chickens (*Gallus domesticus*)	Embryos and 1.5 days posthatching	No sham exposure	Calcium efflux from brain	288 eggs, ~10 V/m, 50 or 60 Hz; 5.9 V/m; 73 nT, brain in vitro	Calculated current density: 0.13 $\mu A/m^2$ in eggs; calcium efflux affected by 50 but not 60 Hz	Not independently confirmed; no indication that studies were blinded
Burack et al. 1984	Sprague-Dawley rats	14-21 days	12 litters; not energized, randomly selected system (room)	Postnatal viability, growth, body weight, developmental landmarks	17 litters; 80 kV/m, 60 Hz	Well-engineered and evaluated system, all required measurements, no shocks while drinking; blind experiment; no change in litter size, viability, or other measures; reduction in percent of exposed males displaying copulatory behavior	Small number of animals; no evaluation of general stress as a confounder; no indication that studies were blinded

Continues on next page

TABLE A4-2 Continued

Study	Species	Developmental Stage	Number of Controls and Exposure Characteristics	Number of Exposed and Exposure Characteristics	End Points Evaluated	Outcome	Comments
Cameron et al. 1985	Medaka fish	Fertilized eggs (2- and 4- cell embryos)	NA	Number of eggs not provided; 100 μT rms, 300 mA/m²; 60 Hz; magnetic only, electric only, and magnetic plus electric	Morphologic defects; developmental delay	No increase in morphologic defects; developmental delay observed with magnetic and magnetic plus electric fields, but not with electric field alone	Not independently confirmed; reported delays did not result in abnormal development or decreased survivability; no indication that studies were blinded
Cox et al. 1993	Chickens (white leghorn)	Last 52 hr of gestation	200 eggs	200 eggs; 10 μT, 50 Hz, plus 17 μT DC	Morphology	No increases in abnormal development	Attempted to confirm Berman studies; analyses blinded
Fam, 1980	Swiss-Webster mice	Parental male and female exposed before and during gestation	Number not provided; system not energized; similar set up to exposure.	23 females and 23 males; 240 kV/m, 60 Hz	Litter growth, blood histology, biochemistry, histology of critical organs	No changes	No information given on evaluation of fields from given dimensions; poor uniformity, 23 animals housed in a single cage; no indication that studies were blinded

Reference	Species	Exposure duration	Control	Exposure	Endpoints	Results	Comments
Free et al. 1981	Sprague-Dawley rats	20-69 days of age	20 animals; not energized, interchangeable with field, all the same.	20 animals; 64, 68, or 80 kV/m; 60 Hz	Testosterone, FSH, LH, corticosterone, prolactin, TSH, GSH, thyroxin	No treatment-related effects	Well-engineered exposure system, described in great detail, all essential parameters required and many desirable given; no indication that studies were blinded
Frolen et al. 1993	CBA/S mice	1-19 days, 2-19 days, 5-19 days, 7-19 days	543 animals; control racks with no coils; stray field at 0.1-0.7 μT	707 animals; 15 μT, 20-kHz saw-toothed, 45-μsec rise and 5-μsec fall	Number of implants, resorptions, living and dead fetuses, malformations, length and weight of live fetuses	Increased resorption rate in all but 7-19 days; fetal body weight and length decreased in 7-19 days; no change in litter size	dc magnetic field measured, 50-Hz ambient 15-52 nT, exposure field not perturbed by cages; no information on vibration and illumination; lack of correlation between increased rates of resorptions and litter size makes it unlikely that detected increase is biologically significant; no indication that studies were blinded
Huuskonen et al. 1993	Wistar rats	1-20 days	72 animals; coil system not energized	144 animals; 35.6 μT (50 Hz) or 15 μT (20 kHz, saw-toothed)	Malformations, resorptions, living and dead fetuses	No increase in malformation or resorption rates; mean number of	6-17% field uniformity, dc and 50-Hz ambient measured; no

Continues on next page

TABLE A4-2 Continued

Study	Species	Developmental Stage	Number of Controls and Exposure Characteristics	Number of Exposed and Exposure Characteristics	End Points Evaluated	Outcome	Comments
						implants and living fetuses increased in 50 Hz; increase in minor skeletal anomalies	indication that studies were blinded
Kowalczuk and Saunders 1990	Mice (male)	Dominant lethal mutation	10 animals; same room, plates not energized	10 animals; 20 kV/m, 50 Hz; 2 wk; positive control: 10 animals	In utero death, litter size, viability	No effects	10% field uniformity; no information on exposure given, except cage position interchanged; females not exposed; no indication that studies were blinded
Margonato and Viola 1982	Rats (male)	Offspring exposed up to 48 days	NA	27 animals; 30 min/day or 8 hr/day; 100 kV/m, 50 Hz	Treated males: fertility, sperm viability, morphology; offspring: number of implantations, percent live/litter, incidence of malformations	No treatment-related effects on male reproduction or offspring	No indication that studies were blinded

Marino et al. 1976	Mice	Three generations	233 animals; either ambient or shielded in same apparatus	331 animals; 10 kV/m, 60 Hz, 3.5 kV/m	Mortality and morbidity during first week postpartum; 8-35 days postpartum	Decreased body weight at 35 days postpartum and increased mortality for three generations	Number of experimental animals approximate; results might be due to grounding microcurrents animals experienced while feeding; no indication that studies were blinded
Marino et al. 1980	Mice	Three generations	519 animals; either ambient or shielded in the same apparatus	497 animals; 3.5 kV/m, 60 Hz, 3.5 kV/m	Mortality and weight of animals for three generations	Increased mortality in each exposed generation and increase in body weight in only the third generation	Ambient electric field 2-12 V/m; rubber foam to prevent unspecified vibrations; no examination of fetuses for birth weight or malformations; insufficient data regarding mortality for independent analysis; number of animals approximated; possible microshocks from drinking apparatus; no indication that studies were blinded

Continues on next page

TABLE A4-2 Continued

Study	Species	Developmental Stage	Number of Controls and Exposure Characteristics	Number of Exposed and Exposure Characteristics	End Points Evaluated	Outcome	Comments
Martin 1992	Chickens (white leghorn)	First 48 hr	100 eggs	100 eggs; 3 μT (peak to peak), 60 Hz; with 2-μsec rise and fall times, 500-μsec duration	Morphology; frequency of malformations	No exposure-related effects	All analyses blinded
McGivern et al. 1990	Sprague-Dawley rats	15-20 days gestation	6 animals; same treatment, but not in the coil; cage control: 6 animals	6 animals; 800 μT (peak intensity), 15 Hz pulsed, 300 μsec duration, 5 μsec fall time	At birth: number live, average weight, anogenital distance; at 120 days postpartum: reproductive morphology, and male testosterone, LH, FSH, testes, accessory sex organ weight, and marking behavior	Calculated internal fields at 0.1-0.5 V/ m; no effect on number live, average weight, anogenital distance; no differences in hormone levels; increased accessory sex organ weight; reduced marking behavior	Not replicated by any other laboratory; no indication that studies were blinded
McRobbie and Foster 1985	Swiss-Webster (CD-1) mice	Not known	NA	Number of animals not given; varying and unspecified periods of exposure; 3.5-12 kT (capacitor	Number of live young and postnatal growth rates	Heat removed by forced air, vibrations eliminated by separate mounting, noise limited but not eliminated; no	Does not conform to scientifically accepted protocols; no indication that studies were blinded

				discharge-MRI simulation)		treatment-related effects	
Persinger et al. 1978	Wistar rat	19 days prepartum to 3 days postpartum	3 animals per group; the same system without magnets	3 animals per group; 5, 100, or 1,000 µT, 0.5-Hz rotating	38 blood, tissue, and consumptive measurements	Random	No indication that studies were blinded
Rommereim et al. 1987	Rats (female)	Adults and offspring exposed 19 hr/day for 4 wk	1,780 animals; not energized, interchangeable with field, all the same	1,831 animals; 100 kV/m; effective field at 65 kV/m, 60 Hz	Copulatory behavior, intrauterine mortality, malformation	No indication of altered mating behavior; effect on fertility not consistent; fetal death lower in one exposed group than in controls	Well-engineered exposure system, described in great detail, all essential parameters required and many desirable given; very large study, found some positive findings but not repeated in duplicate experiments; inconsistent results could be due to random variation or threshold dose; analyses blinded
Rommereim et al. 1989	Rats (female)	Adults through mating, pregnancy parturition, and rearing of young	223 animals; not energized, interchangeable with field, all the same	450 animals; 112, 150 kV/m; 60 Hz; 1-mon exposure of females before mating; exposed 19 hr/day	Litter size, sex ratio, mortality, maternal and fetal weight gain, and growth	No effects on reproduction measurements	Well-engineered exposure system, described in great detail, all essential parameters required and many desirable given; study

Continues on next page

TABLE A4-2 Continued

Study	Species	Developmental Stage	Number of Controls and Exposure Characteristics	Number of Exposed and Exposure Characteristics	End Points Evaluated	Outcome	Comments
							designed to answer question of threshold from 1987 publication; analyses blinded
Rommereim et al. 1990	Sprague-Dawley rats	Adult through pregnancy, parturition, and rearing of the young for two generations	68 animals; not energized, interchangeable with field; all parameters the same	204 animals (3 groups) exposed 19 hr/day; 10, 65, or 130 kV/m; 60 Hz	Percent pregnant, gestational and postnatal weight gain, litter size, neonatal and juvenile mortality, sex ratio, placental weight, number of corpora lutea, implantations, resorptions, malformation	No detrimental effects on survival or growth of the offspring	Well-engineered exposure system, described in great detail, all essential parameters required and many desirable given; analyses blinded
Seto et al. 1984	Rats	Four generations	1,337 animals; not energized, randomly selected system (room)	1,346 animals; 80 kV/m, 60 Hz, 21 hr/day; conceived, born, and raised for four generations	Fertility, litter size at birth and weaning, sex ratio, weight at weaning, frequency of malformations	No effects	No information given on evaluation from given dimensions: poor uniformity, 23 animals housed in one cage, raising

							concerns about shielding; no water available to animals during exposure; no measurements of weight of newborns or careful examinations of those exposed in utero for malformations; analyses blinded
Sikov et al. 1984	Rat	Before mating until term; 8 days from conception; 17-25 days postpartum	128 animals; not energized, interchangeable with field, all the same	337 animals; 100 kV/m; 60 Hz, 20 hr/day; group 1, females exposed 6 days before mating until term; group 2, females exposed 8 days preconception until term; and group 3, same as group 2 except exposure 17-25 days postpartum	Fertility, resorptions, viability, sex ratio, birth weight, postpartum growth, and malformation; postnatal behavioral tests including movement, grooming, standing and righting reflex, and geotropism	No reproductive effects; transient behavioral changes in neonatal period, but not when tested 21 days postpartum	Well-engineered exposure system, described in great detail, all essential parameters required and many desirable given; rats can perceive and respond to 60-Hz field strengths below those used in these studies; no indication that studies were blinded.
Sikov et al. 1987	Hanford miniature swine	F-0 study group for 4 mon before breeding and for first	114 animals; not energized, separate building	261 animals; 30 kV/m, 60-Hz field, 20 hr/day, 7 days/wk. F-0	F-0 and F-1 were birth defect studies; birth weight and litter	No effects on birth weight and litter size; increase in malformations in	Exposure system well engineered and described in detail with all essential

Continues on next page

TABLE A4-2 Continued

Study	Species	Developmental Stage	Number of Controls and Exposure Characteristics	Number of Exposed and Exposure Characteristics	End Points Evaluated	Outcome	Comments
		100 days postpartum		group exposed 4 mon before breeding and for first 100 days postpartum	size, rates of malformation	musculoskeletal and digits; F-1 groups, increase in malformations	parameters given; malformations, and CNS or cardiovascular defects not described; increase in malformations not a consistent finding across generations; no indication that studies were blinded
Stuchly et al. 1988	Rats	2 wk before conception and throughout pregnancy	340 maternal animals	987 maternal animals; 5.7, 23, and 66 µT (alternating field) for 7 hr/day; saw-toothed waveform (18,000 Hz) similar to but higher than a video-display terminal	Maternal weight gain, fetal and placental weight, litter size, live fetuses and resorptions, major and minor malformations	All reproductive results indistinguishable from control results; fewer skeletal variants in higher exposure groups and an increase in minor skeletal anomalies by fetus, but not litter	Appropriate large control and exposed groups; no reproductive effects but a reduction in maternal lymphocyte count, although still within the normal range; no indication that studies were blinded

Reference	Animal	Exposure Period	Controls	Exposure Conditions	Endpoints	Effects	Comments
Wiley et al. 1992	Mice	1-18 days	185 animals; identical system, not energized	558 animals; 3.6, 17, and 200 µT; 20-kHz saw-toothed waveform	Implantations, fetal deaths, resorptions, and body weights; gross external, visceral, and skeletal malformations	No exposure-related effects	Very well characterized, blind, computer monitoring, vibrations, illumination evaluated but not reported; no indication that studies were blinded
Zusman et al. 1990	Rats and mice	Preimplantation embryos through blastocyst	9 animals; sham-exposure conditions not given	34 animals; 1, 20, 50, 70, or 100-Hz, 0.6 V/m; pulse duration 10 msec	Malformations	Cultured rat embryo: abnormal limb development; cultured mouse embryo: developmental retardation; no effects in vivo	Internal fields below 1 µV/m (with 0.6 V/m in air); no indication that studies were blinded

TABLE A4-3 Reports of Special Interest on Electric- and Magnetic-Field Exposure and Neurobehavioral Effects

Study	Species	Developmental Stage	Number per Group	Exposure Characteristics	End Points Evaluated	Outcome	Comments
Wolpaw et al. 1989	Macaque monkeys (male)	4-6 yr	4 sham, 6 exposed	3 kV/m, 10 µT 10 kV/m, 30 µT 30 kV/m, 90 µT	Well being, weight, blood chemistry, simple motor tasks, postmortem	No effects	
Sagan et al. 1987	Sprague-Dawley rats	Adult	32 exposed	0-27 kV/m, 60 Hz	Operant detection threshold	13.3 or 7.9 kV/m detection threshold	
Lovely et al. 1992	Sprague-Dawley rats (male)	Adult 63 and 3 days	8 sham, 32 exposed	3.03 mT, 60 Hz	Avoidance shuttlebox, 1 hr	No effects	
Dowman et al. 1989	Macaque monkeys	Adult, 5-7 kg	6 sham, 4 exposed	3 kV/m, 10 µT 10 kV/m, 30 µT 30 kV/m, 90 µT	Auditory, visual, somatosensory, evoked response 2 times per week	No effects	Somatosensory decrease in amplitude of late response for exposures of 10 kV/m and 30 kV/m
Hong et al. 1988	Sprague-Dawley rats	Exposed 0-14 days; weaned at 21 days; tested at 30 days	46 sham, 50 exposed	0.5 T, 3 exposures per day for 15 min for 2 wk	Repeated reversal of position habit	No effects	
Liboff et al. 1989	Rats	Adult	0 shams, 5 exposed	26.1 µT, 0.139 µT, 60 Hz	Performance on FR/BRL combined operant schedule	DRL baseline disrupted not FR; threshold for effect, 27 µT	

Reference	Species	Age/Weight	N	Exposure	Measure	Results	Comments
Stern and Laties 1989	Rats (male and female)	Adults	5, animals were their own controls	90-100 kV/m, 60 Hz	Press lever to switch exposure off or on	No effect	
Salzinger et al. 1990	Rats	Exposed at 8 days of age, tested as adults at 90 days old	21 exposed, 20 sham	30 kV/m, 100 mT rems, 60 Hz	Complex operant schedule with repeat extinction and releasing	Slower rates of responding on many tasks in exposed groups	Good study
Weigel et al. 1987	Cats	Adult, 2.1-4.6 kg	N/A	245 receptors, 600 kV/m, 60 Hz	Receptor responsive-ness, cats paw stimulation, recorded DRG	Hair removal deceased response; mineral oil decreased response further	
Ossenkopp and Cain 1988	Rats (male)	Adult, 350 g	17 sham, 17 exposed, crossover design	100 μT, 60 Hz	Kindling after discharge duration	Attenuation after discharge in experimental groups	

TABLE A4-4 Electric-Field Exposure and Neurobiologic Effects

Study	Species	Developmental Stage	Number per Group	Exposure Characteristics	End Points Evaluated	Outcome	Comments
Blackwell and Reed 1985	Mice (male)	21-30 g	20 sham, 20 exposed	50-400 V/m, 15-50 Hz	Sleep-time exploration behavior	No effects	
Rosenberg et al. 1981	Deer mice	5-15 mon	34 in plastic cages above ground, 21 grounded	100 kV/m, 10 Hz	Activity, circulation, CO_2, oxygen, temperature	Transient activity and gas increase during inactive phase	
Rosenberg et al. 1983	Deer mice	5-15 mon	8 sham, 60 exposed	10-75 kV/m, 60 Hz	Activity, blood gas	Transient increase with exposures of 50-75 kV/m	
Blackwell 1986	Albino male rats (HMT)	250-400 g	200 recordings, 51 cells	100 V/m, 50, 30, and 45 Hz	Firing rate timed	No effect on rate at 15 and 30 Hz; firing time was dependent on field voltage	
Creim et al. 1984	Sprague-Dawley rats (male)	70 days	3 exposed, 4 groups each	69 kV/m, 133 kV/m, 34 kV/m, 60 Hz	Taste aversion, time drinking saccharine	No effects	
Easley et al. 1991	Baboons	5-6 yr	8 exposed, 8 sham	30 kV/m, 60 Hz; exposed 6 wk, 12 hr/day and 7 days/wk	Social behavior, passive affinity, tension, steroptopy	Social stress only in first 2 wk	
Stern and Laties 1985	Rats (female)	Adult	5 own control	55 kV/m, 60 Hz	Operant detection	3-10 kV/m detection threshold	
Hjeresen et al. 1980	Rats	Adult	8 sham, 32 exposed per experiment	0, 60, and 105 kV/m	Shuttlebox avoidance and general acuity	75 kV/m and greater in long 23.5 hr exposure leads to	

Reference	Species	Age/Weight	Number	Exposure	Measurement	Results
						avoidance of exposed regions during light portion of the day, more activity
Hjeresen et al. 1982	Swine	22-24 mon, 65 kg	15 sham, 7 exposed	30 kV/m, 60 Hz	Shuttlebox avoidance	Spent more time out of electric fields during sleep time
Jaffe et al. 1983	Rats (male)	3-6 wk	Two experiments: 1-14 exposed, 15 sham, 2-46 exposed, 25 sham	100 kV/m, 60 Hz	Synaptic transmission and PNS function SCG (in vitro)	Increased synaptic excitability
Jaffe et al. 1980	Rats (male)	0-30 days	114 sham and exposed	65 kV/m, 60 Hz, 20 hr/day, PD 11-20	Visual-evoked response in cortex	Age effects but no exposure effects
Portet and Cabanes 1988	Rats (male), rabbits	Rats begin prenatally, rabbits 8 wk	Rats: 25 exposed, 25 sham; rabbits: 28 exposed in four equal groups	50 kV/m, 60 Hz	Organ growth, hormone production, many measurements	Only in rabbits, adrenal cortisol decrease, but no decrease in plasma cortisol
Stern and Laties 1989	Rats (male and female)	Adults	5 animals, animal was own control	90-100 kV/m, 60 Hz	Press lever to switch exposure on and off	No effect
Stern et al. 1983	Rats	Adults	19 animals, animal was own control	0-10 kV/m, 60 Hz	Detection threshold, operant response	4-8 kV/m detection threshold

Continues on next page

TABLE A4-4 Continued

Study	Species	Developmental Stage	Number per Group	Exposure Characteristics	End Points Evaluated	Outcome	Comments
Salzinger et al. 1990	Rats	Exposed at 8 days old and tested as adults at 90 days old	21 exposed, 20 sham	30 kV/m, 100 mT rems, 60 Hz	Complex operant schedule with repeat extinction and releasing	Slower rates of responding on many tasks in exposed groups	
Weigel et al. 1987	Cats	Adult, 2.1-4.6 kg		245 receptors, 600 kV/m, 60 Hz	Receptor responsiveness, cats-paw stimulation, record in DRG	Hair removal decreased response; mineral oil decreased response further	Tested hypothesis of direct field effect in neuronal (DRG) membrane; not likely to be membrane action
Hackman and Graves 1981	Mice	Adult, 56-58 days	110 mice, 5-15 per group	5 min/day for 6 wk; 0, 25, 50 kV/m; 60 Hz	Corticosterone levels	Increase immediately after onset	
Cooper et al. 1981	Pigeons	5-12 mon	6	25, 50 kV/m, 60 Hz	Conditioned suppression (detection)	Significant suppression at 50 kV/m	Suppression does not mean aversion
Graves 1981	Pigeons	5-12 mon	6	50 kV/m, 60 Hz	Conditioned suppression	Significant suppression	Not vibration

Reference	Animal	Age	Number	Exposure	Measure	Results	
Graves et al. 1978	Pigeons, leghorn chickens	Adult	60 birds, 20 groups	0, 40, 80 kV/m, 60 Hz	Conditioned suppression EEG heart rate	CS detected at 32 kV/m, EEG increase variance, heart rate increased	
Smith et al. 1979	Rats	Young adult; begins at 30 days	8 sham, 8 exposed	25 kV/m, 60 Hz, for 5 or 6 wk	Body mass, food and water intake, exploratory behavior	No effects	
Sagan et al. 1987	Sprague-Dawley rats	Adult	32	0-27 kV/m, 60 Hz	Operant detection threshold	13.3 or 7.9 kV/m detection threshold	
Jaffe et al. 1981	Rats (male)	Adult males	In 3 experiments, 22 exposed, 14 sham; 13 exposed, 14 sham; 18 exposed, 20 sham	100 kV/m, 60 Hz	Neuromuscular function (FTW), plantor (PTW), soleus muscles (STW), muscle fiber	Fatigue in STW fibers constantly enhanced	
Rosenberg et al. 1981	Deer mice	5-15 mon	34 exposed, 21 sham	100 kV/m rms, 60 Hz	Gross motor CO_2, O_2, and temperature	Activity and gas increased initially then returned to normal	
Hackman and Graves 1981	Mice	70 days	5-15 in all groups	0-50,000 V/m, 60 Hz	Plasma corticosterone	Acute transient increase after exposure to high levels	Perception apparent for only minutes

TABLE A4-5 Magnetic-Field Exposure and Neurobehavioral Effects

Study	Species	Developmental Stage	Number per Group	Exposure Characteristics	End Points Evaluated	Outcome	Comments
Lovely et al. 1992	Rats (male)	Adult, 3 and 63 days	8 sham, 24 exposed	3.03 mT, 60 Hz	Avoidance shuttlebox for 1 hr	No effect	
Thomas et al. 1986	Rats (male)	Adult	5 were own controls	2.6 μT, 60 Hz	FR/BRL performance mixed schedule	Decreased rate of DRL performance only	Cyclotran resonance geomagnetic
Smith and Justesen 1977	mice	Adult	39 in 3 gender groups	1.3 ± 0.3 mT, 60 Hz	Locomotor activity aggression	Field induced increase in activity	
Ossenkopp et al. 1985	Mice (male)	Adult	In 2 experiments: 31 exposed, 20 sham; 10 exposed, 20 sham	147 ± .02 mT, 6.2 Hz (4.7), 8 mT/sec, 10 mT/sec	Analgesia to morphine (latency to respond)	Attenuated analgesia day and night	Theory: pineal gland and calcium bonding
Ossenkopp and Kavaliers 1987	Mice (male)	Adult	15 sham, 10 exposed at 100 μT, 16 exposed at 50 μT twice	2 μT, 100 μT, 150 μT, 30-min exposure	Analgesia to morphine (latency to respond)	Attenuate in linear relation to 100 μT highest at 150 μT	
Ossenkopp and Cain 1988	Rats (male)	Adult, 350 g	17 exposed, 17 sham, cross over design	100 μT, 60 Hz	Kindling after discharge duration	Attenuation of discharge in experimental groups	

Reference	Species	Age/weight	Number	Exposure	Task	Results	
Clarke and Justesen 1979	Leghorn chickens	Exposed 24 hr in utero, tested at 10 wk	2 sham, 2 exposed	4.0 mT dc, 1.7 mT rms, 60 Hz	Conditioned suppression	Detection as evidenced by increased variability, both ac and dc effect	dc movement induced effect
Tucker and Schmitt 1978	Humans	Adult	200	0.75, 1.3, 1.5, and 7.5 mT; 60 Hz	Detection	No detection	
Kavaliers and Ossenkopp 1986a	CF-1 and C57BL mice	1-2 mon, 25-30 g	10 mice per group	30-min exposure 0.15-9 mT, 0.5 Hz	Paw-flick response to 50 ±, 5° C hotplate; 1-min total activity monitor	Exposure for 30 min markedly decreases degree of stress-induced opiod analgesia and hyperactivity	
Rudolph et al. 1985	Wistar rats	Adult, 385 g	2 groups: light and dark; total: 47 rats	50 Hz, 40 μT dc, 4-hr exposure	18-min test of open field activity	40% increase in activity, only at beginning of light field test	
Kavaliers and Ossenkopp 1986b	CF-1 mice	3-4 mon adults, 30-35 g	10 mice per group	0.5 Hz, 0.15-9 mT for 60 min	Paw-flick hotplate response to m, d, k, and S agonists	Significant attenuation of agonist effects for all receptor agonists except SKF-10047 S receptor	
Kavaliers and Ossenkopp 1986c	CF-1 and C57BL mice (male)	3-4 mon, 30-35 g	10 mice per group	0.5 Hz, 0.15-9 mT	1-min activity latency to lift paw in hotplate response to 10 mg/kg morphine	B field inhibited effects of morphine; EGTA blocks inhibition; A21387 (Ca^{2+} ionophore) augments analgesia	Ca^{2+} antagonized morphine effect

Continues on next page

TABLE A4-5 Continued

Study	Species	Developmental Stage	Number per Group	Exposure Characteristics	End Points Evaluated	Outcome	Comments
Ossenkopp and Kavaliers 1987	CF-1 mice (male)	2-3 mon, 25-30 g	27 sham, 50 exposed	Two experiments: 15 sham; 10 exposed at 50 µT, 16 exposed at 100 µT; 12 sham; 12 exposed at 50 µT, 12 exposed at 150 µT; 60 Hz	Analgesia to 10 mg/kg morphine in paw-flick latency response to 50° C	B field inhibited analgesia in dose-dependent manner; biggest effect, nocturnal at 150 µT	
Ossenkopp and Kavaliers 1987	CF-1 mice	1-2 mon, 25-30 g	5 per group	60-min exposure, 0.5 Hz, 0.15-9 mT rotating field	Analgesia to 10 mg/kg morphine effects of Ca^{2+} regulators	Ca^{2+} channel antagonist reduced analgesia; Ca^{2+} channel agonist enhanced inhibition to B field	Ca^{2+} drugs had no effect on morphine-induced analgesia

TABLE A4-6 Magnetic- and Electric-Field Exposure and Neurobehavioral Effects

Study	Species	Developmental Stage	Number per Group	Exposure Characteristics	End Points Evaluated	Outcome	Comments
Wolpaw et al. 1989	Macaque monkeys	4-6 yr	6 exposed, 4 sham	3 kV/m, 10 µT; 10 kV/m, 30 µT; 30 kV/m, 90 µT	Well-being, weight, blood chemistry, simple motor tasks, postmortem	No effects	
Dowman et al. 1989	Macaque monkeys	Adult, 5-7 kg	6 exposed, 4 sham	3 kV/m, 10 µT; 10 kV/m, 30 µT; 30 kV/m, 9 µT	Auditory, visual, somatosensory, evoked response 2 times per week	No effects; only somatosensory decrease in amplitude of late response for 10 kV/m and 30 kV/m	
Davis et al. 1984	CD-1 mice (male), LAF-1 mice (female)	40-70 days	Male mice: >100 exposed, >100 sham; female mice: 10 exposed	1.65 T rms, 60 Hz, exposed 72 hr	Passive avoidance; activity chemical-induced seizures	No effects	
Hong et al. 1988	Sprague-Dawley rats	Exposed 0-14 days, weaned 21 days, tested at 30 days	46 exposed, 50 sham	0.5 T, three exposures of 15 min/day for 2 wk	Repeated reversal of position habit	No effects	
Liboff et al. 1989	Rats	Adult	5 exposed, 0 sham	26.1 µT, 0.139 µT, 60 Hz	Performance on FR/BRL combined operant schedule	DRL baseline disrupted, but not FR; threshold for effect 27 µT	

TABLE A4-7 Effects of Sinusoidal Electric-Field Exposure on Pineal Melatonin Production

Study	Species	Developmental Stage	Number per Group	Exposure Characteristics	End Points Evaluated	Outcome	Comments
Wilson et al. 1981	Sprague-Dawley rats (male)	Adult	5 per group	60 Hz, 0.7-1.9 kV/m for 20 hr/day for 30 days; sham-exposed controls	Pineal NAT activity, pineal melatonin, pineal 5-methoxy-tryptophol	No change in nighttime pineal NAT, reduction in nighttime melatonin, no change in nighttime 5-methoxytryptophol production	Lack of correlation of NAT and melatonin unexpected; unusually low nighttime pineal melatonin
Wilson et al. 1986	Sprague-Dawley rats (male)	Adult	10 per group	60 Hz; 39 kV/m; 20 hr/day for 1, 2, 3, or 4 wk exposure; sham-exposed controls	Pineal NAT activity, pineal melatonin	After 3 and 4 wk exposure, nighttime pineal NAT and melatonin depressed; withdrawal of fields returned nighttime pineal NAT and melatonin to normal	NAT activity and pineal melatonin decreased in parallel
Reiter et al. 1988	Sprague-Dawley rats	Fetal and newborn		60 Hz; 10, 65, or 130 kV/m; exposed in utero and for 23 days after birth; sham-exposed controls	Pineal melatonin	Pineal melatonin depressed by all field strengths	No dose-response relationship

| Grota et al. 1994 | Sprague-Dawley rats | Adult | 60 Hz; 35 kV; exposed 20 hr/day for 30 days; sham-exposed controls | Pineal NAT activity, pineal HIOMT activity, pineal melatonin, blood melatonin | No change in nighttime pineal NAT, HIOMT, or melatonin; depressed nighttime blood melatonin | Exposures conducted with or without concurrent red-light exposure |

TABLE A4-8 Effects of Sinusoidal Magnetic-Field Exposure on the Pineal Gland in Animals in Morphologic Studies

Study	Species	Developmental Stage	Number per Group	Exposure Characteristics	End Points Evaluated	Outcome	Comments
Milin et al. 1988	Wistar rats (male)	Adult	4-6 per group	70-μT exposure for 20 min/day for 14 days; sham-exposed controls	Pineal morphology, ultrastructural studies	Decreased peptidergic activity of light pinealocytes	Very high field strengths used; interpretation of results confounded by stress factors (rats restrained during exposure)
Gimenez-Gonzalez et al. 1991	Wistar rats (male)	Adult	5 per group	50 Hz; 5.2-mT exposure for 1, 3, 7, 15, or 21 days for 30 min/day; room controls	Pineal morphology, light, and ultrastructural studies	After 3 and 7 days, changes most prominent include decreased karyometric index and increased lipid in pineal cells	Very high field strengths used; no quantitation of reported changes; no sham-exposed controls
Martinez-Soriano et al. 1992	Wistar-King rats (male)	Adult	5 per group	50 Hz; 5.2-mT exposure for 1, 3, 7, 15, or 21 days for 30 min/day; room controls	Pineal morphology	After 15 and 21 days, decrease in synaptic ribbon number in pineal cells	Very high field strengths used; no sham controls; little internal consistency in the data
Matsushima et al. 1993	Wistar-King rats (male)	Adult	6 per group	50 Hz; 5-μT circularly polarized; continuous exposure (except for two 2-hr intervals per week) for 42 days; sham-exposed controls	Pineal morphology, light microscopic studies	Slight differences in pinealocyte size, especially the proximal and distal (but not central) portions of gland; changes seasonally dependent	Regional and seasonal differences make interpretation of significance difficult

TABLE A4-9 Effects of Sinusoidal Magnetic-Field Exposure on Pineal Melatonin Production

Study	Species	Developmental Stage	Number per Group	Exposure Characteristics	End Points Evaluated	Outcome	Comments
Martinez-Soriano et al. 1992	Wistar rats (male)	Adult	5 per group	50 Hz; 5.2 mT; exposure for 1, 3, 7, 15, or 21 days; 30 min/day; room controls	Blood melatonin	Depressed daytime blood melatonin at 15 days	Blood melatonin depressed after 15 days but not after 21 days; very high field strengths used
Kato et al. 1993	Wistar-King rats (male)	Adult	8 per group	50 Hz; 0, 0.2, 0.1, 1, 5, 50, or 250 µT, circularly polarized field with continuous exposure for 42 days (except for two 2-hr intervals per week); sham-exposed controls	Pineal melatonin, blood melatonin	Field strength of 1 µT and above suppressed nighttime pineal melatonin and increased daytime melatonin; field strengths of 1 µT and above suppressed day time and nighttime blood melatonin	Rise in daytime pineal melatonin after magnetic-field exposure is unusual
Kato et al. 1994a	Long-Evans rats	Adult	8 per group	50 Hz; 0, 0.2, or 1 µT, circularly polarized field with continuous exposure for 42 days (except for two 2-hr intervals per week);	Pineal melatonin, blood melatonin	Field strengths of 0.2 µT depressed nighttime melatonin; 0.2 and 1 µT depressed daytime and nighttime blood melatonin	Showed that pigmented rats (Long-Evans) respond to magnetic fields as do albino rats (Sprague-Dawley)

Continues on next page

TABLE A4-9 Continued

Study	Species	Developmental Stage	Number per Group	Exposure Characteristics	End Points Evaluated	Outcome	Comments
				sham-exposed controls and room controls			
Kato et al. 1994b	Wistar-King rats	Adult	8 per group	50 Hz; 1 μT; horizontal or vertical; continuous exposure for 42 days (except for two 2-hr intervals per week); sham-exposed controls and room controls	Pineal melatonin, blood melatonin	No effects on pineal or blood melatonin daytime or nighttime	In contrast to circularly polarized 50-Hz, 1-μT fields, horizontal or vertical, 50-Hz, 1-μT fields are without effect on pineal and blood melatonin
Yellon 1994	Djungarian hamsters (male and female)	Adult	4-6 per group	60 Hz; 100 μT; horizontal for 15 min (beginning 2 hr before dark onset); sham-exposed controls	Pineal melatonin, blood melatonin	In two of three experiments, reduced and delayed rise in nighttime pineal and blood melatonin; in one experiment, no effect on nighttime pineal or blood melatonin	Two of three identical experiments showing suppression of nighttime pineal and blood melatonin and one showing no effect; all animals were adults but varied widely in age

TABLE A4-10 Effects of Combined Sinusoidal Electric- and Magnetic-Field Exposure on Pineal Melatonin Production

Study	Species	Developmental Stage	Number per Group	Exposure Characteristics	End Points Evaluated	Outcome	Comments
Lee et al. 1993	Suffolk sheep (female)	Juvenile	10 per group	Exposed under 500-kV power line: continuous 6-kV electric and 4-μT magnetic fields; controls 225 m from power line; continuous <10-V/m electric and <0.03-μT magnetic fields	Blood melatonin between ages of 2 and 10 mon at 8 different times	No effect on 24-hr melatonin rhythms between exposed and controls	Thorough and well-supervised study
Rogers et al. 1995	*Papio cynocephalus* baboons (male)	Adult	3 per group	Random, intermittent (with rapid onset/ offset) 30-kV/m electric and 100-μT magnetic field; animals served as own controls and run at another time in the same facility	Blood melatonin	Nighttime melatonin depressed in experiments	Animals served as own controls; therefore, controls not run simultaneously

TABLE A4-11 Effects of Different Types of Electric- and Magnetic-Field Exposure on Melatonin Metabolism in Humans

Study	Species	Developmental Stage	Number per Group	Exposure Characteristics	End Points Evaluated	Outcome	Comments
Prato et al. 1988-89	Human (male)	Adult	4 per group	MRI exposure: concurrent static magnetic field, time-varying magnetic field, radio frequency fields, at night for 40.5 min; subjects served as own controls and were sham exposed on another night	Blood melatonin	No changes in blood melatonin during exposure	Small number of subjects
Schiffman et al. 1994	Human (male)	Adult	2 per group	MRI exposure: 2.5-T magnetic field for 1 hr; subjects served as own controls and were sham exposed on another night; subjects also used as bright-light positive controls on another night	Blood melatonin	No effect on blood melatonin; bright light slightly reduced blood melatonin	MRI exposure did not alter blood melatonin; bright-light exposure also had surprisingly little effect
Wilson et al. 1990b	Human (male and female)	Adult	28 or 14 per group	Slept under snap safety switch (conventional) or continuous polymer wire (CPW) electric blankets for 6-10 wk	Urinary 6-hydroxy melatonin sulfate	At beginning or discontinuation of CPW blanket use, 7 subjects exhibited change in urinary 6-hydroxy melatonin sulfate; conventional blanket use caused no changes	Implication is that at either beginning or discontinuation of electric-blanket use, nocturnal melatonin synthesis or metabolism might change

TABLE A5-1 Residential Electric- and Magnetic-Field Exposure and Cancer: Study Structure

Study	Geographic Area of Study	Time[a]	Number of Cases[b]	Number of Controls[c]	Exposure Assessment Strategy[d]
		Studies of Childhood Cancer			
Wertheimer and Leeper 1979	United States (Denver, Colorado)	1950-1973	344 cases, 491 residences	344 cases, 472 residences	Wire codes
Fulton et al. 1980	United States (Rhode Island)	1964-1978	119 cases, 209 residences	240 cases, 240 residences	Wire codes
Tomenius 1986	Sweden (Stockholm County)	1958-1973	716 cases, 1,172 residences	716 cases, 1,015 residences	Wire codes, spot field measurements
Savitz et al. 1988	United States (Denver, Colorado)	1976-1983	356	278	Wire codes, spot field measurements
Coleman et al. 1989	United Kingdom (SE London)	1965-1980	811	1,614 cancer controls, 254 population controls	Wire codes
Myers et al. 1990	United Kingdom (Yorkshire health region)	1970-1979	419	656	Distance from overhead lines, calculated fields
London et al. 1991	United States (Los Angeles County)	1980-1987	331	257	Spot field measurements, 24-hr field measurements, wire codes, household-appliance use
Feychting and Ahlbom 1993	Sweden	1960-1985	142	558	Wire codes, spot field measurements

Continues on next page

TABLE A5-1 Continued

Study	Geographic Area of Study	Time[a]	Number of Cases[b]	Number of Controls[c]	Exposure Assessment Strategy[d]
Studies of Childhood Cancer					
Olsen et al. 1993	Denmark	1968-1986	1,707	4,788	Wire codes
Verkasalo et al. 1993	Finland	1970-1989	140	129,800 in cohort	Wire codes
Petridou et al. 1993	Greece				Distance from substations and transmission lines
Fajarado-Gutierrez et al. 1993	Mexico City		81	77	Distance from power lines
Studies of Adult Cancer					
Wertheimer and Leeper 1982	United States (Denver, Boulder, and Longmont, Colorado)	1967-1975 (Boulder and Longmont) 1977 (Denver)	1,179	1,179	Wire codes
McDowall 1986	United Kingdom (East Anglia)	1971-1983	213	7,631 in cohort	Distances from substations and distribution lines
Severson et al. 1988	United States (Washington)	1981-1984	164	204	Wire codes, 24-hr field measurements, spot field measurements

Coleman et al. 1989	United Kingdom (SE London)	1965-1980	811	1,614 cancer controls, 254 population controls	Wire codes
Youngson et al. 1991	United Kingdom (NW and Yorkshire Regional Health Authority Area)	1983-1985 (NW region) 1979-1985 (Yorkshire)	3,276 (1,491 in NW region 1,770 in Yorkshire)	3,144 (1,491 in NW region 1,653 in Yorkshire)	Distance from overhead lines, calculated fields
Schreiber et al. 1993	Netherlands (urban Limmel)	1956-1987	431	3,549 in cohort	Wire codes
Feychting and Ahlbom 1994	Sweden	1960-1985	548	1,091	Wire codes, spot field measurements

[a]Interval over which cancer cases occurred.
[b]Number of eligible cases ascertained for the study.
[c]Number of controls ascertained for the study.
[d]Mode of obtaining data to address residential magnetic-field exposure.

TABLE A5-2 Residential Electric- and Magnetic-Field Exposure and Cancer: Case and Control Selection

Study	Case Definition: Age Range[a]	Type of Cancer[b]	Other Restrictions[c]	Method of Control Selection[d]
Studies of Childhood Cancer				
Wertheimer and Leeper 1979	0-18 yr	All cancer deaths	Denver-area resident, Colorado birth certificate	Birth certificate
Fulton et al. 1980	0-20 yr	Leukemia	Patient in R.I. hospital	Birth certificate
Tomenius 1986	0-18 yr	All tumors	Born and diagnosed in Stockholm county	Birth registry
Savitz et al. 1988	0-14 yr	All cancers	Resident of metropolitan Denver	Random digit dialing
Coleman et al. 1989	All ages	Leukemia	Resident of four borough areas of London	Electoral roll of 1975 for population control, cancer registry for cancer controls
Myers et al. 1990	0-14 yr	Malignant tumors	Born in Yorkshire health region	Birth registry
London et al. 1991	0-10 yr	Leukemia	Resident of Los Angeles County	Friends of controls, random digit dialing
Feychting and Ahlbom 1993	0-15 yr	All cancers	Resident of homes during 1960-85 within 300 m of overhead power lines	Nested case-control study, random selection from study base
Olsen et al. 1993	0-14 yr	Leukemia, malignant lymphoma, tumors of central nervous system	Resident of Denmark	Danish population register
Petridou et al. 1993				
Fajarado-Gutierrez et al. 1993		Leukemia	Referral hospitals	Hospital controls

Studies of Adult Cancer

Reference	Age[a]	Cancer[b]	Eligibility[c]	Controls[d]
Verkasalo et al. 1993	0-19 yr	Leukemia, lymphoma, and tumors of the nervous system	Resident of home during 1970-89 that was within 500 m of power lines of field strength ≥ 0.01 μT	Finland national cancer rates for comparison
Wertheimer and Leeper 1982	>19 yr	All cancer deaths (lung cancer added) and life-threatening forms of cancer diagnosed >5 yr previously in persons living without recurrence in 1979	The sample comprised deaths occurring up to age 62 for all cancers except lung cancer; a fraction of the lung cancer deaths of persons over age 62 were used	For cancer deaths: neighborhood controls and death-certificate controls For cancer survivors: neighborhood controls and random sample from cohort of area residents
Severson et al. 1988	20-79 yr	Acute nonlymphocytic leukemia	Resident of three county areas of Washington state	Random digit dialing
Coleman et al. 1989	All ages	Leukemia	Resident of four borough areas of London	Electoral roll of 1975 for population control, cancer registry for cancer controls
Youngson et al. 1991	>15 yr	Leukemia and lymphoma	Resident of West RHA or Yorkshire RHA	Hospital-discharge records of other cancers and noncancers
Feychting and Ahlbom 1994	>15 yr	Leukemia and brain cancer	Resident of homes during 1960-85 within 300 m of overhead power lines	Nested case-control study, random selection from study base

[a]Age at diagnosis or death to be eligible for study.
[b]Cancer diagnoses eligible for inclusion.
[c]Exclusions from eligibility for reasons other than age or diagnosis.
[d]Sampling frame for generating controls.

TABLE A5-3 Residential Electric- and Magnetic-Field Exposure and Cancer: Exposure Assessment

Study	Operational Definition of Exposure[a]	Blinding of Data Collector[b]	Prevalence of Increased Exposure Among Controls[c]	Residence of Interest[d]
		Studies of Childhood Cancer		
Wertheimer and Leeper 1979	Two exposure categories based on distance of home from substations or overhead power lines and on type of line	No	21.8% of control households	Birth and death residence when available
Fulton et al. 1980	Estimated field exposure for each household, based on the number, type, and distance of the wires from the house; exposure categories divided into quartiles based on controls	Not stated	50% of control households	All residences before diagnosis (analysis based on residences, not individuals)
Tomenius 1986	Magnetic-field measurements: low exposure at <0.3 µT and high exposure at ≥ 0.3 µT; five exposure categories based on type of electric construction and distance from the house	Yes	200 kV wire and ≥0.3 µT: 0.41% all control households; 200 kV wire: 1.34% all control households; ≥ 0.3 µT 1.4% of control households	Residence at birth and at diagnosis
Savitz et al. 1988	Field measurements two categorization methods	Field measurements: no	Field measurements: >0.25 µT (at low power use): 4.3% of controls >0.25 µT (at high power use): 6.5% of controls >0.2 µT (low power use): 7.7% of controls >0.2 µT (high power use): 14.1% of controls	Residence at diagnosis

	Five exposure categories based on wire current configuration and distance from house	Wire codes: yes	Wire codes: Very high (at time of diagnosis): 3.2% of controls Very high (at 2 yr before diagnosis): 1.4% of controls	Residence at diagnosis
Coleman et al. 1989	Four exposure categories based on distance of nearest power source, power line, or substation. Estimated field based on wire codes, peak electric load, and distance of electric source from home.	For cases and cancer controls: yes For population controls: no	Distance of nearest power line: <25 m, 0.1% of cancer controls Distance of nearest substation: <25 m, 3.6% of cancer controls Estimated field for nearest substation: ≥500 kV, 14.4% of controls Sum of all weighted exposure for all substations within 200 m: in the high category, 3.3% of controls	
Myers et al. 1990	Six exposure categories based on distance of closest overhead line of any voltage. Estimated field based on distance of electric source and of maximum electric load	Yes	Distance from overhead line: <25 m, 2.9% of controls Estimated field: ≥ 0.1 µT, 0.69% of controls	Residence at birth
London et al. 1991	24-hr field measurements: four categories based on 50th, 75th, 90th percentiles of all study subjects	Yes	24-hr measurements: ≥0.268 µT, 7.6% of controls	Longest residence occupied during the period 9 mon before birth up to approximately 1 yr before diagnosis

Continues on next page

TABLE A5-3 Continued

Study	Operational Definition of Exposure[a]	Blinding of Data Collector[b]	Prevalence of Increased Exposure Among Controls[c]	Residence of Interest[d]
	Studies of Childhood Cancer			
	Spot field measurements: four categories based on 50th, 75th, 90th percentiles of all study subjects		Spot field measurements: ≥0.125 µT, 10.1% of controls	
	Wire codes: five categories based on distance of home from overhead lines and type of line or electric facility.		Wire codes: very high category, 11.7% of controls	
Fajarado-Gutierrez et al. 1993	Distance from electric facilities	NA	<20 m from distribution lines, 10% of controls	
Feychting and Ahlbom 1993	Estimated field exposure based on distance from electric towers, height of tower, distance between phases, ordering of phases, current load, distance from overhead line; (two categorization methods used)	Yes	Estimated fields: ≥0.3 µT, 32/554 = fraction of controls; ≥ 0.2 µT, 46/554 = fraction of controls	Residence at diagnosis or last home occupied by case within the study area
	Spot field measurements: used same exposure categories as for estimated field exposure method		Spot field measurements: ≥ 0.2 µT, 70/344 = fraction of controls	
	Distance to power lines: high exposure: 0-5 m low exposure: >100 m		Distance from line: 34/554 = fraction of controls	

Olsen et al. 1993	Estimated average field and cumulative field dose based on distance of installation from home, type of line, distance between towers, height of line, distance between phases, current load, duration of exposure	Yes	Estimated fields: ≥0.25 µT, 11/4,788 = % controls; ≥0.40 µT, 3/4,788 = fraction of controls	All residences occupied 9 mon before birth
Verkasalo et al. 1993	Estimated field exposure and cumulative field exposure based on distance of home from power lines, type of line, distance between phases, current load, duration of exposure	Not stated	NA	Residence during 1979-89
Studies of Adult Cancer				
Wertheimer and Leeper 1982	Four levels of exposure based on type of power line and distance of line from the home	Only for 12% of cases and controls	Very high current configuration, 6.3% of controls	Residence of longest duration 3-10 yr before diagnosis
McDowall 1986	Distance of electric installation from the house	NA	NA	Residence occupied in 1971 (0-12 yr before diagnosis)
Severson et al. 1988	Four exposure categories based on type of power line and distance from house	Wire codes: yes	Wire codes: very high (longest residence), 5.5% of controls; very high (for residence closest to reference date), 6.0% of controls	Field measurements: residence at diagnosis
	Estimated field exposure based on type of line, distance from home, and current flow	Field measurements: no	Estimated fields: >0.2 µT (longest residence), 16.4% of controls; >0.2 µT (for residence closest to reference date), 15.7% of controls	Wire codes: residence of longest duration 3-10 yr before diagnosis

Continues on next page

TABLE A5-3 Continued

Study	Operational Definition of Exposure[a]	Blinding of Data Collector[b]	Prevalence of Increased Exposure Among Controls[c]	Residence of Interest[d]
Studies of Adult Cancer				
Youngson et al. 1991	Estimated field strength based on number, type of wire, distance from home, and current flow Five categories based on distance of home from nearest overhead power line	Yes	Estimated field: ≥0.3 µT, 0.25% of controls Distance from overhead line: <25 m, 2.0% of controls	Residence at diagnosis
Schreiber et al. 1993	Exposure categories based on distance of home from power line; high exposure defined as within 100 m; low exposure defined as greater than 100 m from power line	Not stated	NA	Residence occupied during follow-up
Feychting and Ahlbom 1994	Estimated field exposure based on distance from electric towers, height of tower, distance between phases, ordering of phases, current load, distance from overhead line; two categorization methods used Spot field measurements: used same exposure categories as for estimated field exposure method	Yes	Estimated fields: ≥0.2 µT, 8% of controls Cumulative exposure: ≥3 µT, 4% of controls Spot measurements: ≥0.2 µT, 21% of controls	Residence at diagnosis or last home occupied by case within the study area

| Distance to power lines: high exposure, 0-5 m; low exposure, >100 m | Distance from power lines: <50 m, 6% of controls |

[a] The actual definition of exposure used in the analysis that compared cases with controls.
[b] Whether the person ascertaining exposure was aware of whether the residence was occupied by a case or control.
[c] Proportion of controls for leukemia assigned to the high-exposure category.
[d] Time of occupancy for the residences considered in the analysis that compared cases with controls.

TABLE A5-4 Residential Electric- and Magnetic-Field Exposure and Childhood Leukemia: Results

Study	Exposure Category	Number of Cases	Number of Controls	Crude OR[a]	95% CI[b]	Adjusted OR[c]	Adjusted 95% CI[d]	Potential Confounders Addressed[e]
Wertheimer Leeper 1979	Birth addresses:							Age, sex, socioeconomic status (SES), urban residence, family pattern, traffic congestion, onset age
	HCC[f]	52	29	2.3	1.3-3.9			
	LCC[g]	84	107					
	Death addresses:							
	HCC	63	29	3.0	1.8-5.0			
	LCC	92	126					
Fulton et al. 1980	Very high	47.5	56.3	1.0	0.6-1.8			Onset age, SES, age
	High	55.4	56.3	1.2	0.7-2.1			
	Low	49.5	56.3	1.1	0.6-1.9			
	Very low	45.5	56.3					
Tomenius 1986	Total residences:							Age, sex, church district of birth
	≥0.3 μT	4	10	0.3	0.1-1.1			
	<0.3 μT	239	202					
Savitz et al. 1988	Field measurements for low-power conditions:							Maternal age, father's education, family income, maternal smoking in pregnancy, traffic density, age, sex, geographic area of residence
	>0.2 μT	5	16	1.9	0.7-5.6	1.8-2.4		
	<0.2 μT	31	191					
	Field measurements for high-power conditions:							
	>0.2 μT	7	29	1.4	0.6-3.5			
	<0.2 μT	30	175					
	Two-level wire codes:							
	High	27	52	1.5	0.9-2.			
	Low	70	207					
	Wire codes:							
	Very high	7	8	2.8	0.9-8.0			
	Very low	28	88					

Study	Exposure category	Cases	Controls	OR	95% CI	OR	95% CI	Adjustment
Coleman et al. 1989	Distance from substation:							Age, sex, county of residence, year of diagnosis
	0-24 m	3	3	1.6	0.3-8.4			
	25-49 m	11	12	1.5	0.6-3.6			
	50-99 m	22	48	0.7	0.4-1.4			
	≥100 m	48	78					
London et al. 1991	24-hr measurements:							Pesticide use, hair dryer use, black and white TV use, parental occupational exposures, other appliance use, other environmental exposures, residence type, SES
	0-0.067 µT	85	69					
	0.068-0.118 µT	35	42	0.7	0.4-1.2	0.7	0.4-1.2	
	0.119-0.267 µT	24	22	0.9	0.5-1.7	0.9	0.5-1.9	
	≥0.268 µT	20	11	1.5	0.7-3.3	1.7	0.7-4.0	
	Spot measurements:							
	0-0.031 µT	67	56					
	0.032-0.067 µT	34	28	1.0	0.6-1.9			
	0.068-0.124 µT	23	14	1.4	0.7-2.9			
	≥ 0.125 µT	16	11	1.2	0.5-2.8			
	Wire codes:							
	Buried	11	11					
	Very low	20	27					
	Low	58	75	1.0	0.5-1.7	1.0	0.4-1.5	
	High	80	68	1.4	0.8-2.6	1.5	0.8-2.7	
	Very high	42	24	2.2	1.1-4.3	1.7	0.8-3.7	
Feychting and Ahlbom 1993	Estimated fields:							Sex, age, county, residence type, diagnosis year, SES, NO₂
	≥ 0.3 µT	7	32	3.8	(1.4-9.3)	—	—	
	0.1-0.29 µT	4	47	1.5	(0.4-4.2)	—	—	
	≥ 0.2 µT	7	46	2.7	(1.0-6.3)	3.1	(1.1-8.6)	
	0.1-0.19 µT	4	33	2.1	(0.6-6.1)	1.5	(0.3-7.4)	
	<0.09 µT	27	475	—				
	Distance to power line:							
	<51 m	6	34	2.9	(1.0-7.3)	—		
	51 m-100 m	6	89	1.1	(0.4-2.7)	—		
	≥101 m	26	431	—				

Continues on next page

TABLE A5-4 Continued

Study	Exposure Category	Number of Cases	Number of Controls	Crude OR[a]	95% CI[b]	Adjusted OR[c]	Adjusted 95% CI[d]	Potential Confounders Addressed[e]
	Spot measurements:							
	≥0.2 µT	4	70	0.6	(0.2-1.8)	—		
	0.1-0.19 µT	1	67	0.2	(0.0-0.9)	—		
	<0.1 µT	19	207	—				
Olsen et al. 1993	Estimated Fields:							Sex, onset age, age
	≥ 0.4 µT	3	1	—	—	6.0	(0.8-44)	
	0.1-0.39 µT	1	7	—	—	0.3	(0.0-2.0)	
	≥ 0.25 µT	5	4	—	—	1.5	(0.3-6.7)	
	0.1-0.24 µT	1	4	—	—	0.5	(0.1-4.3)	
	≥0.1 µT	4	8	—	—	1.0	(0.3-3.3)	
	<0.1 µT	829	1658	—	—	—		

Study	Exposure	Cases observed	Cases Expected[h]	Standardized Incidence Ratio	95% CI	Potential Confounders Addressed
Verkasalo et al. 1993	≥ 0.2 µT	3	1.93	1.6	0.3-4.5	Age, sex
	0.01-0.19 µT	32	36.1	0.9	0.6-1.3	
	≥ 0.4 µT-yr	3		1.2	0.3-3.6	
	0.01-0.39 µT-yr	32		0.9	0.6-1.3	

[a]Odds ratio calculated without consideration of possible confounders (ratio of exposed cases divided by the ratio of exposed to unexposed controls).

[b]95% confidence intervals for the odds ratio calculated without consideration of possible confounders.

[c]Odds radio adjusted statistically for possible confounding factors.

[d]Corresponding confidence intervals calculated for the odds ratio adjusted for possible confounding factors.

[e]Includes all factors considered to be potential confounders whether or not statistical adjustments were made for them.

[f]HCC, high current configuration.

[g]LCC, low current configuration.

[h]Cases expected on the basis of incidence data for the disease in the general population.

TABLE A5-5 Residential Electric- and Magnetic-Field Exposure and Childhood Brain Tumors: Results

Study	Exposure Category	Number of Cases	Number of Controls	Crude OR[a]	95% CI[b]	Adjusted OR[c]	Adjusted 95% CI[d]	Potential Confounders Addressed[e]
Wertheimer and Leeper 1979	Birth addresses:							Age of onset, sex, socioeconomic status (SES), urban residence, family pattern, traffic congestion
	HCC[f]	22	12	2.4	1.0-5.4			
	LCC[g]	35	45					
	Death addresses:							
	HCC	30	17	2.4	1.2-5.0			
	LCC	36	49					
Tomenius 1986	Total residences:							Age, sex, church district of birth
	≥0.3 μT	13	3	3.9	1.1-13.7			
	<0.3 μT	281	250					
Savitz et al. 1988	Field measurements for low-power conditions:							Maternal age, father's education, family income, maternal smoking in pregnancy, traffic density, age, sex, geographic area of residence
	>0.2 μT	2	16	1.0	0.2-4.8			
	<0.2 μT	23	191					
	Field measurements for high-power conditions:							
	>0.2 μT	3	29	0.8	0.2-2.9			
	<0.2 μT	22	175					
	Two-level wire codes:							
	High	20	52	2.0	1.1-3.8			
	Low	39	207					
	Wire codes:							
	Very high	3	8	1.9	0.5-8.0			
	Very low	17	88					

Continues on next page

TABLE A5-5 Continued

Study	Exposure Category	Number of Cases	Number of Controls	Crude OR[a]	95% CI[b]	Adjusted OR[c]	Adjusted 95% CI[d]	Potential Confounders Addressed[e]
Olsen et al. 1993	Estimated fields:							Sex, onset age, age
	≥0.4 µT	2	1			6.0	0.7-44	
	0.1-0.39 µT	1	8			0.4	0.1-2.8	
	≥0.25 µT	2	6			1.0	0.2-5.0	
	0.1-0.24 µT	1	3			1.0	0.1-9.6	
	≥0.1 µT	3	9			1.0	0.3-3.7	
	<0.1 µT	621	1,863					
Feychting and Ahlbom 1993	Estimated fields:							Sex, age, county, residence type, diagnosis year, SES, NO$_2$
	≥0.3 µT	2	32	1.0	(0.2-3.9)			
	0.1-0.29 µT	2	47	0.7	(0.1-2.6)			
	≥0.2 µT	2	46	0.7	(0.1-2.7)			
	0.1-0.19 µT	2	33	1.0	(0.2-3.8)			
	<0.1 µT	29	475	—	—			
	Distance to power line:							
	<50 m	1	34	0.5	(0.0-2.8)			
	50 m-100 m	7	89	1.4	(0.5-3.1)			
	>100 m	25	431	—	—			
	Spot measurements:							
	>0.2 µT	5	70	1.5	(0.4-4.9)			
	0.1-0.2 µT	8	67	2.5	(0.9-6.6)			
	<0.1 µT	10	207	—	—			

Study	Exposure	Cases Observed	Cases Expected[h]	Standardized Incidence Ratio	95% CI	Potential Confounders Addressed
Verkasalo et al. 1993	Estimated fields:					Age, sex
	≥0.2 μT	4	2.16	2.3	0.8-5.4	
	0.01-0.19 μT	34	39.82	0.9	0.6-1.2	
	≥0.4 μT-yr	7		2.3	0.9-4.8	
	0.01-0.39 μT-yr	32		0.8	0.6-1.2	

[a] Odds ratio calculated without consideration of possible confounders (ratio of exposed to unexposed cases divided by the ratio of exposed to unexposed controls).
[b] 95% confidence intervals for the odds ratio calculated without consideration of possible confounders.
[c] Odds radio adjusted statistically for possible confounding factors.
[d] Corresponding confidence intervals calculated for the odds ratio adjusted for possible confounding factors.
[e] Includes all factors considered to be potential confounders whether or not statistical adjustments were made for them.
[f] HCC, high current configuration.
[g] LCC, low current configuration.
[h] Cases expected on the basis of incidence data for the disease in the general population.

TABLE A5-6 Residential Electric- and Magnetic-Field Exposure and Childhood Lymphoma: Results

Study	Exposure Category	Number of Cases	Number of Controls	Crude OR[a]	95% CI[b]	Adjusted OR[c]	Adjusted 95% CI[d]	Potential Confounders Addressed[e]
Wertheimer and Leeper 1979	Birth addresses:							Age of onset, sex, socioeconomic status (SES), urban residence, family pattern, traffic congestion
	HCC[f]	10	5	2.5	0.7-8.4			
	LCC[g]	21	26					
	Death addresses:							
	HCC	18	11	2.1	0.8-5.2			
	LCC	26	33					
Tomenius 1986	Total residences:							Age, sex, church district of birth
	≥0.3 μT	2	1	1.8	0.2-19.8			
	<0.3 μT	130	115					
Savitz et al. 1988	Field measurements for low-power conditions:							Maternal age, father's education, family income, maternal smoking in pregnancy, traffic density, age, sex, geographic area of residence
	>0.2 μT	2	16	2.2	0.5-10.3	3.2		
	<0.2 μT	11	191					
	Field measurements for high-power conditions:							
	>0.2 μT	3	29	1.8	0.5-6.9			
	<0.2 μT	10	175					
	Two-level wire codes:							
	High	5	52	0.8	0.3-2.2			
	Low	25	207					
	Wire codes:							
	Very high	3	8	3.3	0.8-13.7			
	Very low	10	88					

Study	Exposure	Cases Observed	Cases Expected[h]	Standardized Incidence Ratio	95% CI	Potential Confounders Addressed
Olsen et al. 1993	Estimated field:					Sex, onset age, age
	≥0.4 μT	1	1	5.0	0.3-82	
	0.1-0.39 μT	2	2	5.0	0.7-36	
	≥0.25 μT	1	1	5.0	0.3-82	
	0.1-0.24 μT	2	2	5.0	0.7-36	
	≥0.1 μT	3	3	5.0	1.0-25	
	<0.1 μT	247	1,247			
Feychting and Ahlbom 1993	Estimated fields:					sex, age, county, residence type, diagnosis year, SES, NO_2
	≥0.3 μT	1	32	0.9	0.0-5.4	
	0.1-0.29 μT	2	47	1.3	0.2-5.0	
	≥0.2 μT	2	46	1.3	0.2-5.1	
	0.1-0.19 μT	1	33	0.9	0.0-5.2	
	<0.09 μT	16	475			
Verkasalo et al. 1993	≥0.2 μT	0	0.88		0-4.2	Age, sex
	0.01-0.19 μT-yr	15	16.55	0.9	0.5-1.5	
	≥0.4 μT-yr	1		0.6	0.02-3.6	
	0.01-0.39 μT	14		0.9	0.5-1.5	

[a] Odds ratio calculated without consideration of possible confounders (ratio of exposed to unexposed cases divided by the ratio of exposed to unexposed controls).

[b] 95% confidence intervals for the odds ratio calculated without consideration of possible confounders.

[c] Odds radio adjusted statistically for possible confounding factors.

[d] Corresponding confidence intervals calculated for the odds ratio adjusted for possible confounding factors.

[e] Includes all factors considered to be potential confounders whether or not statistical adjustments were made for them.

[f] HCC, high current configuration.

[g] LCC, low current configuration.

[h] Cases expected on the basis of incidence data for the disease in the general population.

TABLE A5-7 Electric- and Magnetic-Field Exposure and Childhood Cancers Other than Leukemia and Brain Cancer: Results

Study	Exposure Category	Number of Cases	Number of Controls	Crude OR[a]	95% CI[b]	Potential Confounders Addressed[c]
Wertheimer and Leeper 1979	Birth addresses:					Age of onset, sex, socioeconomic status, urban residence, family pattern, traffic congestion
	HCC[d]	17	9	2.4	0.9-6.1	
	LCC[e]	31	39			
	Death addresses:					
	HCC	18	17	1.1	0.5-2.4	
	LCC	45	46			
Tomenius 1986	Total residences:					Age, sex, church district of birth
	≥0.3 µT	11	0			
	<0.3 µT	352	309			
Savitz et al. 1988	Field measurements for low-power conditions:					Maternal age, father's education, family income, maternal smoking in pregnancy, traffic density, age, sex, geographic area of residence
	>0.2 µT	1	16	0.3	0.4-2.1	
	<0.2 µT	39	191			
	Field measurements for high-power conditions:					
	>0.2 µT	3	29	0.5	0.1-1.7	
	<0.2 µT	37	175			
	Two-level wire codes:					
	High	28	52	1.5	0.9-2.6	
	Low	74	207			
	Wire codes:					
	Very high	4	8	1.6	0.5-5.8	
	Very low	27	88			

Study	Exposure	Cases Observed	Cases Expected[f]	Crude OR[a]	95% CI[b]	Potential Confounders Addressed[c]
Verkasalo et al. 1993	≥ 0.2 μT	3	2.42	1.2	0.3-3.6	Age, sex
	0.01-0.19	48	44.7	1.1	0.8-1.4	
	≥0.4 μT-yr	4		1.0	0.3-2.6	
	0.01-0.39	47		1.1	0.8-1.4	

[a]Odds ratio calculated without consideration of possible confounders (ratio of exposed to unexposed cases divided by the ratio of exposed to unexposed controls).

[b]95% confidence interval for the odds ratio calculated without consideration of possible confounders.

[c]Includes all factors considered to be potential confounders whether or not statistical adjustments were made for them.

[d]HCC, high current configuration.

[e]LCC, low current configuration.

[f]Cases expected on the basis of incidence data for the disease in the general population.

TABLE A5-8 Residential Electric- and Magnetic-Field Exposure and All Childhood Cancers: Results

Study	Exposure Category	Number of Cases	Number of Controls	Crude OR[a]	95% CI[b]	Adjusted OR[c]	Potential Confounders Addressed[d]
Wertheimer and Leeper 1979	Birth address:						Age of onset, sex, traffic congestion, socioeconomic status (SES), urban residence, family pattern
	HCC[e]	101	55	2.3	1.6-3.4		
	LCC[f]	171	217				
	Death address:						
	HCC	129	74	2.2	1.6-3.1		
	LCC	199	254				
Tomenius 1986	Total residences:						Age, sex, church district of birth, permanent vs. transient residence
	≥0.3 μT	34	14	2.1	1.1-4.0		
	<0.3 μT	1,095	955				
Savitz et al. 1988	Field measurements for low-power conditions:						Maternal age, father's education, family income, maternal smoking in pregnancy, traffic density, age, sex, geographic area of residence
	>0.2 μT	13	16	1.4	0.6-2.9	(1.2-1.5)	
	<0.2 μT	115	191				
	Field measurements for high-power conditions:						
	>0.2 μT	19	29	1.0	0.6-2.0		
	<0.2 μT	110	175				
	Two-level wire codes:						
	High	89	52	1.5	1.0-2.3		
	Low	231	207				
	Wire codes:						
	Very high	19	8	2.2	1.0-5.2		
	Very low	95	88				

Study	Exposure					Adjustment factors
Myers et al. 1990	Distance to power line:					Age, sex, residence type, county of birth
	<25 m	13	17	1.1	0.5-2.6	
	≥25 m <50 m	7	15	0.7	0.3-2.0	
	≥50 m <75 m	10	17	1.0	0.5-2.2	
	≥75 m <100 m	8	9	1.5	0.6-4.0	
	<100 m	38	58	1.0	0.6-1.7	
	≥100 m	336	530			
	Estimated field:					
	≥0.1 μT	1	4	0.4	0.04-4.1	
	≥0.03 μT <0.1 μT	8	4	2.6	0.8-9.0	
	≥0.01 μT <0.03 μT	7	13	1.0	0.4-2.5	
	≥0.01 μT <0.1 μT	15	17	1.4	0.6-3.0	
	≥0.01 μT	16	21	1.2	0.6-2.6	
	<0.01 μT	358	567			
Feychting and Ahlbom 1993	Estimated field:					Sex, age, county, residence type, diagnosis year, SES, NO₂
	≥0.3 μT	10	32	1.3	0.6-2.7	
	0.1-0.29 μT	14	47	1.2	0.6-2.3	
	≥0.2 μT	12	46	1.1	0.5-2.1	
	0.1-0.19 μT	12	33	1.5	0.7-2.9	
	<0.1 μT	117	475			
Olsen et al. 1993	Estimated field:					Age, sex, age at diagnosis
	≥0.4 μT	6	3	5.6	1.6-19	
	0.1-0.39 μT	4	17	0.7	0.2-2.0	
	≥0.25 μT	6	11	1.5	0.6-4.1	
	0.1-0.24 μT	4	9	1.3	0.4-4.1	
	≥0.1 μT	10	20	1.4	0.7-3.0	
	<0.1 μT	4	21	0.6	0.2-17	
	Not exposed, distant	16	49	0.9	0.5-1.6	
	Not exposed	1,677	4,698			

Continues on next page

TABLE A5-8 Continued

Study	Exposure Category	Cases Observed	Cases Expected[g]	Standardized Incidence Ratio	95% CI[b]	Potential Confounders Addressed[d]
Verkasalo et al. 1993	≥ 0.2 μT	11	7.39	1.5	0.7-2.7	Age, sex
	0.01-0.19 μT	129	137.17	0.9	0.8-1.1	
	≥ 0.4 μT-yr	15		1.4	0.8-2.3	
	0.01-0.39 μT-yr	125		0.9	0.8-1.1	

[a]Odds ratio calculated without consideration of possible confounders (ratio of exposed to unexposed cases divided by the ratio of exposed to unexposed controls).
[b]95% confidence intervals for the odds ratio calculated without consideration of possible confounders.
[c]Odds radio adjusted statistically for possible confounding factors.
[d]Includes all factors considered to be potential confounders whether or not statistical adjustments were made for them.
[e]HCC, high current configuration.
[f]LCC, low current configuration.
[g]Cases expected on the basis of incidence data for the disease in the general population.

TABLE A5-9 Residential Electric- and Magnetic-Field Exposure and Cancer: Results of Cohort Studies Including Subjects of All Ages

Study	Exposure Description	Number of Cases Observed	SMR[a]	95% CI[b]	Potential Confounders Addressed[c]
McDowall 1986	Distance from power line for				Age, sex, calendar time
	Leukemia:				
	0-14 m	1	1.4	0.0-8.0	
	15-34 m	2	0.8	0.1-2.8	
	35-50 m	3	1.2	0.3-3.5	
	Lymphoma:				
	0-14 m	3	3.3	0.7-9.7	
	15-34 m	2	0.6	0.1-2.1	
	35-50 m	5	1.5	0.5-3.4	
	All cancers:				
	0-14 m	27	1.0	0.7-1.5	
	15-34 m	97	1.1	0.9-1.3	
	35-50 m	89	1.0	0.8-1.2	
Schreiber et al. 1993	Wire codes:				Age, sex
	High exposure	0			
	Low exposure	3	1.3	0.3-3.9	
	Hodgkin's disease:				
	High exposure	2	4.7	0.5-17.0	
	Low exposure	0			
	Non-Hodgkin's lymphoma:				
	High exposure	2	1.8	0.2-6.4	
	Low exposure	0			
	Brain tumors:				
	High exposure	0			
	Low exposure	3	2.0	0.4-5.7	

Continues on next page

TABLE A5-9 Continued

Study	Exposure Description	Number of Cases Observed	SMR[a]	95% CI[b]	Potential Confounders Addressed[c]
	All cancers:				
	High exposure	46	0.9	0.6-1.1	
	Low exposure	65	0.9	0.7-1.2	

[a]Standard mortality ratio, ratio of observed number of deaths to the number expected based on mortality in the general population.
[b]95% confidence intervals for the odds ratio calculated without consideration of possible confounders.
[c]Includes all factors considered to be potential confounders whether or not statistical adjustments were made for them.

TABLE A5-10 Residential Electric- and Magnetic-Field Exposure and Adult Leukemia: Results

Study	Exposure Description	Number of Cases	Number of Controls	Crude OR[a]	95% CI[b]	Potential Confounders Addressed[c]
	Wire codes at time of longest residence:					Age, sex, family income, race, cigarette smoking
	Very high	5	6	0.8	0.2-2.9	
	High	21	23	0.8	0.4-1.7	
	Low	21	37	0.6	0.3-1.2	
	Very low	42	44			
	Wire codes at residence closest to reference date:					
	Very high	5	7	0.8	0.2-2.9	
	High	24	19	1.4	0.6-3.0	
	Low	26	38	0.8	0.4-1.6	
	Very low	42	52			
	Estimated field at longest residence:					
	>0.2 μT	14	18	0.8	0.3-1.8	
	0.05-0.199 μT	46	64	0.7	0.4-1.3	
	0.0-0.05 μT	29	28			
	Estimated field at residence closest to reference date:					
	>0.2 μT	23	25	1.0	0.5-2.0	
	0.05-0.1992 μT	70	92	0.8	0.5-1.4	
	0.0-0.052 μT	40	42			

Continues on next page

TABLE A5-10 Continued

Study	Exposure Description	Number of Cases	Number of Controls	Crude OR[a]	95% CI[b]	Potential Confounders Addressed[c]
	Field measurement mean exposure for low power:					
	>0.2 μT			1.5	0.5-4.7	
	0.05-0.1992 μT			1.2	0.5-2.6	
	0.0-0.052 μT					
	Field measurement mean exposure; for low power:					
	>0.2 μT			1.6	0.5-5.0	
	0.05-0.1992 μT			0.6	0.3-1.2	
	0.0-0.052 μT					
Coleman et al. 1989	Distance to substation, using population controls:					
	0-24 m	4	4	1.3	(0.3-5.3)	
	25-49 m	11	13	1.1	(0.5-2.5)	
	50-99 m	63	69	1.2	(0.8-1.8)	
	≥100 m	112	145	—	—	
Youngson et al 1991	Distance from power line:					
	<25 m	77	62	1.3	(0.9-1.8)	
	≥25 m <50 m	60	47	1.3	(0.9-1.9)	
	≥50 m <75 m	52	50	1.1	(0.7-1.6)	
	≥75 m <100 m	47	53	0.9	(0.6-1.3)	
	<100 m	236	212	1.1	(0.9-1.4)	
	≥100 m	2,908	2,932	—	—	

Estimated magnetic field:					
≥0.3 µT	15	8	1.9	(0.8-4.4)	Age, sex, residence type
≥0.1 µT	129	125	1.0	(0.7-1.5)	
Feychting and Ahlbom 1994					Age, sex
Estimated magnetic field:					
≥ 0.2 µT	26	83	1.0	0.7-1.7	
0.1-0.19 µT	20	76	0.9	0.5-1.5	
<0.09 µT	278	924	—	—	

[a] Odds ratio calculated without consideration of possible confounders (ratio of exposed to unexposed cases divided by the ratio of exposed to unexposed controls).

[b] 95% confidence intervals for the odds ratio calculated without consideration of the possible confounders.

[c] Includes all factors considered to be potential confounders whether or not statistical adjustments were made for them.

TABLE A5-11 Residential Electric- and Magnetic-Field Exposure and Adult Cancer: Results

Study	Exposure Description	Number of Cases	Number of Controls	Crude OR[a]	Adjusted 95% CI[b]	Potential Confounders Addressed[c]
Wertheimer and Leeper 1982	Wire codes:					Sex, age, socioeconomic status, onset age, urban exposure
	VHCC[d]	108	74	2.2	1.5-3.2	
	OHCC[e]	330	298	1.7	1.2-2.2	
	OLCC[f]	642	659	1.5	1.1-1.9	
	End pole[g]	99	148			

[a]Odds ratio calculated without consideration of possible confounders (ratio of exposed to unexposed cases divided by the ratio of exposed to unexposed controls).
[b]95% confidence intervals for the odds ratio calculated without consideration of possible confounders.
[c]Includes all factors considered as potential confounders whether or not statistical adjustments were made for them.
[d]VHCC, very high current configuration.
[e]OHCC, ordinary high current configuration.
[f]OLCC, ordinary low current configuration.
[g]End pole, very low current configuration.

Appendix B

Exposure Assessment in Residential Studies

Wire Codes

Whether developed as a stroke of serendipity or as a work of expert insight, wire codes and their relatively consistent association with disease end points have persisted and are largely responsible for continued interest in this avenue of research. Wire codes (or wiring configurations) were first used by Wertheimer and Leeper (1979). In that study, houses of cases and controls were classified as either high current configuration (HCC) or low current configuration (LCC), depending on the proximity of the house to several "wiring classes." In a subsequent paper (Wertheimer and Leeper, 1982), high current configuration was further divided into very high current configuration (VHCC) and ordinary high current configuration (OHCC), and low current configuration was divided into ordinary low current configuration (OLCC) and end pole (VLCC).

The basic components of the wiring codes used by Wertheimer-Leeper are depicted in Figure B-1. As defined by Wertheimer and Leeper, Class 1 wiring comprises high-voltage transmission lines, distribution lines with six or more wires (more than one distribution circuit), or a single three-phase distribution circuit with thick wires. These power line configurations are shown schematically in Figure 2-1 (Chapter 2). Class 1 is represented by the two power-line configurations at the top of the figure. If a Class 1 wiring configuration lies within 50 feet of a home, that home is classified VHCC. If the home is 50 to 130 feet from the line it is OHCC.

Class 2 wiring refers to single-circuit three-phase distribution lines with thin wires. A Class 2 wiring configuration must lie within 25 feet of a home for that home to be classified VHCC. From 25 to 50 feet from Class 2 wiring, the home is classified OHCC, and from 50 to 130 feet, the home is classified OLCC.

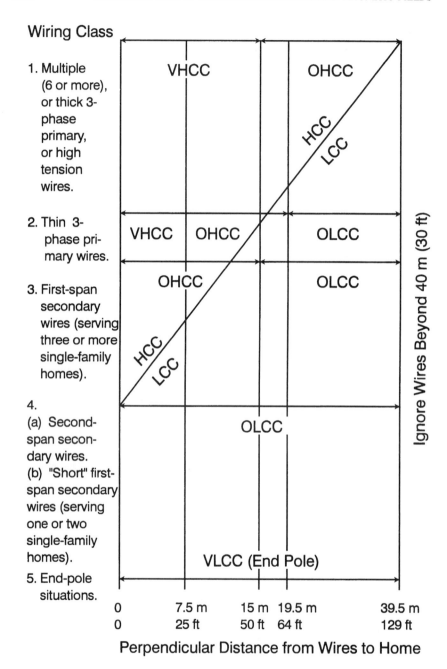

FIGURE B-1   Definition of the wiring codes in relation to the distances from wires to homes. Source: Adapted from Jones 1993.

Errata: Note shaded area.

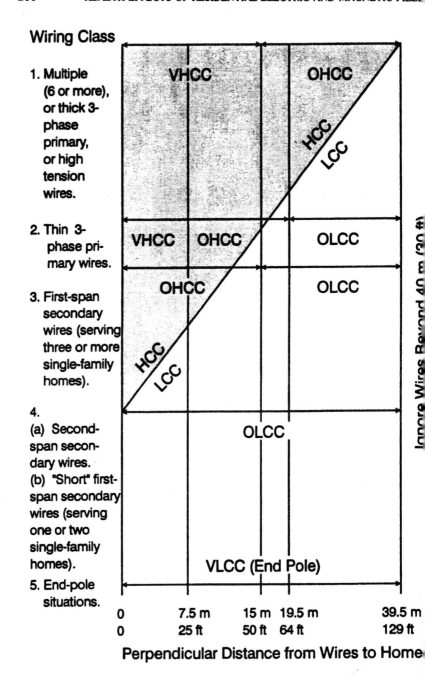

FIGURE B-1   Definition of the wiring codes in relation to the distances from wires to homes. Source: Adapted from Jones 1993.

Distribution transformers, usually pole-mounted cylindrical devices, reduce the voltage from relatively high-voltage distribution lines that can carry thousands of volts to low-voltage wiring that normally carries 120 or 220 volts for use directly in homes. Classes 3, 4, and 5 wiring comprise only secondary wires from transformers at 120 and 220 volts.

Class 3 wiring configurations are only those secondary wires that are connected to the transformer on one end (first-span wires), and serve three or more homes. If a Class 3 wiring configuration is within 50 feet of a home, that home is coded OHCC. If the home is between 50 and 130 feet of the line, that home is classified OLCC.

Class 4 wiring configurations are second-span secondary wires that serve three or more homes or first-span secondary wires that serve one or two homes. Homes within 130 feet of Class 4 wiring configurations are coded as OLCC.

Class 5, or end-pole configurations, are secondary wiring configurations that serve only one or two homes and that are not attached directly to the transformer.

It should be noted that the first wire-coding system used by Wertheimer and Leeper (1979) had only two classifications, HCC and LCC. In that scheme the HCC homes comprise both VHCC and OHCC homes as defined above. Those classifications are represented by the shaded area in Figure B-1. The LCC homes include those described above as OLCC and end-pole and are represented in Figure B-1 as the unshaded area.

Various modifications have subsequently been made to the original Wertheimer-Leeper wire codes. Savitz et al. (1988) expanded them to include a separate "buried wire" category, and changed the name "end-pole configuration" to "very low current configuration;" this is the wire code shown in Figure 2-1 (Chapter 2). They found that the coding was very reliable, with agreement over 95% when the code was rescored by independent judges.

The Wertheimer-Leeper wire codes were developed as a nonintrusive way to assess in-home exposure to power-line magnetic flux density. Magnetic flux density is proportional to power-line current (or load) and decreases with distance, factors that were hypothesized to be related to cancer incidence. Magnetic fields also are not shielded by the walls of houses, so they are more likely to reflect residents' actual exposure to power lines than would electric fields, which are easily shielded. Wertheimer and Leeper (1979) also believed that "relatively continuous exposure" to low fields is "more effective than relatively brief but strong exposure" and that the wire codes, which remain the same for years, reflect relatively continuous exposure. Additionally, wire codes do not have the time variability of spot measurements and can be gathered without access to homes. Wertheimer and Leeper (1979) reported taking some magnetic-field measurements and found that wire codes correlated with those measurements.

Some studies have examined the stability of wire codes. In the "Back-to-Denver" study (Dovan et al. 1993), 81 homes were reexamined, comparing 1985 and 1990 coding. In one case, the wire configuration had been changed; for 91%

of the remaining, the two codings were in agreement. Differences arose from errors in estimating distance (four cases), thickness (two cases), and first span (one case).

MAGNETIC-FIELD MEASUREMENTS

As an alternative approach to the classification of magnetic fields in homes, direct measurements of the fields have been incorporated into epidemiologic studies (Tomenius 1986; Savitz et al. 1988; Severson et al. 1988; London et al. 1991). Because these studies are concerned with historical rather than contemporaneous exposures, the measurements should not be viewed as an inherently valid indicator of the magnetic-field exposure of the home occupants but as another potential marker for such exposure.

Measurements typically consist of "spot measurements," in which the magnetic field is measured at a point in time and a predetermined protocol is used to choose the location of the measurement in the home. For example, Savitz et al. (1988) obtained magnetic-field measurements in three standard locations (front door, parents' bedroom, and child's bedroom) and in rooms reported to be occupied by the child for 1 or more hours per day. In those locations, a measurement was taken near the center of the room, 1 m above the floor, but away from local field sources.

Other studies have incorporated longer time periods of measurement, often 24 hr, to incorporate short-term fluctuations. London et al. (1991) monitored exposure for a full day. In principle, the 24-hr measurements should yield a more accurate depiction of long-term exposures of interest and possibly yield a more accurate indication of historical exposures of etiologic interest.

WIRE CODES AND RESIDENTIAL MAGNETIC FIELDS

Prediction of Mean Fields and Variance in Fields

A few studies have examined the degree to which wire codes are associated with magnetic-field measurements. The largest and most recent investigation was based on the national study presented by the Electric Power Research Institute (EPRI),[1] Survey of Residential Magnetic Field Sources, also known as the 1,000-home study (EPRI 1993a,b). For the underground wiring configuration, VLCC,

---

[1]The Electric Power Research Institute has funded many of the studies of electric and magnetic fields important to the interpretation of biologic and epidemiologic research. Although their reports are not peer reviewed in the same manner as journal articles, thus falling somewhat short of the selection criteria for this review, their reports are subject to both internal and external review that is of comparable rigor to most journal reviews, and their material is sufficiently important to be considered here.

OLCC, OHCC, and VHCC, the median spot magnetic-field measurements were 0.05, 0.04, 0.06, 0.08, and 0.12 μT, and the field exceeded at 5% of the residences was 0.17, 0.19, 0.36, 0.61, and 0.64 μT.

Similar results were obtained in epidemiologic studies with smaller numbers of measurements in Denver (Wertheimer and Leeper 1982; Savitz et al. 1988), Seattle (Kaune et al. 1987), and Los Angeles (London et al. 1991), although the absolute values reported for the high-wire codes from Los Angeles were lower (Table B-1). Measurements in the first four studies listed in Table B-1 were generally made near the centers of rooms, away from obvious local magnetic-field sources. Because people spend some time near such local sources, somewhat higher values might be expected for personal exposure measurements, such as the EMDEX data in Table B-1, than for spot measurements (EPRI 1993c). The spot measurements were similar in VLCC and OLCC homes, increased in OHCC homes, and increased further in VHCC homes.

The magnetic fields used in Table B-1 from the Wertheimer and Leeper (1982), Savitz et al. (1988), and the EPRI (1993a,b) studies were all spot measurements. For the London et al (1991) study, 24-hr measurements were used. Although spot measurements were reported in the London study, they were inconsistent with the 24-hr measurements. The spot measurements, which were made near the centers of rooms, were much lower than the 24-hr measurements, which were made using a different type of instrument near the child's bed location. The bed was more likely to be near sources of magnetic fields. The 24-hr OHCC and VHCC measurements from the Los Angeles study are low in comparison to those from the Denver and Seattle studies (and the underground configuration, VLCC, and OLCC seem high), but they lie within a range that is consistent with the 1,000 home study data. The differences might be related to the comparatively larger proportion of homes in the OHCC and VHCC categories in the Los Angeles study or to the differences in distribution-wiring practices in Denver and Los Angeles, as noted by the authors.

The median field measurements of 0.05, 0.04, and 0.06 μT (0.5, 0.4 and 0.6

TABLE B-1   Measured Magnetic Fields and the Associated Wire-Code Current Configurations

	Median Measured Field (μT)			
Study	VLCC	OLCC	OHCC	VHCC
Wertheimer and Leeper 1982	<0.05	<0.05	0.12	0.25
Savitz et al. 1988	0.03	0.051	0.09	0.216
London et al. 1991	0.043	0.058	0.066	0.107
1000 Homes Study (EPRI 1993a,b)	0.04	0.06	0.08	0.12
EMDEX Study (EPRI 1993c)	0.063	0.09	0.124	0.205
Average	0.044	0.065	0.096	0.180

mG) from the 1,000 homes study for the three lowest configurations (underground, VLCC, and OLCC) are very similar because they reflect sources within the residences (appliances and the grounding systems) as well as the power lines that determined the wire codes. The fields for the two highest wire codes (0.08 and 0.12 μT for OHCC and VHCC, respectively) reflect the power-line fields to a greater extent. The local sources within the home appear to add a random component that is more evident in low background fields (EPRI 1993c). The 1,000-homes study mathematically separated the power-line fields from other fields measured within the home, producing a clearer relationship between wire codes and power-line-derived magnetic fields; for underground, VLCC, OLCC, OHCC and VHCC, the median power-line 24-hr average field calculations were 0.03, 0.02, 0.04, 0.06, and 0.10 μT, respectively. On average, sources within residences appear to contribute about 0.02 μT for each of the wire codes.

Although the rank ordering of fields in homes is predicted reasonably well by wire codes, the total variance in fields is not accurately characterized by wire codes. The suggestion that local sources add a random component to residential measurements might help to explain why studies that examine the correlation between wire codes and spot measurements (or personal exposure) generally show low explained variance. In a number of studies, most of which are relatively small, the proportion of variance in measured fields accounted for by wire codes is reported. Kaune and co-workers reported that the Wertheimer-Leeper wire code accounts for about 15-20% of the measured low-power magnetic field in bedrooms in the Seattle (Kaune et al. 1987) and the Denver study (Kaune and Savitz 1994). Similarly, Bracken (1992) reported that wire-code categories accounted for little of the variance (10-16%) in magnetic-field measurements. Yost et al. (1992) found that wire codes were very weakly associated with the average magnetic field in the bedroom, having an explained variance of only 7%, but an alternative multivariate wire code reportedly yielded much higher explained variance (approximately 70%).

The Geomet study (EPRI 1992a), which involved 28 subjects in Frederick City, Maryland, found that wire codes explained 49-56% of the variance in various spot measurements. For a subset of 12 subjects, they found that 76% and 81% for winter and spring personal exposure measurements, respectively, were accounted for by wire codes. The authors of the Geomet study also suggested that the wire codes correlated more strongly with spot measurements that were averaged over two seasons than with season-specific measurements, suggesting that reduction in the random variations due to seasons enhances the predictiveness of wire codes. The study of adult leukemia by Severson et al. (1988) reported 17% explained variance in 24-hr magnetic-field measurements in a "limited sample of the residences." The Enertech study, in which two visits were made to 35 homes in Massachusetts and northern California, reported 36% and 37% explained variance for log-transformed personal exposures based on wire codes (EPRI 1992b).

The EMDEX residential study, a much larger, multicity study involving 993 visits to 380 residences served by 39 different electric utility companies, found that wire codes explained 16% and 17% of the variance in inside and outside spot measurements, respectively. Explained variance in long-term (at least 24-hr) stationary measurements in homes was 7% and that in personal exposure measurements in homes was 9% (EPRI 1993c). Analysis of the data gathered for the 1,000-homes study shows an approximate 10% explained variance of fields measured in bedrooms. The generally lower explained variance in the large multicity studies suggests that some other source of variance, such as differences in distribution-wiring practices among utilities, is involved (EPRI 1993c). That possibility suggests that wire codes might not be useful in multicity epidemiologic studies, if the average magnetic field is the exposure of interest.

Prediction of Categories of Fields

Because random magnetic-field sources in homes seem to dominate measurements for the three lowest wire-code categories (underground, VLCC, and OLCC), an alternative approach to assessing the usefulness of wire codes is to examine the extent to which the original dichotomous wire code (LCC and HCC) defines two exposure groups. That approach was used on the Enertech study data in a recent report by Kaune and Zaffanella (1994). In spite of the modest predictiveness of wire codes for fields overall, they found that 80% of the subjects with personal exposures above 0.2 μT were correctly classified based on their living in HCC homes, and 76% of the subjects with personal exposures below 0.2 μT were correctly classified based on living in LCC homes. A similar result was found in the Geomet study in which it was reported that wire codes correctly classified children's average exposure (above or below 0.2 μT) in about 77% of the cases (87% when only EMDEX-measured fields were considered).

The authors in the Enertech study pointed out that the homes in the study were not selected randomly but were chosen to have various combinations of ground-current and power-line magnetic-field sources. Because of this wire-code sampling bias, they explained, the estimates "may be different (most likely smaller) in a true random sample." The analysis can be improved by normalizing for the number of homes in the various categories and looking at ratios. Using Table B-2, four of five HCC houses in the Enertech study have fields above 0.2

TABLE B-2   Personal-Exposure Data from Kaune and Zaffanella (1994)

| Field | Wire Code | | |
	LCC	HCC	Total Houses
<0.2 μT	16	5	21
>0.2 μT	1	4	5
Total houses	17	9	26

μT and 5 of 21 have fields below 0.2 μT, so that if a person lives in an HCC house, that person is 3.4 times more likely than a person in an LCC house (a 77% chance) to have an exposure over 0.2 μT. Similarly, if a person lives in an LCC house, that person is 3.8 times more likely than a person in an HCC house (a 79% chance) to have a personal exposure of 0.2 μT or less.

The Enertech and Geomet studies were small, so the analyses are statistically unreliable regardless of how the calculations are done. The same analysis can be done with data from the much larger EMDEX residential study. As seen in Table B-3, if a person lives in an HCC house, that person is 1.8 times more likely than a person in an LCC house to have a personal exposure over 0.2 μT (a 64% chance). If a person lives in an LCC house, that person is 1.7 times more likely than a person in an HCC house to have a personal exposure under 0.2 μT (a 63% chance).

Data from the 1,000-homes study permits the same type of calculation for average spot measurements at the centers of bedrooms instead of personal exposure (Table B-4). An HCC house is 2.3 times more likely than an LCC house to have a bedroom field over 0.2 μT (a 70% chance). A LCC house is 1.8 times more likely than an HCC house to have a bedroom field under 0.2 μT (a 64% chance).

The EMDEX residential study data permits the same calculation. Using a long-term mean magnetic field (Table B-5), an HCC house is 1.9 times more likely than an LCC house to have a bedroom field over 0.2 μT (a 66% chance). An LCC house is 1.5 times more likely than an HCC house to have a bedroom field under 0.2 μT (a 60% chance).

TABLE B-3 Personal-Exposure Data from the EMDEX Study (EPRI 1993c)

| Field | Wire Code | | |
	LCC	HCC	Total Houses
<0.2 μT	357	190	547
>0.2 μT	83	137	220
Total houses	440	327	767

TABLE B-4 Spot-Measurement Data from the 1,000-Homes Study (EPRI 1993a,b)

| Field | Wire Code | | |
	LCC	HCC	Total Houses
<0.2 μT	667	223	890
>0.2 μT	39	54	93
Total houses	706	277	983

TABLE B-5   Long-Term Measurements from the EMDEX Study (EPRI 1993c)

Field	Wire Code		
	LCC	HCC	Total Houses
<0.2 μT	476	267	743
>0.2 μT	58	134	192
Total houses	534	401	935

WIRE CODES AND PERSONAL EXPOSURE TO MAGNETIC FIELDS

The ultimate inferences from wire codes pertain not to characterizing the home's magnetic fields but rather characterizing the home occupant's magnetic-field exposure. A few studies mentioned above have obtained data pertinent to addressing the question of the extent to which wire codes provide valid estimates of individual exposure.

Wire codes reflect only power-line fields, which, as noted above, are only part of the residential exposure. Spot and long-term-average measurements are generally made near the centers of rooms away from obvious sources of magnetic fields, and personal-exposure measurements are made while the subjects move around the house. Because the subjects occasionally use or are close to appliances, personal-exposure measurements are generally higher than the spot or long-term-average magnetic-field measurements. Based on the measurements reported above, it appears that the average grounding system contributes about 0.02 μT (0.2 mG) to residential personal exposure (regardless of wire code); appliances contribute between 0.02 and 0.03 μT, and power lines contribute the remainder (0.02, 0.04, 0.06, and 0.10 μT for VLCC, OLCC, OHCC, and VHCC wire codes, respectively).

It should not be surprising, therefore, that wire codes account for only a relatively small fraction of the variance in residential background fields and residential personal exposure (approximately 18%). Most of the variance between people in personal exposure is a result of variation in exposure to residential sources (appliances and grounding system); in a recent study of children, the correlation coefficient between log-transformed 24-hr residential background fields and total personal exposure was 0.97 (Kaune et al. 1994).

Although the literature on wire codes, residential measurements, and personal exposure to magnetic fields is still quite incomplete, some tentative conclusions can be drawn. The replicated association of wire codes with residential magnetic fields extends to an association between wire codes and personal time-integrated exposure. As shown in the last section, use of wire codes to classify personal exposure into broad magnetic field categories (e.g., greater than 0.2 μT and less than 0.2 μT) appears to be somewhat more accurate than estimation of actual levels of exposure, for which the proportion of magnetic fields accounted for is typically, but not always, under 20%. Wire codes (or residential measurements) might more accurately predict exposures of children, who spend a higher proportion of time in their home, than those of adults.

Refinements in Wire Codes

A number of attempts have been made to improve and simplify the Wertheimer-Leeper wire codes by using residential magnetic-field measurements as the standard. Using selected wiring characteristics to optimize the prediction of magnetic fields, Kaune et al. (1987) developed an approach that accounted for 52% of the variance, but that approach has not been shown to be generalizable to other settings. Leeper et al. (1991) reported that spun (wrapped) secondary wires are not good predictors of increased field strengths, but that first-span open secondary wires are. Using this information for the Denver study of Savitz et al. (1988) increases the odds ratio from 1.5 to 1.8, a modest increase because spun secondary wires were rare in the area (Leeper et al. 1991). A three-category wire code proposed more recently is reported to account for slightly more (21% vs. 18%) of the variance in homes in low-power magnetic-field measurements in bedrooms than the Wertheimer-Leeper code (Kaune and Savitz 1994). A recent report from Melbourne, Australia, optimized a wire code to suit their particular electric-distribution system and found an extremely strong relationship between their optimized wire code and 24-hr field measurements. Wire codes accounted for 72% of the variance in field strengths.

Efforts have been made to incorporate additional information on power loads or appliance use to refine the predictiveness of wire codes for measured fields. The Enertech study (EPRI 1992b) examined whether residential power use and utility load-flow records would help in extrapolating exposure predictions by using data from the date of the first visit to the date of the second visit and found that it was not helpful. Calculations using load-flow records are accurate when a transmission line is the sole outdoor magnetic-field source, but calculations are more difficult for distribution lines. Neighborhood homes were also found to contribute substantially to ground currents, and therefore to the fields, in many homes. The Geomet study used an appliance survey for a similar purpose. The survey conducted immediately after exposure monitoring had little predictive power (EPRI 1992a).

Overall, the evidence is fairly consistent that wire codes are weakly predictive of magnetic fields measured in the home. When statistical procedures to optimize the relationship are applied to a specific data set, the strength of the prediction improves markedly. However, unless such approaches can be generalized, they are of little value to subsequent investigators.

Historical Versus Contemporary Measurements

Studies to date are consistent in showing only a weak relationship between wire codes and measured fields, although the use of wire-code categories is more predictive of disease than measurements of present-day field strengths. The evidence linking contemporary magnetic-field spot measurements to cancer incidence is not as strong as the evidence for linking cancer to wire-code categories. However, if magnetic fields are the causal exposure, it is the historical, prediagnosis fields that are of interest, not the contemporary levels. Therefore, contemporary fields are of interest only to the extent that they are predictive of historical fields.

The study by Feychting and Ahlbom (1993) addressed that point in relation to fields from transmission lines in Sweden. They found that average low-power measurements of the magnetic field correlated better with contemporary calculations (0.70) than with historical calculations (0.52); that finding is illustrated in Figures B-2 and B-3. According to the authors, the result was taken "as an

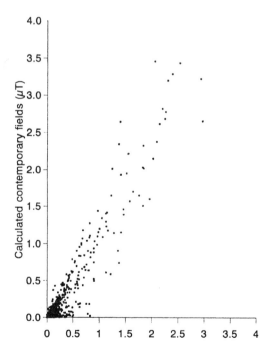

FIGURE B-2   Plot of calculated contemporary fields by low-power measurements: Children living close to 220- or 440-kV power lines in Sweden, 1960-1985. Source: Feychting and Ahlbom 1993. Reprinted with permission; copyright 1993, *American Journal of Epidemiology*, The Johns Hopkins University School of Hygiene and Public Health.

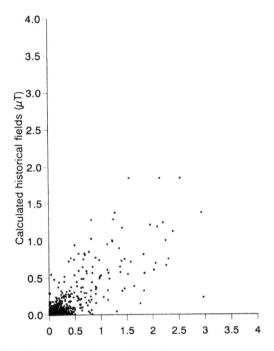

Figure B-3   Plot of calculated historical fields by low-power measurements: Children living close to 220- or 440-kV power lines in Sweden, 1960-1985. Source: Feychting and Ahlbom 1993. Reprinted with permission; copyright 1993, *American Journal of Epidemiology,* The Johns Hopkins University School of Hygiene and Public Health.

indication that contemporary fields would not be accurate as predictors of past fields.''

In the Enertech study (EPRI 1992b), two sets of personal-exposure measurements were made (spring and winter, about 7 or 8 months apart). The correlation between the first-visit and the second-visit personal-exposure levels was ''weak to nonexistent.'' The authors state, ''The fact that personal exposures measured months apart in this study were weakly correlated, at best, raises questions as to the precision with which any set of contemporary measurements can be used to assess historical exposures to power-frequency magnetic-fields'' (EPRI 1992b).

In contrast, in the Back-to-Denver study (Dovan et al. 1993), the overall correlation between log-transformed spot-measurements of the magnetic field made in 1985 and again in 1990 was 0.70. According to Kaune et al. (1994), that result provides ''some evidence for the usefulness of contemporaneous measurements as estimators of historical exposures.''

Kaune et al. (1994) examined the correlation in bedroom spot measurements over periods of up to 24 hr. The correlation in spot measurements was generally greater than 0.9 for measurements taken within 3 hr; the correlations dropped to 0.7-0.8 when the time difference was 6-24 hr.

In the EMDEX residential project (EPRI 1993c), multiple measurements (up to five times over a 17-month period) were made at 380 homes; most homes were measured two or three times. The authors did not examine the correlation between successive measurements, but they did plot the difference between personal-exposure (and stationary measurement) means for pairs of visits as a function of the time between visits (which varied from 2 to 493 days, with a median of 148 days). There was no apparent pattern in the differences between means over time, as might be expected if the 24-hr trend of decreasing correlations (Kaune et al. 1994) continued over a period of weeks or months.

Unfortunately, the available studies do not provide a clear answer to the question of how predictive current magnetic fields are of past exposure levels. This issue is particularly challenging for researchers to address because the period of interest extends over years rather than days or months. At present, the overall impression is that present-day magnetic fields are somewhat predictive of historical fields, with correlations possibly as high as 0.5-0.8.

Variability in the Wire-Code and Magnetic-Field Association

Examination of the current literature on wire codes and residential magnetic-field strengths reveals several anomalous results. In principle, at least in the United States, a specific Wertheimer-Leeper wire-code category would be expected to correspond to approximately the same magnetic-field strength. Although that is the case in Denver (Wertheimer and Leeper 1979; Savitz et al. 1988) and Seattle (Kaune et al. 1987), it is not so clearly true in Los Angeles (London et al. 1991). Because the patterns of cancer risk in relation to the wire code are similar in Denver and Los Angeles, the inconsistency in the corresponding magnetic fields is particularly anomalous and warrants additional evaluation.

The study by Feychting and Ahlbom (1993) presents another apparent anomaly. For leukemia in children and exposure defined from calculated historical fields, the study shows an increase in estimated relative risks, which increase with level of exposure. However, when the data are stratified by type of home, the results seem to be confined to one-family homes; the results disappear for apartments. One possible explanation mentioned by the authors is that the small numbers make the relative risk estimated unstable.

Another possibility is that the calculation method was unsuitable for apartments (which frequently have hidden distribution wiring). When the Feychting and Ahlbom (1993) cross-tabulated contemporary calculated fields with spot measurements for one-family homes and for apartments, the agreement was considerably better for the one-family homes; the analysis also showed that the spot measurements (but not the calculated fields) were higher in apartments than in one-family homes.

The pattern of changes in field strengths in relation to power lines over time observed in Sweden (Feychting and Ahlbom 1993) raises the question of whether wire codes, which predict fields over time more generally, are accurate in predicting temporal changes. For example, when a given power line is constructed, it will be used at far less than its capacity, but over time the use will rise to its intended capacity and perhaps somewhat beyond. Thus, early in its life, the magnetic field predicted from the line will be too low and later it might be too high. Knowledge of such a pattern would be valuable and possibly could result in refinements in estimating magnetic-field exposure.

ALTERNATIVE MAGNETIC-FIELD INDICES

The relation of wire codes to measured fields depends, in part, on the aspect of magnetic field of interest. Much of the literature is concerned with determining the exact value of the field at each point in time (a series of spot measurements), implicitly assuming that all variability in field strengths (from one point in time to the next) is potentially important. An alternative assumption might be that the only important attribute of the magnetic field is whether it falls above or below some critical value. If that were the case, then the apparent predictiveness of the wire code would differ.

When magnetic fields are measured over an extended period, such as 24 or 48 hr, then additional decisions are required in establishing the index of interest. Time-integrated values have typically been the focus of attention, but a broad range of parameters based on percentiles, time above or below thresholds, and indicators of temporal variability might be as appropriate. An approach to better understanding wire codes might be to ask what aspects of magnetic fields they do predict well. For example, they might accurately predict time above a low threshold (e.g., 0.1 μT), yet poorly predict time-integrated residential field strengths. The same concerns apply to personal exposures, which are collected over some extended time.

Kaune et al. (1994) examined the relation between wire codes and other magnetic-field characteristics (such as peak field, fields above a minimum threshold, and occasional or frequent changes in field strength), but found that none was more strongly correlated with the wire-code category than it was with the arithmetic mean of the 24-hr bedroom magnetic field. The EMDEX residential project found no relation between wire-code categories and maximum field strengths, nor was it found between wire-code categories and first measured differences in personal exposures (an estimate of field variance) (EPRI 1993c).

WIRE CODES AND TRANSIENT AND VARIABLE FIELDS

In a study of residential transient magnetic fields (EPRI 1994a), the authors found that transients within the home might be (1) random, originating from manual and automatic switching devices, such as those associated with lights and appliances, or (2) repetitive, originating from electronic controllers, such as dimmer switches, that produce transients twice per power-frequency cycle. They also found that the rate of occurrence of transients increases from residential locations to business and industrial locations. Transients might propagate from a residence or business through secondary distribution wiring to neighboring residences and along ground pathways (conductive plumbing pipes). Switching operations in the primary distribution lines can also generate transients within the residences.

Personal exposure to transient fields is extremely complex because of their strong spatial dependence. Exposure to high-frequency fields is typically due to sources within the residence, but exposure to low-frequency fields can also be due to external sources in the neighborhood.

The authors suggest that ''higher current configurations may correlate locally with higher demand indicating either increased population density or individual customers with large current usage such as industries or businesses. Thus, a residence classified as being in a 'higher' current configuration would more likely be in a locale where the rate of transient occurrence is higher, because of the likelihood of having more neighbors or being closer to businesses and industries. Furthermore, such a residence would be physically closer to neighboring transient sources, implying shorter propagation distances and an increased probability of 'receiving' a transient generated at a neighboring residence'' (EPRI 1994a).

On the other hand, current flows are generally more variable at points in the power-delivery system that serve fewer customers; therefore, the most uniform power flows should be found on the multiple or thick three-phase primary lines and high-tension lines near high current configuration (HCC) homes. So, if the causal exposure is related to long-period variability, it would probably not corre- late with the average field strength or with wire codes.

References

Abou-Samra, A.-B., H. Jüppner, T. Force, M.W. Freeman, X.-F. Kong, E. Schipani, P. Urena, J. Richards, J.V. Bonventre, J.T. Potts, Jr., H.M. Kronenberg, and G.V. Segre. 1992. Expression cloning of a common receptor for parathyroid hormone and parathyroid hormone-related peptide from rat osteoblast-like cells: A single receptor stimulates intracellular accumulation of both cAMP and inositol trisphosphates and increases intracellular free calcium. Proc. Natl. Acad. Sci. USA 89:2732-2736.

Adair, R.K. 1992. Criticism of Lednev's mechanism for the influence of weak magnetic fields on biological systems. Bioelectromagnetics 13:231-235.

Adey, W.R. 1983. Some fundamental aspects of biological effects of extremely low frequency (ELF). Pp. 561-580 in Biological Effects and Dosimetry of Nonionizing Electromagnetic Fields, M. Grandolfo and S.M. Michaelson, eds. New York: Plenum.

Adey, W.R. 1992a. ELF magnetic fields and promotion of cancer experimental studies. Pp. 23-46 in Interaction Mechanisms of Low-Level Electromagnetic Fields in Living Systems, B. Norden and C. Ramel, eds. New York: Oxford University Press.

Adey, W.R. 1992b. Collective properties of cell membranes. Pp. 47-77 in Interaction Mechanisms of Low-Level Electromagnetic Fields in Living Systems, B. Norden and C. Ramel, eds. New York: Oxford University Press.

Ahlbom, A., M. Feychting, M. Koskenvuo, J.H. Olsen, E. Pukkala, G. Schulgen, and P. Verkasalo. 1993. Electromagnetic fields and childhood cancer [letter]. Lancet 342(8882):1295-1296.

Albert, E.N., F. Slaby, J. Roche, and J. Loftus. 1987. Effect of amplitude-modulated 147 MHz radiofrequency radiation on calcium ion efflux from avian brain tissue. Radiat. Res. 109:19-27.

Algers, G., and J. Hultgren. 1987. Effects of long-term exposure to a 400-kV, 50-Hz transmission line on estrousand fertility in cows. Prev. Vet. Med. 5:21-36.

Anderson, J.C., and C. Eriksson. 1970. Piezoelectric properties of dry and wet bone. Nature 227:491-492.

Andrienko, L.G. 1977. The effect of an electromagnetic field of industrial frequency on the generative function in an experiment [in Russian]. Gig. Sanit. 6:22-25; Engl. Trans. Gig. Sanit. 7:27-31.

ANSI/IEEE. 1987. Procedures for Measurement of Power Frequency Electric and Magnetic Fields from AC Power Lines. ANSI/IEEE 644-1987. New York: IEEE.

Auerbach, G.D., S.J. Marx, and A.M. Spiegel. 1985. Parathyroid hormone, calcitonin and the calciferols. Pp. 1137-1217 in Williams Textbook of Endocrinology, 7th Ed., J.D. Wilson and D.W. Foster, eds. Philadelphia: Saunders.

Balcer-Kubiczek, E.K., and G.H. Harrison. 1991. Neoplastic transformation of C3H/10T1/2 cells following exposure to 120-Hz modulated 2.45-GHz microwaves and phorbol tumor promoter. Radiat. Res. 126:65-72.

Banks, R. 1988. Magnetic fields and illness. Public Health 102:393-394.

Baris, D., and B. Armstrong. 1990. Suicide among electric utility workers in England and Wales [letter]. Br. J. Ind. Med. 47:788-789.

Baroncelli, P., S. Battisti, A. Checcucci, P. Comba, M. Grandolfo, A. Serio, and P. Vecchia. 1986. A health examination of railway high-voltage substation workers exposed to ELF electromagnetic fields. Am. J. Ind. Med. 10:45-55.

Bassett, C.A.L. 1982. Pulsing electromagnetic fields: A new method to modify cell behavior in calcified and noncalcified tissues. Calcif. Tissue Int. 34:1-8.

Bassett, C.A.L. 1983. Biomedical implications of pulsing electromagnetic fields. Surg. Rounds Jan.:22-31.

Bassett, C.A.L. 1989. Fundamental and practical aspects of therapeutic uses of pulsed electromagnetic fields (PEMFs). CRC Crit. Rev. Biomed. Eng. 17:451-529.

Bassett, C.A.L. 1990. Premature alarm over electromagnetic fields. Issues Sci. Technol. 6:37-39.

Bassett, C.A.L., and R.O. Becker. 1962. Generation of electric potentials in bone in response to mechanical stress. Science 137:1063-1064.

Bassett, C.A.L., R.J. Pawluk, and R.O. Becker. 1964. Effect of electric currents on bone in vivo. Nature 204:652-654.

Bassett, C.A.L., R.J. Pawluk, and A.A. Pilla. 1974a. Augmentation of bone repair by inductively coupled electromagnetic fields. Science 184:575-577.

Bassett, C.A.L., R.J. Pawluk, and A.A. Pilla. 1974b. Acceleration of fracture repair by electromagnetic fields. A surgically noninvasive method. Ann. N.Y. Acad. Sci. 238:242-262.

Bassett, C.A.L., A.A. Pilla, and R.J. Pawluk. 1977. A non-operative salvage of surgically-resistant pseudarthrodeses and non-unions by pulsing elecromagnetic fields. Clin. Orthop. Rel. Res. 124:128-143.

Bassett, C.A.L., S.N. Mitchell, and S.R. Gaston. 1982. Pulsing electromagnetic field treatment in ununited fractures and failed arthrodeses. JAMA 247:623-628.

Baum, A., M. Mevissen, K. Kamino, U. Mohr, and W. Löscher. 1995. A histopathological study on alterations in DMBA-induced mammary carcinogenesis in rats with 50 Hz, 100 μT magnetic field exposure. Carcinogensis 16:119-125.

Bawin, S.M., and W.R. Adey. 1976. Sensitivity of calcium binding in cerebral tissue to weak environmental electric fields oscillating at low frequency. Proc. Natl. Acad. Sci. USA 73:1999-2003.

Becker, R.O. 1974. The significance of bioelectric potentials. Bioelectrochem. Bioenerget. 1:187-199.

Beniashvili, D.Sh., V.G. Bilanishvili, and M.Z. Menabde. 1991. Low-frequency electromagnetic radiation enhances the induction of rat mammary tumors by nitrosomethyl urea. Cancer Lett. 61:75-79.

Berman, E., L. Chacon, D. House, B.A. Koch, W.E. Koch, J. Leal, S. Lovtrup, E. Mantiply, A.H. Martin, G.I. Martucci, K. Hansson Mild, J.C. Monahan, M. Sandstrom, K. Shamsaifar, R. Tell, M.A. Trillo, and P. Wagner. 1990. Development of chicken embryos in a pulsed magnetic field. Bioelectromagnetics 11:169-187.

Bersani, F., A. Cossarizza, and C. Franceschi. 1989. Effects of extremely low frequency (ELF) pulsed electromagnetic fields on immunocompetent cells: In vitro studies. Alta Freq. 58:375-380.

Blackman, C.F., J.A. Elder, C.M. Weil, S.G. Benane, D.C. Eichinger, and D.E. House. 1979. Induction of calcium-ion efflux from brain tissue by radiofrequency radiation: Effects of modulation frequency and field strength. Radio Sci. 14:93-98.

Blackman, C.F., S.G. Benane, J.A. Elder, D.E. House. J.A. Lampe, and J.J. Faulk. 1980a. Induction of calcium-ion efflux from brain tissue by radiofrequency radiation: Effect of sample number and modulation frequency on the power-density window. Bioelectromagnetics 1:35-43.

Blackman, C.F., S.G. Benane, W.T. Joines, M.A. Hollis, and D.E. House. 1980b. Calcium-ion efflux from brain tissue: Power-density versus internal field-intensity dependencies at 50 MHz RF radiation. Bioelectromagnetics 1:277-283.

Blackman, C.F., S.G. Benane, L.S. Kinney, W.T. Joines, and D.E. House. 1982. Effects of ELF fields on calcium-ion efflux from brain tissue in vitro. Radiat. Res. 92:510-520.

Blackman, C.F., S.G. Benane, D.E. House, and W.T. Joines. 1985a. Effects of

ELF (1-120 Hz) and modulated (50 Hz) RF fields on the efflux of calcium ions from brain tissue in vitro. Bioelectromagnetics 6:1-11.

Blackman, C.F., S.G. Benane, J.R. Rabinowitz, D.E. House, and W.T. Joines. 1985b. A role for the magnetic field in the radiation-induced efflux of calcium ions from brain tissue in vitro. Bioelectromagnetics 6:327-337 [Erratum (1986) 7:347].

Blackman, C.F., S.G. Benane, D.J. Elliott, D.E. House, and M.M. Pollock. 1988a. Influence of electromagnetic fields on the efflux of calcium ions from brain tissue in vitro: A three-model analysis consistent with the frequency response up to 510 Hz. Bioelectromagnetics 9:215-227.

Blackman, C.F., D.E. House, S.G. Benane, W.T. Joines, and R.J. Spiegel. 1988b. Effect of ambient levels of power-line-frequency electric fields on a developing vertebrate. Bioelectromagnetics 9:129-140.

Blackman, C.F., L.S. Kinney, D.E. House, and W.T. Joines. 1989. Multiple power-density windows and their possible origin. Bioelectromagnetics 10:115-128.

Blackman, C.F., S.G. Benane, and D.E. House. 1991. The influence of temperature during electric- and magnetic-field-induced alteration of calcium-ion release from in vitro brain tissue. Bioelectromagnetics 12:173-182.

Blackwell, R.P. 1986. Effects of extremely-low-frequency electric fields on neuronal activity in rat brain. Bioelectromagnetics 7:425-434. Erratum 1987 8:213.

Blackwell, R.P., and A.L. Reed. 1985. Effects of electric field exposure on some indices of CNS arousal in the mouse. Bioelectromagnetics 6:105-107.

Blair, A., J. Burg, J. Foran, H. Gibb, S. Greenland, R. Morris, G. Raabe, D. Savitz, J. Teta, D. Wartenberg, O. Wong, and R. Zimmerman. 1995. Guidelines for application of meta-analysis in environmental epidemiology. Regul. Toxicol. Pharmacol. 22:189-197.

Blank, M. 1992. Na,K-ATPase function in alternating electric fields. FASEB J. 6:2434-2438.

Blank, M., and L. Soo. 1992. Threshold for inhibition of Na,K-ATPase by ELF alternating currents. Bioelectromagnetics 13:329-333.

Blank, M., L. Soo, H. Lin, and R. Goodman. 1993. Stimulation of transcription in HL-60 cells by alternating currents from electric fields. Pp. 516-519 in Electricity and Magnetism in Biology and Medicine: Review and Research Papers Presented at the First World Congress for Electricity and Magnetism in Biology and Medicine, M. Blank, ed. San Francisco: San Francisco Press.

Blasiak, J., M. Zmyslony, Z. Jozwiak, J. Rsoin, and S. Szmigielski. 1990. Enhancement or reduction of calcium-ion efflux from brain tissues in vitro following exposure to ELF fields dependent on intensity of local geomagnetic field. J. Bioelectr. 9:55-60.

Blask, D.E. 1993. Melatonin in oncology. Pp. 447-476 in Melatonin: Biosynthesis,

Physiological Effects, and Clinical Applications, H.S. Yu and R.J. Reiter, eds. Boca Raton, Fla.: CRC.

Bonnell, J., W. Norris, J. Pickles, J. Male, and R. Cartwright. 1983. Comments on "Environmental power-frequency magnetic fields and suicide." Health Phys. 44:698-699.

Borgens, R.B. 1984. Endogenous ionic currents traverse intact and damaged bone. Science 225:478-482.

Borle, A.B. 1990. An overview of techniques for the measurement of calcium distribution, calcium fluxes, and cytosolic free calcium in mammalian cells. Environ. Health. Perspect. 84:45-56.

Bracken, T.D. 1992. Experimental macroscopic dosimetry of extremely-low-frequency electric and magnetic fields. Bioelectromagnetics Suppl. 1:15-26.

Bracken, T.D., R.F. Rankin, R.S. Senior, and J.R. Alldredge. 1992. Relationships within and between residential magnetic field exposure characterization methods. P. A-47 in Project Abstracts: The Annual Review of Research on Biological Effects of Electric and Magnetic Fields. Frederick, Md.: W/L Associates.

Bracken, T.D., L.I. Kheifets, and S.S. Sussman. 1993. Exposure assessment for power frequency electric and magnetic fields (EMF) and its application to epidemiologic studies. J. Expo. Anal. Environ. Epidemiol. 3:1-22.

Bracken, M.B., K. Belanger, K. Hellenbrand, L. Dlogoza, T.R. Holford, J.E. McSharry, K. Addeso, and B. Leaderer. 1995. Exposure to electromagnetic fields during pregnancy with emphasis on electrically heated beds: Association with birth weight and intrauterine growth retardation. Epidemiology 6:263-270.

Brainard, G.C., J.P. Gaddy, F.M. Barker, J.P. Hanifan, and M.D. Rollag. 1993. Mechanisms in the eye that mediate the biological and therapeutic effects of light. Pp. 29-54 in Light and Biological Rhythms in Man, L. Wetterberg, ed. New York: Pergamon.

Brighton, C.T., and Z.B. Friedenberg. 1974. Electrical stimulation and oxygen tension. Ann. N.Y. Acad. Sci. 238:314-320.

Brighton, C.T., and W.P. McCluskey. 1986. Cellular response and mechanisms of action of electrically induced osteogenesis. Pp. 213-254 in Bone and Mineral Research, Vol. 4, W.A. Peck, ed. Amsterdam: Elsevier.

Brighton, C.T., Z.B. Friedenberg, and J. Black. 1979. Evaluation of the use of constant direct current in the treatment of non-union. Pp. 519-545 in Electrical Properties of Bone and Cartilage: Experimental Effects and Clinical Applications, C.T. Brighton, J. Black, and S.R. Pollack, eds. New York: Grune & Stratton.

Broadbent, D.E., M.H. Broadbent, J.C. Male, and M.R. Jones. 1985. Health of workers exposed to electric fields. Br. J. Ind. Health 42:75-84.

Burack, G.D., Y.J. Seto, S.T. Hsieh, and J.L. Dunlap. 1984. The effects of prenatal exposure to a 60-Hz high-intensity electric field on postnatal development and sexual differentiation. J. Bioelectr. 3:451-467.

Byus, C.V., S.E. Pieper, and W.R. Adey. 1987. The effects of low-energy 60-Hz environmental electromagnetic fields upon the growth-related enzyme ornithine decarboxylase. Carcinogenesis 8:1385-1389.

Cain, C.D., and R.A. Luben. 1987. Pulsed EMF effects on PTH stimulated cAMP accumulation and bone resorption in mouse calvariae. Pp. 269-278 in Interaction of Biological Systems with ELF, L.E. Anderson, B.J. Kelman, and R.J. Weigel, eds. Conf. Publ. No. 24. Columbus, Ohio: Battelle Press.

Cain, C.D., W.R. Adey, and R.A. Luben. 1987. Evidence that pulsed electromagnetic fields inhibit coupling of adenylate cyclase by parathyroid hormone in bone cells. J. Bone Miner. Res. 2:437-441.

Cameron, I.L., K.E. Hunter, and W.D. Winters. 1985. Retardation of embryogenesis by extremely-low-frequency 60-Hz electromagnetic fields. Physiol. Chem. Phys. Med. NMR 17:135-138.

Caputa, K., and M.A. Stuchly. 1996. Computer controlled system for producing uniform magnetic fields and its application in biomedical research. IEEE Trans. Instrum. Meas. 45(3): 701-709.

Cardossi, R., F. Bersani, A. Cossarizza, P. Zucchini, G. Emilia, G. Torelli, and C. Franceschi. 1992. Lymphocytes and low-frequency electromagnetic fields. FASEB J. 6:2667-2674.

Carson, J.J.L., F.S. Prato, D.J. Drost, L.D. Diesbourg, and S.J. Dixon. 1990. Time-varying magnetic fields increase cytosolic free Ca^{2+} in HL-60 cells. Am. J. Physiol. 259:687-692.

Chahal, R., D.Q.M. Craig, and R.J. Pinney. 1993. Investigation of potential genotoxic effects of low frequency electromagnetic fields on *Escherichia coli*. J. Pharm. Pharmacol. 45:30-33.

Champney, T.H., A.P. Holtorf, R.W. Steger, and R.J. Reiter. 1984. Concurrent determination of enzymatic activities and substrate concentrations in the melatonin synthetic pathway within the same rat pineal gland. J. Neurosci. Res. 11:59-66.

Chen, K.M., H.R. Chuang, and C.J. Lin. 1986. Quantification of interaction between ELF-LF electric fields and human bodies. IEEE Trans. Biomed. Eng. 33:746-755.

Cheng, J., M.A. Stuchly, C. DeWagter, and L. Martens. 1995. Magnetic field induced currents in a human head from use of portable appliances. Phys. Med. Biol. 40:495-510.

Chernoff, N., J.M. Rogers, and R. Kavet. 1992. A review of the literature on potential reproductive and developmental toxicity of electric and magnetic fields. Toxicology 74:91-126.

Chiban, A., K. Isaka, M Kitagawa, T. Matsuo, M. Nagata, and Y. Yokoi. 1984. Application of finite element method to analysis of induced current densities inside human models exposed to 60-Hz electric fields. IEEE Trans. Power Appar. Syst. 103:1895-1902.

Clarke, R.L., and D.R. Justesen. 1979. Behavioral sensitivity of a domestic bird to 60-Hz AC and DC magnetic fields. Radio Sci. 14:209-216.

Cleary, S.F. 1993. A review of in vitro studies: Low-frequency electromagnetic fields. J. Am. Ind. Hyg. Assoc. 54:178-185.

Cochran, G.V.B., R.J. Pawluk, and C.A.L. Bassett. 1968. Electromechanical characteristics of bone under physiologic moisture conditions. Clin. Orthop. Relat. Res. 58:249-270.

Cohen, M.M. 1986. In Vitro Genetic Effects of Electromagnetic Fields. Final Report to New York State Power Lines Project. Wadsworth Center, Albany, N.Y.

Cohen, M.M., A Kunska, J.A. Astemborski, and D. McCulloch. 1986a. The effect of low-level 60-Hz electromagnetic fields on human lymphoid cells: II. Sister-chromatid exchanges in peripheral lymphocytes and lymphoblastoid cell lines. Mutat. Res. 172:177-184.

Cohen, M.M., A. Kunska, J.A. Astemborski, D. McCulloch, and D.A. Paskewitz. 1986b. Effect of low-level 60-Hz electromagnetic fields on human lymphoid cells: I. Mitotic rate and chromosome breakage in human peripheral lymphocytes. Bioelectromagnetics 7:415-423.

Colacicco, G., and A.A. Pilla. 1983. Chemical, physical, and biological correlations in the Ca-uptake by embryonal chick tibia in vitro. Biochem. Bioenerg. 10:119-131.

Coleman, M., C.M.J. Bell, H.L. Taylor, and M. Primic-Zakelj. 1989. Leukaemia and residence near electricity transmission equipment: A case-control study. Br. J. Cancer 60:793-798.

Compere, C.L. 1982. Electromagnetic fields and bones [editorial]. JAMA 247:669.

Cone, C.D., Jr. 1971. Unified theory on the basic mechanism of normal mitotic control and oncogenesis. J. Theor. Biol. 30:151-181.

Cook, M., C. Graham, H. Cohen, and M. Gerkovich. 1992. A replication study of human exposure to 60-Hz fields: Effects on neurobehavioral measures. Bioelectromagnetics 13:261-285.

Cooke, P., and P.G. Morris. 1981. The effects of NMR exposure in living organisms. II. A genetic study of human lymphocytes. Br. J. Radiol. 54:622-625.

Cooper, L.J., H.B. Graves, J.C. Smith, D. Poznaniak, and A.H. Madjid. 1981. Behavioral responses of pigeons to high-intensity 60-Hz electric fields. Behav. Neural Biol. 32:214-228.

Cossarizza, A., D. Monti, P. Sola, G. Moschini, R. Cadossi, F. Bersani, and C. Franceschi. 1989. DNA repair after gamma irradiation in lymphocytes exposed to low-frequency pulsed electromagnetic fields. Radiat. Res. 118:161-168.

Cox, C.F., L.J. Brewer, C.H. Raeman, C.A. Schryver, S.Z. Child, and E.L. Cartensen. 1993. A test for teratological effects of power frequency magnetic fields on chick embryos. IEEE Trans. Biomed. Eng. 40:605-610.

Creim, J.A., R.H. Lovely, W.T. Kaune, and R.D. Phillips. 1984. Attempts to produce taste-aversion learning in rats exposed to 60-Hz electric fields. Bioelectromagnetics 5:271-282.

Czerska, E., J. Casamento, J. Ning, M. Swicord, H. Al-Abrazi, C. Davis, and E. Elson. 1992. Comparison of the effect of ELF on c-*myc* oncogene expression in normal and transformed human cells. Ann. N.Y. Acad. Sci. 649:340-342.

D'Ambrosio, G., A. Scaglione, D. Di Berardino, M.B. Lioi, L. Iannuzzi, E. Mostacciuolo, and M.R. Scarfi. 1985. Chromosomal aberrations induced by extremely low frequency electric fields. J. Bioelectr. 4:279-284.

D'Ambrosio, G., R. Massa, D. Di Berardino, M.B. Lioi, A. Scaglione, and M.R. Scarfi. 1988-1989. Chromosomal aberrations in bovine lymphocytes exposed to 50-Hz electric currents. J. Bioelectr. 7:239-245.

Davis, H.P., S.J.Y Mizumori, H. Allen, M.R. Rosenzweig, E.L. Bennet, and T.S. Tenforde. 1984. Behavioral studies with mice exposed to DC and 60-Hz magnetic fields. Bioelectromagnetics 5:147-164.

Dealler, S.F. 1981. Electrical phenomena associated with bones and fractures and the therapeutic use of electricity in fracture healing. J. Med. Eng. Technol. 5:73-79.

Deguchi, T., and J. Axelrod. 1972. Sensitive assay for serotonin N-acetyltransferase activity in rat pineal. Anal. Biochem. 50:174-179.

Delpizzo, V. 1990. A model to assess personal exposure to ELF magnetic fields from common household sources. Bioelectromagnetics 11:139-147.

Delpizzo, V. 1993. Misclassification of ELF occupational exposure resulting from spatial variation of the magnetic field. Bioelectromagnetics 14:117-130.

Delpizzo, V. 1994. Epidemiological studies of work with video display terminals and adverse pregnancy outcomes (1984-1992). Am. J. Ind. Med. 26:465-480.

Demers, P.A., D.B. Thomas, K.A. Rosenblatt, and L.M. Jimenez. 1991. Occupational exposure to electromagnetic fields and breast cancer in men. Am. J. Epidemiol. 134:340-347.

Deno, D.W. 1977. Currents induced in the human body by high voltage transmission line electric field—Measurement and calculation of distribution and dose. IEEE Trans. Power Appar. Syst. 96:1517-1527.

DerSimonian, R., and N. Laird. 1986. Meta-analysis in clinical trials. Controlled Clin. Trials 7:177-188.

Dickersin, K., and J. Berlin. 1992. Meta-analysis: State of the science. Epidemiol. Rev. 14:154-176.

Dimbylow, P.J. 1987. Finite difference calculations of current densities in a homogeneous model of a man exposed to extremely low frequency electric fields. Bioelectromagnetics 8:355-375.

Dimbylow, P.J. 1988. The calculation of induced currents and absorbed power in a realistic, heterogeneous model of the lower leg for applied electric fields from 60 Hz to 30 MHz. Phys. Med. Biol. 33:1453-1468.

Dlugosz, L., J. Vena, T. Byers, L. Sever, M. Bracken, and E. Marshall. 1992. Congenital defects and electric bed heating in New York State: A registry-based case-control study. Am. J. Epidemiol. 135:1000-1011.

Dosemeci, M., S. Wacholder, and J.H. Lubin. 1990. Does nondifferential misclassification of exposure always bias a true effect toward the null value? Am. J. Epidemiol. 132:746-748.

Dovan, T., W.T. Kaune, and D.A. Savitz. 1993. Repeatability of measurements of residential magnetic fields and wire codes. Bioelectromagnetics 14:145-159.

Dowman, R., J.R. Wolpaw, R.F. Seegal, and S. Satya-Murti. 1989. Chronic exposure of primates to 60-Hz electric and magnetic fields: III. Neurophysiologic effects. Bioelectromagnetics 10:303-317.

Dowson, D.I., G.T. Lewith, M. Campbell, M.A. Mullee, and L.A. Brewster. 1988. Overhead high-voltage cables and recurrent headache and depressions. Practitioner 232:435-436.

Dumanskii, Y.D., Iu.M. Popovich, and I.P. Kozyiarin. 1977. Effect of a low-frequency (50 Hz) electromagnetic field on the functional state of the human body [in Russian]. Gig. Sanit. 12:32-35.

Dutta, S.K., K.P. Das, B. Ghosh, and C.F. Blackman. 1992a. Dose dependence of acetylcholineesterase activity in neuroblastoma cells exposed to modulated radio-frequency electromagnetic radiation. Bioelectromagnetics 13:317-322.

Dutta, S.K., B. Watson, Jr., and K.P. Das. 1992b. Intensity dependence of enolase activity by modulated radiofrequency radiation. Bioelectrochem. Bioenerg. 27:179-189.

Easley, S.P., A.M. Coelho, Jr., and W.R. Rogers. 1991. Effects of exposure to a 60-kV/m, 60-Hz electric field on the social behavior of baboons. Bioelectromagnetics 12:361-375.

Eichwald, C., and F. Kaiser. 1993. Model for receptor-controlled cystolic calcium oscillations and for external influences on the signal pathway. Biophys. J. 65:2047-2058.

EPA (U.S. Environmental Protection Agency). 1990. Evaluation of the Potential Carcinogenicity of Electromagnetic Fields. Review Draft. EPA/600/6-90/005B. U.S. Environmental Protection Agency, Washington, D.C.

EPA (U.S. Environmental Protection Agency). 1992. EMF in Your Environment: Magnetic Field Measurements of Everyday Electrical Devices. EPA/402/R-92/008. Office of Radiation and Indoor Air, U.S. Environmental Protection Agency, Washington, D.C.

EPA (U.S. Environmental Protection Agency). 1993. EMF Exposure Environments Summary Report. Contract 68D20185. Prepared by the Volpe National Transportation Systems Center for the Office of Radiational and Indoor Air, Radiation Studies Division, U.S. Environmental Protection Agency, Washington, D.C.

EPRI (Electric Power Research Institute). 1990. The EMDEX Project: Technology Transfer and Occupational Measurements. Project RP2966-1. Rep. EN-7048-V1, -V2, and -V3. Prepared by T. Dan Bracken, Inc., for Electric Power Research Institute, Palo Alto, Calif.

EPRI (Electric Power Research Institute). 1992a. Assessment of Children's Long-Term Exposure to Magnetic Fields (the Geomet Study). Project RP2966-04. Rep. TR-101406. Prepared by Geomet Technologies, Inc., for Electric Power Research Institute, Palo Alto, Calif.

EPRI (Electric Power Research Institute). 1992b. Assessment of Children's Long-Term Exposure to Magnetic Fields (the Enertech Study). Project RP2966-06. Rep. TR-101407. Prepared by Enertech Consultants and High Voltage Transmission Research Center for Electric Power Research Institute, Palo Alto, Calif.

EPRI (Electric Power Research Institute). 1993a. Survey of Residential Magnetic Field Sources: Goals, Results, and Conclusions, Vol. 1. Project RP3335-02. Rep. TR-102759-V1. Prepared by High Voltage Transmission Research Center for Electric Power Research Institute, Palo Alto, Calif.

EPRI (Electric Power Research Institute). 1993b. The EMDEX Project Residential Study: Interim Report. Project RP2966-01. Rep. TR-102011. Prepared by T. Dan Bracken, Inc., for Electric Power Research Institute, Palo Alto, Calif.

EPRI (Electric Power Research Institute). 1993c. Survey of Residential Magnetic Field Sources: Protocol, Data Analysis, and Management, Vol. 2. Project RP3335-02. Rep. TR-102759-V2. Prepared by High Voltage Transmission Research Center for Electric Power Research Institute, Palo Alto, Calif.

EPRI (Electric Power Research Institute). 1994a. Residential Transient Magnetic Field Research: Interim Report. Project RP2966-07. Rep. TR-103470. Electric Power Research Institute, Palo Alto, Calif.

EPRI (Electric Power Research Institute). 1994b. Association of Wire Code Configuration with Long-Term Average 60-Hz Magnetic Fields: Interim Report. Project RP3533-01. Rep. TR-104656-V2. Electric Power Research Institute, Palo Alto, Calif.

Fairbairn, D.W., and K.L. O'Neill. 1994. The effect of electromagnetic field exposure on the formation of DNA single strand breaks in human cells. Cell. Mol. Biol. 40:561-567.

Fajardo-Gutierrez, A, J. Garduno-Espinosa, L. Yamamoto-Kimura, D.M. Hernandez-Hernandez, A. Gomez-Delgado, M. Mrjia-Arangure, A. Cartagena-Sandoval, and M.C. Martinez-Garcia. 1993. Residence close to high-tension electric power lines and its association with leukemia in children (in Spanish). Biol. Med. Hosp. Infant Mex. 50:32-38.

Falugi, C., M. Grattorola, and G. Prestipino. 1987. Effects of low-intensity pulsed electromagnetic fields on the early development of sea urchins. Biophys. J. 51:999-1003.

Fam, W.Z. 1980. Long-term biological effects of very intense 60 Hz electric field on mice. IEEE Trans. Biomed. Eng. 27:376-381.

Fewtrell, C. 1993. Ca^{2+} oscillations in non-excitable cells. Annu. Rev. Physiol. 55:427-454.

Feychting, M., and A. Ahlbom. 1993. Magnetic fields and cancer in children residing near Swedish high-voltage power lines. Am. J. Epidemiol. 138:467-481.

Feychting, M., and A. Ahlbom. 1994. Magnetic fields, leukemia, and central nervous system tumors in Swedish adults residing near high-voltage power lines. Epidemiology 5:501-509.

Fiorani, M., O. Cantoni, P. Sestili, R. Conti, P. Nicolini, F. Vetrano, and M. Dachà. 1992. Electric and/or magnetic field effects on DNA structure and function in cultured human cells. Mutat. Res. 282:25-29.

Fiorio, R., E. Morichetti, R. Vellosi, and G. Bronzetti. 1993. Mutagenicity and toxicity of electromagnetic fields. J. Environ. Pathol. Toxicol. Oncol. 12(3):139-142.

Fitton-Jackson, S., and C.A.L. Bassett. 1980. The response of skeletal tissues to pulsed magnetic fields. Pp. 21-28 in Use of Tissue Culture in Medical Research II, R.J. Richards and K.T. Rajan, eds. Oxford, U.K.: Pergamon.

Fitzsimmons, R.J., J.R. Farley, W.R. Adey, and D.J. Baylink. 1989. Frequency dependence of increased cell proliferation, in vitro, in exposures to a low-amplitude, low-frequency electric field: Evidence for dependence on increased mitogen activity released into culture medium. J. Cell. Physiol. 139:586-591.

Fitzsimmons, R.J., D.D. Strong, S. Mohan, and D.J. Baylink. 1992. Low-amplitude, low-frequency electric field-stimulated bone cell proliferation may in part be mediated by increased IGF-II release. J. Cell. Physiol. 150:84-89.

Fleiss J., and A.J. Gross. 1991. Meta-analysis in epidemiology, with special reference to studies of the association between exposure to environmental tobacco smoke and lung cancer: A critique. J. Clin. Epidemiol. 44:127-139.

Floderus, B., T. Persson, C. Stenlund, A. Wennberg, A. Ost, and B. Knave. 1993. Occupational exposure to electromagnetic fields in relation to leukemia and brain tumors: A case-control study in Sweden. Cancer Causes Control 4:465-476.

Floderus, B., S. Tornqvist, and C. Stenlund. 1994. Incidence of selected cancers in Swedish railway workers, 1961-79. Cancer Causes Control 5:189-194.

Foster, K.R., and H.P. Schwan. 1986. Dielectric permittivity and electrical conductivity of biological materials. Pp. 27-96 in Handbook of Biological Effects of Electromagnetic Fields, C. Polk and E. Postow, eds. Boca Raton, Fla.: CRC.

Frazier, M.E., J.A. Reese, J.E. Morris, R.F. Jostes, and D.L. Miller. 1990. Exposure of mammalian cells to 60-Hz magnetic or electric fields: Analysis of DNA repair induced, single-strand breaks. Bioelectromagnetics 11:229-234.

Free, M.J., W.T. Kaune, R.D. Phillips, and H.C. Cheng. 1981. Endocrinological effects of strong 60-Hz electric fields on rats. Bioelectromagnetics 2:105-121.

Friedenberg, Z.B., and C.T. Brighton. 1966. Bioelectric potentials in bone. J. Bone Jt. Surg. 48-A:915-923.

Friedenberg, Z.B., and C.T. Brighton. 1981. Bioelectricity and fracture healing. Plast. Reconstr. Surg. 68:435-443.

Friedenberg, Z.B., E.T. Andrews, B.I. Smolenski, B.W. Pearl, and C.T. Brighton. 1970. Bone reaction to varying amounts of direct current. Surg. Gynecol. Obstet. 131:894-899.

Friedenberg, Z.B., P.G. Roberts, Jr., N.H. Didizian, and C.T. Brighton. 1971a. Stimulation of fracture healing by direct current in the rabbit fibula. J. Bone Jt. Surg. 53-A:1400-1408.

Friedenberg, Z.B., M.C. Harlow, and C.T. Brighton. 1971b. Healing of nonunion of the medial malleolus by means of direct current: A case report. J. Trauma 11:883-885.

Friedenberg, Z.B., M.C. Harlow, R.B. Heppenstall, and C.T. Brighton. 1973. The cellular origin of bioelectric potentials in bone. Calcif. Tissue Res. 13:53-62.

Friedenberg, Z.B., L.M. Zemsky, R.P. Pollis, and C.T. Brighton. 1974. The response of non-traumatized bone to direct current. J. Bone Jt. Surg. 56-A:1023-1030.

Frolen, H., B.-M. Svedenstal, and L.-E. Paulsson. 1993. Effects of pulsed magnetic fields on the developing mouse embryo. Bioelectromagnetics 14:197-204.

Fukuda, E., and I. Yasuda. 1957. On the piezoelectric effect of bone. J. Phys. Soc. Jpn. 10:1158-1162.

Fulton, J.P., S. Cobb, L. Preble, L. Leone, and E. Forman. 1980. Electrical wiring configurations and childhood leukemia in Rhode Island. Am. J. Epidemiol. 111:292-296.

Gamberale, F., B.A. Olson, P. Eneroth, T. Lindh, and A. Wennberg. 1989. Acute effects of ELF electromagnetic fields: A field study of linesmen working with 400 kV power lines. Br. J. Ind. Med. 46:729-737.

Gandhi, O.P., and J.-Y. Chen. 1992. Numerical dosimetry at power-line frequencies using anatomically based models. Bioelectromagnetics Suppl. 1:43-60.

Garciá-Sagredo, J.M., and J.L. Monteagudo. 1991. Effect of low-level pulsed electromagnetic fields on human chromosomes in vitro: Analysis of chromosomal aberrations. Hereditas 115:9-11.

Garciá-Sagredo, J.M., A.L. Parada, and J.L. Monteagudo. 1990. Effect on SCE in human chromosomes in vitro of low-level pulsed magnetic field. Environ. Mol. Mutagen. 16:185-188.

Gimenez-Gonzalez, M., F. Martinez-Soriano, E. Armanazas, and A. Ruiz-Torner. 1991. Morphometric and structural study of the pineal gland of the Wistar rat subjected to the pulse action of 52 gauss (50 Hz) magnetic field. Evaluative analysis over 21 days. J. Hirnforsch. 6:779-786.

Goodman, R., and A. Shirley-Henderson. 1991. Transcription and translation in cells exposed to extremely low frequency electromagnetic fields. Bioelectrochem. Bioenerg. 25:335-355.

Goodman, R., J. Bumann, L.-X. Wei, and A. Shirley-Henderson. 1992. Exposure of human cells to electromagnetic fields: Effect of time and field strength on transcript levels. Electro- Magnetobiology 11:19-28.

Graves, H.B. 1981. Detection of a 60-Hz electric field by pigeons. Behav. Neural Biol. 32:229-234.

Graves, H.B., J.H. Carter, D. Kellmel, L. Cooper, D.T. Poznaniak, and J.W. Bankoske. 1978. Perceptibility and electrophysiological response of small birds to intense 60-Hz electric fields. IEEE Trans. Power Appar. Syst. PAS-97:1070-1073.

Greenland, S. 1994. Invited commentary: A critical look at some popular meta-analytic methods. Am. J. Epidemiol. 140:290-296.

Grota, L.J., R.J. Reiter, P. Keng, and S. Michaelson. 1994. Electric field exposure alters serum melatonin but not pineal melatonin synthesis in male rats. Bioelectromagnetics 15:427-437.

Grundeler, W., F. Kaiser, F. Keilmann, and J. Wolleczek. 1992. Mechanisms of electromagnetic field interaction with cellular systems. Naturwissenschaften 79:551-559.

Hackman, R.M., and H.B. Graves. 1981. Corticosterone levels in mice exposed to high-intensity electric fields. Behav. Neural Biol. 32:201-213.

Halle, B. 1988. On the cyclotron resonance mechanism for magnetic field effects on transmembrane ion conductivity. Bioelectromagnetics 9:381-385.

Harkins, T.T., and C.B. Grissom. 1994. Magnetic field effects on B12 ethanolamine ammonia lyase: Evidence for a radical mechanism. Science 263:958-960.

Harkness, J.E., and M.D. Ridgway. 1980. Chromodacryorrhea in laboratory rats (*Rattus norvegicus*): Etiologic considerations. Lab. Anim. Sci. 30:841-844.

Hart, F.X. 1990. Use of a spread sheet to calculate the current-density distribution produced in human and rat models by low-frequency electric fields. Bioelectromagnetics 11:213-228.

Hart, F.X. 1992a. Numerical and analytical methods to determine the current density distributions produced in human and rat models by electric and magnetic fields. Bioelectromagnetics Suppl. 1:27-42.

Hart, F.X. 1992b. Electric fields induced in rat and human models by 60-Hz magnetic fields: Comparison of calculated and measured values. Bioelectromagnetics 13:313-316.

Hart, F.X., K. Evely, and C.D. Finch. 1993. Use of a spreadsheet program to calculate the electric field/current density distributions induced in irregularly shaped, inhomogeneous biological structures by low-frequency magnetic fields. Bioelectromagnetics 14:161-172.

Harvey, S.M. 1987. Magnetic field exposure apparatus for small animal studies: Phase II. Rep. No. 87-118-K. Ontario Hydro Research Division.

Hassler, C.R., E.F. Rybicki, R.B. Diegle, and L.C. Clark. 1977. Studies of enhanced bone healing via electrical stimuli. Clin. Orthop. Relat. Res. 124:9-19.

Hatch, M. 1992. The epidemiology of electric and magnetic field exposures in the power frequency range and reproductive outcomes. Paediatr. Perinat. Epidemiol. 6:198-214.

Haysom, C., D. Dowson, and M. Campbell. 1990. The relevance of headaches and migraine in populations resident near overhead power lines—An epidemiological study. Computers Med. Res. 4:12-15.

Hedges, L.V., and I. Olkin. 1985. Statistical Methods for Meta-Analysis. Orlando, Fla.: Academic.

Hill, A. Bradford. 1961. The environment and disease: Association or causation? Proc. R. Soc. Med. 58:295-300.

Hilton, D.I., and R.D. Phillips. 1980. Cardiovascular response of rats exposed to 60-Hz electric fields. Bioelectromagnetics 1:55-64.

Hiraki, Y., N. Endo, M. Takigawa, A. Asada, H. Takahashi, and F. Suzuki. 1987. Enhanced responsiveness to parathyroid hormone and induction of functional differentiation of cultured rabbit costal chondrocytes by a pulsed electromagnetic field. Biochim. Biophys. Acta 931:94-100.

Hjeresen, D.L., W.T Kaune, J.R. Decker. and R.D. Phillips. 1980. Effects of 60-Hz electric fields on avoidance behavior and activity of rats. Bioelectromagnetics 1:299-312.

Hjeresen, D.L., M.C. Miller, W.T. Kaune, and R.D. Phillips. 1982. A behavioral response of swine to a 60-Hz electric field. Bioelectromagnetics 3:443-451.

Hong, C.-Z., P. Huestis, R. Thompson, and J. Yu. 1988. Learning ability of young rats is unaffected by repeated exposure to static electromagnetic field in early life. Bioelectromagnetics 9:269-273.

Huang, T.-S., J. Duyster, and J.Y.J. Wang. 1995. Biological response to phorbol ester determined by alternative G_1 pathways. Proc. Natl. Acad. Sci. USA 92:4793-4797.

Hungate, F.P., M.P. Fujihara, and S.R. Strankman. 1979. Mutagenic effects of high-strength electric fields. Pp. 530-537 in Biological Effects of Extremely Low Frequency Magnetic Fields, Proceedings of the 18th Annual Hanford Life Sciences Symposium, R.D. Phillips, M.F. Gillis, W.T. Kaune, and D.D. Malham, eds. Technical Information Center, U.S. Department of Energy, Oak Ridge, Tenn.

Huuskonen, H., J. Juutilainen, and H. Kamulainen. 1993. Effects of low-frequency magnetic fields on fetal development in rats. Bioelectromagnetics 14:205-213.

Ikeda, K., T. Sugimoto, M. Fukase, and T. Fujita. 1991. Protein kinase C is involved in PTH-induced homologous desensitization by directly affecting PTH receptor in the osteoblastic osteosarcoma cells. Endocrinology 128:2901-2906.

Iskander, M.F. 1993. Electromagnetic Fields and Waves. Englewood Cliffs, N.J.: Prentice Hall.

Jackson, J.D. 1992. Are the stray 60-Hz electromagnetic fields associated with the distribution and use of electric power a significant cause of cancer? Proc. Natl. Acad. Sci. USA 89:3508-3510.

Jaffe, R.A., B.L. Laszewski, D.B. Carr, and R.D. Phillips. 1980. Chronic exposure to a 60-Hz electric field: Effects on synaptic transmission and the peripheral nerve function in the rat. Bioelectromagnetics 1:131-147.

Jaffe, R.A., B.L. Laszewski, and D.B. Carr. 1981. Chronic exposure to a 60-Hz electric field: Effects on neuromuscular function in the rat. Bioelectromagnetics 2:277-239.

Jaffe, R.A., C.A. Lopresti, D.B. Carr, and R.D Phillips. 1983. Perinatal exposure to 60-Hz electric fields: Effects on the development of the visual-evoked response in rats. Bioelectromagnetics 4:327-339.

Jahn, T.L. 1968. A possible mechanism for the effect of electrical potentials on apatite formation in bone. Clin. Orthop. Relat. Res. 56:261-273.

John, E.M., D.A. Savitz, and D.P. Sandler. 1991. Prenatal exposure to parents' smoking and childhood cancer. Am. J. Epidemiol. 133:123-132.

Joines, W.T., and C.F. Blackman. 1980. Power density, field intensity, and carrier frequency determinants of RF-energy-induced calcium-ion efflux from brain tissue. Bioelectromagnetics 1:271-275.

Joines, W.T., C.F. Blackman, and M.A. Hollis. 1981. Broadening of the RF power-density window for calcium-ion efflux from brain tissue. IEEE Trans. Biomed. Eng. 28:568-573.

Jones, T. 1993. EMF wire code research. IEEE Power Eng. Rev. 13:10-12.

Jones, T.L., C.H. Shih, D.H. Thurston, B.J. Ware, and P. Cole. 1993. Selection bias from differential residential mobility as an explanation for associations of wire codes with childhood cancer. J. Clin. Epidemiol. 46:545-548.

Juutilainen, J., and A. Liimatainen. 1986. Mutation frequency in Salmonella exposed to weak 100-Hz magnetic fields. Hereditas 104:145-147.

Juutilainen, J., P. Matilainen, S. Saarikoski, E. Laara, and S. Suonio. 1993. Early pregnancy loss and exposure to 50-Hz magnetic fields. Bioelectromagnetics 14:229-236.

Karabakhtsian, R., N. Broude, N. Shalts, S. Kochlatyi, R. Goodman, and A.S. Henderson. 1994. Calcium is necessary in the cell response to EM fields. FEBS Lett. 349:1-6.

Kato, M., K. Honma, T. Shigemitsu, and Y. Shiga. 1993. Effects of exposure to circularly polarized 50-Hz magnetic field on plasma and pineal melatonin levels in rats. Bioelectromagnetics 14:97-106.

Kato, M., K. Honma, T. Shigemitsu, and Y. Shiga. 1994a. Circularly polarized 50-Hz magnetic field exposure reduces pineal gland and blood melatonin concentrations of Long-Evans rats. Neurosci. Lett. 166:59-62.

Kato, M., K. Honma, T. Shigemitsu, and Y. Shiga. 1994b. Horizontal or vertical 50-Hz, 1-μT magnetic fields have no effect on pineal gland or plasma melatonin concentration of albino rats. Neurosci. Lett. 168:205-208.

Kaune, W.T. 1981a. Interactive effects in 60-Hz electric field exposure systems. Bioelectromagnetics 2:33-50.

Kaune, W.T. 1981b. Power-frequency electric fields averaged over the body surfaces of grounded humans and animals. Bioelectromagnetics 2:403-406.

Kaune, W.T., and W.C. Forsythe. 1985. Current densities measured in human models exposed to 60-Hz electric fields. Bioelectromagnetics 6:13-32.

Kaune, W.T., and W.C. Forsythe. 1988. Current densities induced in swine and rat models by power-frequency electric fields. Bioelectromagnetics 9:1-24.

Kaune, W.T., and M.L. Gillis. 1981. General properties of the interaction between animals and ELF electric fields. Bioelectromagnetics 2:1-11.

Kaune, W.T., and R.D. Phillips. 1980. Comparison of coupling of grounded humans, swine and rats to vertical 60-Hz electric fields. Bioelectromagnetics 1:117-129.

Kaune, W.T., and D.A. Savitz. 1994. Simplification of the Wertheimer-Leeper wire code. Bioelectromagnetics 15:275-282.

Kaune, W.T., and L.E. Zaffanella. 1994. Assessing historical exposures of children to power-frequency magnetic fields. J. Expo. Anal. Environ. Epidemiol. 4:149-170.

Kaune, W.T., R.G. Stevens, N.J. Callahan, R.K. Severson, and D.B. Thomas. 1987. Residential magnetic and electric fields. Bioelectromagnetics 8:315-335.

Kaune, W.T., S.D. Darby, S.N. Gardner, Z. Hrubec, R.H. Iriye, and M.S. Linet. 1994. Development of a protocol for assessing time-weighted average exposures of young children to power-frequency magnetic fields. Bioelectromagnetics 15:33-51.

Kavaliers, M., and K.P. Ossenkopp. 1986a. Stress-induced opioid analgesia and activity in mice: Inhibitory influences of exposure to magnetic fields. Psychopharmacology 89:440-443.

Kavaliers, M., and K.P. Ossenkopp. 1986b. Magnetic fields differentially inhibit mu, delta, kappa and sigma opiate-induced analgesia in mice. Peptides 7:449-453.

Kavaliers, M., and K.P. Ossenkopp. 1986c. Magnetic field inhibition of morphine-induced analgesia and behavioral activity in mice: Evidence for involvement of calcium ions. Brain Res. 379:30-38.

Kavet, R. 1991. An alternate hypothesis for the association between electrical wiring configurations and cancer. Epidemiology 2:224-229.

Kavet, R.I., J.M. Silva, and D. Thornton. 1992. Magnetic-field exposure assessment for adult residents of Maine who live near and far away from overhead transmission lines. Bioelectromagnetics 13:35-55.

Kelsey, J.L., W.D. Thompson, and A.S. Evans. 1986. Methods in Observational Epidemiology. New York: Oxford University Press.

Kempf, E., P. Mandel, A. Oliverio, and S. Puglisi-Allegra. 1982. Circadian variations of noradrenaline, 5-hydroxytryptamine and dopamine in specific brain areas of C57BL/6 and BALB/c mice. Brain Res. 232:472-478.

Khalil, A.M., and W. Qassem. 1991. Cytogenic effects of pulsing electromagnetic field on human lymphocytes in vitro: Chromosome aberrations, sister-chromatid exchanges and cell kinetics. Mutat. Res. 247:141-146.

Kikkawa, U., A. Kishimoto, and Y. Nishizuka. 1989. The protein kinase C family: Heterogeneity and its implications. Annu. Rev. Biochem. 58:31-44.

Kinlen, L.J., and S.M. John. 1994. Wartime evacuation and mortality from childhood leukemia in England and Wales in 1945-9. Br. Med. J. 390:197-202.

Kirschvink, J.L. 1992. Uniform magnetic fields and double-wrapped coil systems: Improved techniques for the design of bioelectromagnetic experiments. Bioelectromagnetics 13:401-411.

Klaassen, C.D., and D.L. Eaton. 1991. Principles of toxicology. Pp. 12-49 in Casarett and Doull's Toxicology: The Basic Science of Poisons, 4th Ed., M.O. Amdur, J. Doull, and C.D. Klaassen, eds. New York: Pergamon.

Klavinsh, E.E., Y.Y. Galvanovsky, and A.P. Kreimanis. 1991. Low-frequency electromagnetic pulses enhance calcium intake by chick small intestine in vitro. Bioelectrochem. Bioenerg. 25:437-446.

Knave, B., F. Gamberale, S. Bergstrom, E. Birke, A. Iregrin, B. Kolmodin-Hedman, and A. Wennberg. 1979. Long-term exposure to electric fields: A cross-sectional epidemiologic investigation of occupationally exposed workers in high-voltage substations. Scand. J. Work Environ. Health 5:115-125.

Kowalczuk, C.I., and R.D. Saunders. 1990. Dominant lethal studies in male mice after exposure to a 50-Hz electric field. Bioelectromagnetics 11:129-137.

Kraus, J.D. 1992. Electromagnetics, 4th Ed. New York: McGraw-Hill.

Lacy-Hulbert, A., R.C. Wilkins, T.R. Hesketh, and J.C. Metcalfe. 1995. No effect of 60-Hz electromagnetic fields on *myc* or beta-actin in human leukemic cells. Radiat. Res. 144:9-17.

LaPorte, R., L. Kus, R.A. Wisniewski, M.M. Perchel, B. Azar-Kia, and J.A. McNulty. 1990. Magnetic resonance imaging (MRI) effects on rat pineal neuroendocrine function. Brain Res. 506:294-296.

Lednev, V.V. 1991. Possible mechanism for the influence of weak magnetic fields on biological systems. Bioelectromagnetics 12:71-75.

Lee, Q.P., A.W. Guy, H. Lai, and A. Horita. 1987. The Effects of Modulated Radio-Frequency Radiation on Calcium Efflux from Chick Brains In Vitro. Abstract, p. 10. Ninth Annual Bioelectromagnetics Society Meeting, June 21-25, 1987, Portland Oreg.

Lee, J.M., Jr., F. Stromshak, J.M. Thompson, P. Thiessen, L.J. Painter, E.G.

Olenchels, D.C. Hess, R. Forbes, and D.L. Foster. 1993. Melatonin secretion and puberty in female lambs exposed to environmental electric and magnetic fields. Biol. Reprod. 49:857-864.

Leeper, E., N. Wertheimer, D. Savitz, F. Barnes, and H. Wachtel. 1991. Modification of the 1979 "Denver Wire Code" for different wire or plumbing types. Bioelectromagnetics 12:315-318.

Leung, F.C., D.N. Rommereim, R.A. Miller, and L.E. Andersen. 1990. Brown-colored deposits on the hair of female rats chronically exposed to 60-Hz electric fields. Bioelectromagnetics 11:257-259.

Levy, D.D., and B. Rubin. 1972. Inducing bone growth in vivo by pulse stimulation. Clin. Orthop. Relat. Res. 88:218-222.

Libertin, C.R., J. Panozzo, K.R. Groh, C.M. Chang-Liu, S. Schreck, and G.E. Woloschak. 1994. Effects of gamma rays, ultraviolet radiation, sunlight, microwaves, and electromagnetic fields on gene expression mediated by human immunodeficiency virus promoter. Radiat. Res. 140:91-96.

Liboff, A.R., J.R. Thomas, and J. Scrot. 1989. Intensity threshold for 60-Hz magnetically induced behavioral changes in rats. Bioelectromagnetics 10:111-113.

Liboff, A.R., B.R. McLeod, and S.D. Smith. 1990. Ion cyclotron resonance effects in biological systems. Pp. 251-289 in Extremely Low Frequency Fields: The Question of Cancer, B.W. Wilson, R.G. Stevens, and L.E. Anderson, eds. Columbus, Ohio: Battelle Press.

Liburdy, R.P. 1992a. ELF fields and the immune system: Signal transduction, calcium metabolism, and mitogenesis in lyrnphocytes with relevance to carcinogenesis. Pp. 217-239 in Interaction Mechanisms of Low-Level Electromagnetic Fields in Living Systems, B. Norden and C. Ramel, eds. New York: Oxford University Press.

Liburdy, R.P. 1992b. Calcium signaling in lymphocytes and ELF field: Evidence for an electric field metric and a site of interaction involving the calcium ion channel. FEBS Lett. 301:53-59.

Liburdy, R.P., D.E. Callahan, and J.D. Harland. 1993a. Protein shedding and ELF magnetic fields: Antibody binding at the CD3 and CD20 receptor sites of human lymphocytes. Pp. 651-653 in Electricity and Magnetism in Biology and Medicine: Review and Research Papers Presented at the First World Congress for Electricity and Magnetism in Biology and Medicine, M. Blank, ed. San Francisco: San Francisco Press.

Liburdy, R.P., T.R. Sloma, R. Sokolic, and P. Yaswen. 1993b. ELF magnetic fields, breast cancer, and melatonin: 60-Hz fields block melatonin's oncostatic action on ER-positive breast cancer cell proliferation. J. Pineal Res. 14:89-97.

Liburdy, R.P., D.E. Callahan, T.R. Sloma, and P. Yaswen. 1993c. Intracellular calcium, calcium transport, and c-*myc* mRNA induction in lymphocytes exposed to 60-Hz magnetic fields: The cell membrane and the signal transduction pathway. Pp. 311-314 in Electricity and Magnetism in Biology and

Medicine: Review and Research Papers Presented at the First World Congress for Electricity and Magnetism in Biology and Medicine, M. Blank ed. San Francisco: San Francisco Press.

Lin, R., and P. Lu. 1989. An epidemiologic study of childhood cancer in relation to residential exposure to electromagnetic fields. P. A-40 in Project Abstracts: The Annual Review of Research on Biological Effects of Electric and Magnetic Fields. DOE/EPRI Contractor's Review Meeting. Bioelectromagnetics Society, Federick, Md.

Lin, H., R. Goodman, and A. Shirley-Henderson. 1994. Specific region of the c-myc promoter is responsive to electric and magnetic fields. J. Cell. Biochem. 54:281-288.

Lindbohm, M., L. Hietanen, P. Kyyronen, M. Sallmen, P. Von Mandelstadh, H. Taskinen, M. Pekkarinen, M. Ylikoski, and K. Hemminki. 1992. Magnetic fields of video display terminals and spontaneous abortion. Am. J. Epidemiol. 136:1041-1051.

Litovitz, T.A., D. Krauss, and J.M. Mullins. 1991. Effect of coherence time of the applied magnetic field on ornithine decarboxylase activity. Biochem. Biophys. Res. Commun. 178:862-865.

Livingston, G.K., O.P. Gandhi, I. Chatterjee, K. Witt, and J.L. Roti Roti. 1986. Reproductive integrity of mammalian cells exposed to 60-Hz electromagnetic fields. Final Report. New York State Power Lines Project Contract No. 218209. Wadsworth Center for Laboratories and Research, New York.

Livingston, G.K., K.L. Witt, O.P. Gandhi, I. Chatterjee, and J.L. Roti Roti. 1991. Reproductive integrity of mammalian cells exposed to power frequency electromagnetic fields. Environ. Mol. Mutagen. 17:49-58.

Loew, L.M. 1992. Voltage-sensitive dyes: Measurement of membrane potentials induced by DC and AC electric fields. Bioelectromagnetics Suppl. 1:179-189.

London, S.J., D.C. Thomas, J.D. Bowman, E. Sobel, T.-C. Cheng, and J.M. Peters. 1991. Exposure to residential electric and magnetic fields and risk of childhood leukemia. Am. J. Epidemiol. 134:923-937.

London, S.J., J.D. Bowman, E. Sobel, D.C. Thomas, D.H. Garabrant, N. Pearce, L. Bernstein, and J.M. Peters. 1994. Exposure to magnetic fields among electrical workers in relation to leukemia risk in Los Angeles County. Am. J. Ind. Med. 26:47-60.

Loomis, D.P., D.A. Savitz, and C.V. Ananth. 1994. Breast cancer mortality among female electrical workers in the United States. J. Natl. Cancer Inst. 86:921-925.

Löscher, W., M. Mevissen, W. Lehmacher, and A. Stamm. 1993. Tumor promotion in a breast cancer model by exposure to a weak alternating magnetic field. Cancer Lett. 71:75-81.

Löscher, W., U. Wahnschaffe, M. Mevissen, A. Lerchl, and A. Stamm. 1994. Effects of weak alternating magnetic fields on nocturnal melatonin production and mammary carcinogenesis in rats. Oncology 51:288-295.

Lovely, R.H., J.A. Creim, W.T. Kaune, M.C. Miller, R.D. Phillips. and L.E. Anderson. 1992. Rats are not aversive when exposed to 60-Hz magnetic fields at 3.03 mT. Bioelectromagnetics 13:351-362.

Lowenthal, R., J. Panton, M. Baikie, and J. Lickiss. 1991. Exposure to high tension power lines and childhood leukemia: A pilot study [letter]. Med. J. Aust. 155:347.

Luben, R.A. 1991. Effects of low-energy electromagnetic fields (pulsed and DC) on membrane signal transduction processes in biological systems. Health Phys. 61:15-28.

Luben, R.A. 1993. Effects of low-energy electromagnetic fields (EMF) on signal transduction by G protein-linked receptors. Pp. 57-62 in Electricity and Magnetism in Biology and Medicine: Review and Research Papers Presented at the First World Congress for Electricity and Magnetism in Biology and Medicine, M. Blank, ed. San Francisco: San Francisco Press.

Luben, R.A. 1994. In vitro systems for study of electromagnetic effects on bone and connective tissue. In Biological Effects of Electric and Magnetic Fields, 2 vol., D.O. Carpenter and S. Ayrapetyan, eds. San Diego, Calif.: Academic.

Luben, R.A., C.D. Cain, M.Y.C. Chen, D.M. Rosen, and W.R. Adey. 1982. Inhibition of parathyroid hormone actions on bone cells in culture by induced low energy electromagnetic fields. Proc. Natl. Acad. Sci. USA 79:4180-4184.

Lymangrover, L.R., E. Keku, and J.J.Seto. 1983. 60-Hz electric field alters the steroidogenic response of rat adrenal tissue in vitro. Life Sci. 32:691-696.

MacGregor, D.G., P. Slovic, and M.G. Morgan. 1994. Perception of risks from electromagnetic fields: A psychometric evaluation of a risk-communication approach. Risk Anal. 14:815-828.

MacMahon, B. 1992. Is acute lymphoblastic leukemia in children virus-related? Am. J. Epidemiol. 136:916-924.

MacMahon B., and V.A. Newell. 1962. Birth characteristics of children dying of malignant neoplasms. J. Natl. Cancer Inst. 28:231-244.

Mader, D.L., and S.B. Peralta. 1992. Residential exposure to 60-Hz magnetic fields from appliances. Bioelectromagnetics 13:287-301.

Manolagas, S.C., and R.L. Jilka. 1995. Bone marrow, cytokines, and bone remodeling: Emerging insights into the pathophysiology of osteoporosis. New Engl. J. Med. 332:305-311.

Margonato, V., and D. Viola. 1982. Gonadal function in the rat exposed to high-intensity fields of 50 Hz [in Italian]. Boll. Soc. Ital. Biol. Sper. 58:275-281.

Marino, A.A., and R.O. Becker. 1977. Biological effect of extremely low frequency electric and magnetic fields: A review. Physiol. Chem. Phys. 9:131-139.

Marino, A.A., R.O. Becker, and B. Ullrich. 1976. The effect of continuous exposure to low frequency electric fields on three generations of mice: A pilot study. Experientia 32:565-566.

Marino, A.A., T.J. Berger. B.P. Austin, R.O. Becker, and F.X. Hart. 1977. In vivo bioelectrochemical changes associated with exposure to extremely low frequency electric fields. Physiol. Chem. Phys. 9:433-441.

Marino, A.A., M. Reichmanis, R.O. Becker, B. Ullrich, and J.M. Cullen. 1980. Power frequency electric fields induces biological changes in successive generations of mice. Experientia 36:309-311.

Martin, A. 1992. Development of chicken embryos following exposure to 60-Hz magnetic fields with differing waveforms. Bioelectromagnetics 13:223-230.

Martin, R.B., and W. Gutman. 1978. The effect of electric fields on osteoporosis of disuse. Calcif. Tissue Res. 25:23-27.

Martinez-Soriano, F., M. Gimenez-Gonzalez, E. Armanazas, and A. Ruiz-Torner. 1992. Pineal "synaptic ribbons" and serum melatonin levels in the rat following the pulse action of 52-Gs (50 Hz) magnetic fields: An evolutive analysis over 21 days. Acta Anat. 143:289-293.

Matanoski, G.M., P.N. Breysse, and E.A. Elliott. 1991. Electromagnetic field exposure and male breast cancer [letter]. Lancet 337(8743):737.

Matanoski, G.M., E.A. Elliott, P.N. Breysse, and M.C. Lynberg. 1993. Leukemia in telephone linemen. Am. J. Epidemiol. 137:609-619.

Matsushima, S., Y. Sakai, Y. Hira, M. Kato, T. Shigemitsu, and Y. Shiga. 1993. Effect of magnetic field on pineal gland volume and pinealocyte size in rat. J. Pineal Res. 14:145-150.

McCann, J.E., F. Dietrich, C. Rafferty, and A.O. Martin. 1993. A critical review of the genotoxic potential of electric and magnetic fields. Mutat. Res. 297:61-95.

McClanahan, B.J., and R.D. Phillips. 1983. The influence of electric field exposure on bone growth and fracture repair in rats. Bioelectromagnetics 4:11-19.

McDowall, M.E. 1986. Mortality of persons resident in the vicinity of electricity transmission facilities. Br. J. Cancer 53:271-279.

McGivern, R.F., R.A. Sokol, and W.R. Adey. 1990. Prenatal exposure to a low-frequency electromagnetic field demasculinizes adult scent marking behavior and increases accessory sex organ weights in rats. Teratology 41:1-8.

McLean, J.R.N., M.A. Stuchly, R.E.J. Mitchel, D. Wilkinson, H. Yang, M. Goddard, D.W. Lecuyer, M. Schunk, E. Callary, and S.D. Morrison. 1991. Cancer promotion in a mouse-skin model by a 60-Hz magnetic field: II. Tumor development and immune response. Bioelectromagnetics 12:273-287.

McLeod, B.R., A.A. Pilla, and M.W. Sampsel. 1983. Electromagnetic fields induced by Helmholtz aiding coils inside saline-filled boundaries. Bioelectromagnetics 4:357-370.

McLeod, K.J. 1992. Microelectrode measurements of low frequency electric field effects in cells and tissues. Bioelectromagnetics Suppl. 1:161-178.

McLeod, K.J., and C.T. Rubin. 1990. Frequency specific modulation of bone adaptation by induced electric fields. J. Theor. Biol. 145:385-396.

McLeod, K.J., and C.T. Rubin. 1992. The effect of low-frequency electrical fields on osteogenesis. J. Bone Jt. Surg. 74-A:920-929.

McLeod, K.J., R.C. Lee, and H.P. Ehrlich. 1987. Frequency dependence of electric field modulation of fibroblast protein synthesis. Science 236:1465-1468.

McLeod, K.J., H.J. Donahue, P.E. Levin, and C.T. Rubin. 1991. Low-frequency sinusoidal electric fields alter calcium fluctuations in osteoblast-like cells. Pp. 111-115 in Electromagnetics in Biology and Medicine, C.T. Brighton and S.R. Pollack, eds. San Francisco: San Francisco Press.

McLeod, K.J., H.J. Donahue, P.E. Levin, M.A. Fontaine, and C.T. Rubin. 1993. Electric fields modulate bone cell function in a density-dependent manner. J. Bone Min. Res. 8:977-984.

McMahan, S., J. Ericson, and J. Meyer. 1994. Depressive symptomatology in women and residential proximity to high-voltage transmission lines. Am. J. Epidemiol. 139:58-63.

McRobbie, D., and M.A. Foster. 1985. Pulsed magnetic field exposure during pregnancy and implications for NMR feotal imaging: A study with mice. Magn. Reson. Imaging 3:231-234.

Merritt, R., C. Purcell, and G. Stroink. 1983. Uniform magnetic fields produced by three, four and five square coils. Rev. Sci. Instrum. 54:879-882.

Mevissen, M., A. Stamm, S. Buntenkotter, R. Zwingelberg, U. Wahnschaffe, and W. Löscher. 1993. Effects of magnetic fields on mammary tumor development induced by 7,12-dimethylbenz[a]anthracene in rats. Bioelectromagnetics 14:131-143.

Meyer, T., and L. Stryer. 1991. Calcium spiking. Annu. Rev. Biophys. Chem. 20:153-174.

Mileva, M. 1982. Effect of a permanent magnetic field on human lymphocytes. Kosm. Biol. Aviakosm. Med. 16:86-87.

Milham, S. 1982. Mortality from leukemia in workers exposed to electrical and magnetic fields [letter]. New Engl. J. Med. 307:249.

Milin, J., M. Bajic, and V. Borkus. 1988. Morphodynamic response of the pineal gland of rats chronically exposed to stable strong magnetic field. NeuroScience 26:1083-1092.

Miller, D.L. 1991. Miniature-probe measurements of electric fields and currents induced by a 60-Hz magnetic field in rat and human models. Bioelectromagnetics 1:157-171.

Miller, D.L., M.C. Miller, and W.T. Kaune. 1989. Addition of magnetic field capability to existing extremely-low-frequency electric field exposure systems. Bioelectromagnetics 10:85-98.

Misakian, M. 1991. In vitro exposure parameters with linearly and circularly polarized ELF magnetic fields. Bioelectromagnetics 12:377-381.

Misakian, M., and W.T. Kaune. 1990. Optimal experimental design for in vitro studies with ELF magnetic fields. Bioelectromagnetics 11:251-255.

Misakian, M., A.R. Sheppard, D. Krause, M.E. Frazier, and D.L. Miller. 1993. Biological, physical, and electrical parameters for in-vitro studies with ELF magnetic and electric fields: A primer. Bioelectromagnetics Suppl. 2:1-73.

Moore, R.L. 1979. Biological effects of magnetic fields: Studies with microorganisms. Can. J. Microbiol. 25:1145-1151.

Morgan, G. 1995. Fields from Electric Power. Part 1: What Are Fields? Part 2: Do 60-Hz Fields Pose Health Risks? Part 3: What Can and Should Be Done? Department of Engineering and Public Policy, Carnegie Mellon University, Pittsburgh, Pa.

Morgan, W.W., L.S. McFaddin, and L.Y. Harvey. 1973. A daily rhythm of norepinephrine content in regions of the hamster brain. Comp. Gen. Pharmacol. 4:47-52.

Morris, J.A. 1991. The age of incidence of childhood acute lymphoblastic leukaemia. Med. Hypotheses 35:4-10.

Moses, G.C., and A.H. Martin. 1992. Effects of extremely low-frequency electromagnetic fields on three plasma membrane-associated enzymes in early chicken embryos. Biochem. Int. 28:659-664.

Murphy, J.C., D.A. Kaden, J. Warren, and A. Sivak (International Commission for Protection Against Environmental Mutagens and Carcinogens). 1993. Power frequency electric and magnetic fields: A review of genetic toxicology. Mutat. Res. 296:221-240.

Myers, A., A.D. Clayden, R.A. Cartwright, and S.C. Cartwright. 1990. Childhood cancer and overhead power lines: A case-control study. Br. J. Cancer 62:1008-1014.

Nair, I., M. Granger-Morgan, and H.K. Florig. 1989. Biological Effects of Power Frequency Electric and Magnetic Fields. Publ. No. OTA-BP-E-53. Office of Technology Assessment, Washington, D.C.

Nordenson, I., K. Hansson Mild, S. Nordstrom, A. Sweins, and E. Birke. 1984. Clastogenic effects in human lymphocytes of power frequency electric fields: In vivo and in vitro studies. Radiat. Environ. Biophys. 23:191-201.

Nordenson, I., K. Hansson Mild, G. Anderson, and M. Sandstrom. 1994. Chromosomal aberrations in human amniotic cells after intermittent exposure to fifty hertz magnetic fields. Bioelectromagnetics 15:293-301.

Nordstrom, S., E. Birke, and L. Gustavsson. 1983. Reproductive hazards among workers at high voltage substations. Bioelectromagnetics 4:91-101.

Norton, L.A., G.A. Rodan, and L.A. Bourret. 1977. Epiphyseal cartilage cAMP changes produced by electrical and mechanical perturbations. Clin. Orthop. Relat. Res. 124:59-68.

Nossol, B., G. Buse, and J. Silney. 1993. Influence of weal static and 50-Hz magnetic fields of the redox activity of cytochrome-C oxidose. Bioelectromagnetics 14:361-372.

Novelli, G., M. Gennarelli, L. Potenza, P. Angelongi, and B. Ballagiccola. 1991. Study of the effects of DNA of electromagnetic fields using clamped homogenous electric field gel electrophoresis. Biomed. Pharmacother. 45:451-454.

NRC (National Research Council). 1983. Risk Assessment in the Federal Government: Managing the Process. Washington, D.C.: National Academy Press.

NRPB (National Radiological Protection Board). 1992. Electromagnetic Fields and the Risk of Cancer, Vol. 3, pp. 1-138. National Radiological Protection Board, Chilton, Didcot, U.K.

NRPB (National Radiological Protection Board). 1994. Electromagnetic Fields and the Risk of Cancer, Vol. 5, pp. 77-81. National Radiological Protection Board, Chilton, Didcot, U.K.

Olcese, J., and S. Reuss. 1986. Magnetic field effects on pineal gland melatonin synthesis: Comparative studies on albino and pigmented rodents. Brain Res. 369:365-368.

Olkin, I. 1994. Invited commentary: A critical look at some popular meta-analytic methods. Am. J. Epidemiol. 140:297-299.

Olsen, J.H., A. Nielsen, and G. Schulgen. 1993. Residence near high-voltage facilities and the risk of cancer in children. Br. Med. J. 307:891-895.

ORAU (Oak Ridge Associated Universities). 1992. Health Effects of Low Frequency Electric and Magnetic Fields. Publ. No. ORAU 92/F8. Committee in Interagency Radiation Research and Policy Coordination, Oak Ridge Associated Universities, Oak Ridge, Tenn.

Ossenkopp, K.P., and D.P. Cain. 1988. Inhibitory effects of acute exposure to low-intensity 60-Hz magnetic fields on electrically kindled seizures in rats. Brain Res. 442:255-260.

Ossenkopp, K.P., and M. Kavaliers. 1987. Morphine-induced analgesia and exposure to low-intensity 60-Hz magnetic fields: Inhibition of nocturnal analgesia in mice is a function of magnetic field intensity. Brain Res. 418:356-360.

Ossenkopp, K.P., M. Kavaliers, F.S. Prato, G.C. Teskey, E. Sestini, and M. Hirst. 1985. Exposure to nuclear magnetic resonance imaging procedure attenuates morphine-induced analgesia in mice. Life Sci. 37:1507-1514.

Ozawa, H., E. Abe, Y. Shibasaki, T. Fukuhara, and T. Suda. 1989. Electric fields stimulate DNA synthesis of mouse osteoblast-like cells (MC3T3-E1) by a mechanism involving calcium ions. J. Cell. Physiol. 138:477-483.

Paneth, N. 1993. Neurological effects of power-frequency electromagnetic fields. Environ. Health Perspect. 101:101-106.

Parkinson, W.C., and C.T. Hanks. 1989a. Search for cyclotron resonance in cells in vitro. Bioelectromagnetics 10:129-145.

Parkinson, W.C., and C.T. Hanks. 1989b. Experiments on the interaction of electromagnetic fields with mammalian systems. Biol. Bull. Suppl. 176:170-178.

Parkinson, W.C., and G.L. Sulik. 1992. Diatom response to extremely low-frequency magnetic fields. Radiat. Res. 130:319-330.

Peach, H.G., W.J. Bonwick, and T. Wyse. 1992. Report of the Panel on Electromagnetic Fields and Health. The Victorian Government, Melbourne, Australia.

Peck, E.R. 1953. Electricity and Magnetism. New York: McGraw-Hill.

Perry, S., and L. Pearl. 1988. Power frequency magnetic field and illness in multi-storey blocks. Public Health 102:11-18.

Perry, S., L. Pearl, and R. Binns. 1989. Power frequency magnetic field: Depressive illness and myocardial infarction. Public Health 103:177-180.

Perry, F.S., M. Reichmanis, A.A. Marino, and R.R. Becker. 1981. Environmental power-frequency magnetic fields and suicide. Health Phys. 41:267-277.

Persinger, M.A., N.C. Carrey, G.F. Lafreniere, and A. Mazzuchin. 1978. Thirty-eight blood, tissue and consumptive measures from rats exposed perinatally and as adults to 0.5 Hz magnetic fields. Int. J. Biometeorol. 22:213-226.

Peteiro-Cartelle, F.J., and J. Cabezas-Cerrato. 1989. Absence of kinetic and cytogenic effects on human lymphocytes exposed to static magnetic fields. J. Bioelect. 8:11-19.

Peters, J.M., S. Preston-Martin, S.J. London, J.D. Bowman, J.D. Budkley, and D.C. Thomas. 1994. Processed meats and risk of childhood leukemia (California, USA). Cancer Causes Control 5:195-202.

Petitti, D. 1994. Meta-Analysis, Decision Analysis, and Cost-Effectiveness Analysis. New York: Oxford University Press. 246 pp.

Petridou E., D. Kassimos, M. Kalmanti, H. Kosmidis, S. Haidas, V. Flytzani, D. Tong, and D. Trichopoulos. 1993. Age of exposure to infections and risk of childhood leukaemia. Br. Med. J. 307:774.

Phillips, J.L., W. Hagren, W.J. Thomas, T. Ishida-Jones, and W.R. Adey. 1992. Magnetic field induced changes in specific gene transcription. Biochim. Biophys. Acta 1132:140-144.

Pienkowski, D., and S.R. Pollack. 1983. The origin of stress generated potentials in fluid saturated bone. J. Orthop. Res. 1:30-41.

Pilla, A.A. 1993. State of the art in electromagnetic therapeutics. Pp. 17-22 in Electricity and Magnetism in Biology and Medicine: Review and Research Papers Presented at the First World Congress for Electricity and Magnetism in Biology and Medicine, M. Blank, ed. San Francisco: San Francisco Press.

Polk, C. 1986. Introduction. Pp. 1-24 in Handbook of Biological Effects of Electromagnetic Fields, C. Polk and E. Postwa, eds. Boca Raton, Fla.: CRC.

Polk, C. 1990. Electric fields and surface charges due to ELF magnetic fields. Bioelectromagnetics 11:189-201.

Polk, C. 1992a. Dosimetric extrapolations of extremely-low-frequency electric and magnetic fields across biological systems. Bioelectromagnetics Suppl. 1:205-208.

Polk, C. 1992b. Dosimetry of extremely-low-frequency magnetic fields. Bioelectromagnetics Suppl. 1:209-235.

Polk, C. 1993. Therapeutic applications of low frequency electric and magnetic fields. Pp. 129-148 in Advances in Electromagnetic Fields in Living Systems, J.C. Lin, ed. New York: Plenum.

Polk, C., and J.H. Song. 1990. Electric fields induced by low frequency magnetic fields in homogeneous biological structures that are surrounded by an electric insulator. Bioelectromagnetics 11:235-249.

Pollack, S.R. 1984. Bioelectrical properties of bone: Endogenous electrical signals. Orthop. Clin. North Am. 15:3-14.

Poole, C., and R. Kavet. 1993. Depressive symptoms and headaches in relation to proximity of residence to an althernating-currect transmission line right-of-way. Am. J. Epidemiol. 1373:318-330.

Poole, C., and D. Trichopoulos. 1991. Extremely low-frequency electric and magnetic fields and cancer. Cancer Causes Control 2:267-276.

Portet, R., and J. Cabanes. 1988. Development of young rats and rabbits exposed to a strong electric field. Bioelectromagnetics 9:95-104.

Prato, F.S., K.P. Ossenkopp, M. Kavaliers, P. Uksik, R.L. Nicholson, D. Droat, and E.A. Sestini. 1988-1989. Effects of exposure to magnetic resonance imaging on nocturnal serum melatonin and other hormone levels in adult males: Preliminary findings. J. Bioelectr. 7:169-180.

Preston-Martin, S., J.M. Peters, M.C. Yu, D.H. Garabrant, and J.D. Bowman. 1988. Myelogenous leukemia and electric blanket use. Bioelectromagnetics 9:207-213.

Quinlan, W.J., D. Petronda, N. Lebda, S. Pettit, and S.M. Michaelson. 1985. Neuroendocrine parameter in the rat exposed to 60-Hz electric fields. Bioelectromagnetics 6:381-389.

Raisz, L.G. 1977. Bone metabolism and calcium regulation. Metab. Bone Dis. 1:1-48.

Rannug, A., T. Ekstrom, K. Hansson Mild, B. Holmberg, I. Gimenez-Conti, and T.J. Slaga. 1993a. A study on skin tumor formation in mice with 50-Hz magnetic field exposure. Carcinogenesis 14:573-578.

Rannug, A., B. Holmberg, T. Ekstrom, and K. Hansson Mild. 1993b. Rat liver foci study on coexposure with 50-Hz magnetic fields and known carcinogens. Bioelectromagnetics 14:17-27.

Rannug, A., B. Holmberg, and K. Hansson Mild. 1993c. A rat liver foci promotion study with 50-Hz alternating magnetic fields. Environ. Res. 62:223-229.

Rasmussen, H., and P.Q. Barrett. 1984. Calcium messenger system: An integrated view. Physiol. Rev. 64:938-984.

Rea, W.J., Y. Pan, E.J. Fenyes, I. Sujisava, H. Suyana, N. Samadi, and G.D. Ross. 1991. Electromagnetic field sensitivity. J. Bioelectr. 10:241-246.

Reese, J.A., R.F. Jostes, and M.E. Frazier. 1988. Exposure of mammalian cells to 60-Hz magnetic or electric fields: Analysis for DNA single-strand breaks. Bioelectromagnetics 9:237-247.

Reese, J.A., M.E. Frazier, J.E. Morris, R.L. Buschborn, and D.L. Miller. 1991. Evaluation of changes in diatom mobility after exposure to 16 Hz electromagnetic fields. Bioelectromagnetics 12:21-26.

Reichmanis, M., F. Perry, A. Marino, and R. Becker. 1979. Relation between suicide and the electromagnetic field of overhead power lines. Physiol. Chem. Phys. 11:395-403.

Reiter, R.J. 1980. The pineal and its hormones in the control of reproduction in mammals. Endocr. Rev. 1:109-131.

Reiter, R.J. 1985. Action spectra, dose-response relationships, and temporal aspects of light's effects on the pineal gland. Ann. N.Y. Acad. Sci. 453:215-230.

Reiter, R.J. 1991. Pineal melatonin: Cell biology of its synthesis and of its physiological interactions. Endocr. Rev. 12:151-180.

Reiter, R.J. 1992. Alterations of the circadian melatonin rhythm by the electromagnetic spectrum: A study in environmental toxicology. Regul. Toxicol. 15:226-244.

Reiter, R.J. 1993a. The mammalian pineal glands as an end organ of the visual system. Pp. 145-160 in Light and Biological Rhythms in Man, L. Wetterberg, ed. New York: Pergamon.

Reiter, R.J. 1993b. Static and extremely low frequency electromagnetic field exposure: Reported effects of the circadian production of melatonin. J. Cell. Biochem. 51:394-403.

Reiter, R.J. 1993c. Electromagnetic fields and melatonin production. Biomed. Pharmacol. Ther. 47:439-444.

Reiter, R.J., and B.A. Richardson. 1992. Magnetic field effects on pineal indoleamine metabolism and possible biological consequences. FASEB J. 6:2283-2287.

Reiter, R.J., L.E. Anderson, R.I. Buschbom, and B.W. Wilson. 1988. Reduction of the nocturnal rise in pineal melatonin levels in rats exposed to 60-Hz electric fields in utero and for 23 days after birth. Life Sci. 42:2203-2206.

Robert, E. 1993. Birth defects and high voltage power lines: An exploratory study based on registry data. Reprod. Toxicol. 7:283-287.

Robison, L.L., and A. Daigle. 1984. Control selection using random digit dialing for cases of childhood cancer. Am. J. Epidemiol. 120:164-166.

Robison, L.L., A. Mertens, and J.P. Neglia. 1991. Epidemiology and etiology of childhood cancer. Pp. 11-28 in Clinical Pediatric Oncology, 4th Ed., D.J. Fernbach and T.J. Vietti, eds. St. Louis: Mosby Year Book.

Rodan, G.A., and T.J. Martin. 1981. Role of osteoblasts in the hormonal regulation of bone resorption—A hypothesis. Calcif. Tissue Int. 33:349-351.

Rodan, G.A., L.A. Bourret, and L.A. Norton. 1978. DNA synthesis in cartilage cells is stimulated by oscillating electric fields. Science 199:690-692.

Rogers, W., R.J. Reiter, H.D. Smith, and L. Barlow-Walden. 1995. Rapid onset/offset, variably scheduled 60-Hz electric and magnetic field exposure reduces nocturnal serum melatonin concentration in nonhuman primates. Bioelectromagnetics Suppl. 3:119-122.

Rollag, M.D., and G.D. Niswender. 1976. Radioimmunoassay of serum concentrations of melatonin in sheep exposed to different lighting regimes. Endocrinology 98:482-489.

Rommereim, D.N., W.T. Kaune, R.L. Buschbom, R.D. Phillips, and M.R. Sikov. 1987. Reproduction and development in rats chronologically exposed to 60-Hz electric fields. Bioelectromagnetics 8:243-258.

Rommereim, D.N., W.T. Kaune, L.E. Anderson, and M.R. Sikov. 1989. Rats reproduce and rear litters during chronic exposure to 150-kV/m, 60-Hz electric fields. Bioelectromagnetics 10:385-389.

Rommereim, D.N., R.L. Rommereim, M.R. Sikov, R.L. Buschbom, and L.E. Anderson. 1990. Reproduction, growth, and development of rats during chronic exposure to multiple field strengths of 60-Hz electric fields. Fundam. Appl. Toxicol. 14:608-621.

Rosen, D.M., and R.A. Luben. 1983. Multiple hormonal mechanisms for the control of collagen synthesis in an osteoblast-like cell line, MMB-1. Endocrinology 112:992-999.

Rosenbaum, P.F., J.E. Vena, M.A. Zielezny, and A.M. Michalek. 1994. Occupational exposures associated with male breast cancer. Am. J. Epidemiol. 139:30-36.

Rosenberg, R.S., P.H. Duffy, and G.A. Sacher. 1981. Effects of intermittent 60-Hz high voltage electric fields in metabolism, activity, and temperature in mice. Bioelectromagnetics 2:291-303.

Rosenberg, R.S., P.H. Duffy, G.A. Sacher, and C.F. Ehret. 1983. Relationship between field strength and arousal response in mice exposed to 60-Hz electric fields. Bioelectromagnetics 4:181-191.

Rosenthal, M., and G. Obe. 1989. Effects of 50-hertz electromagnetic fields on proliferation and on chromosomal alterations in human peripheral lymphocytes untreated or pretreated with chemical mutagens. Mutat. Res. 210:329-335.

Rothman, K.J. 1986. Modern Epidemiology. Boston: Little, Brown.

Rubin, P., ed. 1983. Clinical Oncology: A Multidisciplinary Approach. New York: American Cancer Society.

Rubin, C.T., K.J. McLeod, and L.E. Lanyon. 1989. Prevention of osteoporosis by pulsed electromagnetic fields. J. Bone Jt. Surg. 71-A:411-417.

Rudolph, K., K. Kräuchi, A. Wirz-Justice, and H Feer. 1985. Weak 50-Hz electromagnetic fields activate rat open field behavior. Physiol. Behav. 35:505-508.

Saffer, J.D., and S.J. Thurston. 1995. Short exposures to 60 Hz magnetic fields do not alter *myc* expression in HL60 or Daudi cells. Radiat. Res. 144:18-25.

Sagan, P.M., M.E. Stell, G.K. Bryan, and W.R. Adey. 1987. Detection of 60-Hz vertical electric fields by rats. Bioelectromagnetics 8:303-313.

Sahl, J.D., M.A. Kelsh, and S. Greenland. 1993. Cohort and nested case-control studies of hematopoietic cancers and brain cancer among electric utility workers. Epidemiology 4:104-114.

Sakai, T. 1981. The mammalian Harderian gland: Morphology, biochemistry, funciton and physiology. Arch. Histol. Jpn. 44:299-333.

Salzinger, K., S. Freimark, M. McCullough, D. Phillips. and L. Birenbaum. 1990. Altered operant behavior of adult rats after perinatal exposure to a 60-Hz electromagnetic field. Bioelectromagnetics 11:105-116.

Savitz, D.A., and A. Ahlbom. 1994. Epidemiologic evidence on cancer in relation to residential and occupational exposures. Pp. 233-261 in Biologic Effects of Electric and Magnetic Fields: Beneficial and Harmful Effects, D.O. Carpenter and S. Ayrapetyan, eds. San Diego, Calif.: Academic.

Savitz, D.A., and E.E. Calle. 1987. Leukemia and occupational exposure to electromagnetic fields: Review of epidemiologic surveys. J. Occup. Med. 29:47-51.

Savitz, D.A., and J. Chen. 1990. Parental occupation and childhood cancer: Review of epidemiologic studies. Environ. Health Perspect. 88:325-337.

Savitz, D.A., and L. Feingold. 1989. Association of childhood cancer with residential traffic density. Scand. J. Work Environ. Health 15:360-363.

Savitz, D.A., and W.T. Kaune. 1993. Childhood cancer in relation to a modified residential wire code. Environ. Health Perspect. 101:76-80.

Savitz, D.A., and D.P. Loomis. 1995. Magnetic field exposure in relation to leukemia and brain cancer mortality among electric utility workers. Am. J. Epidemiol. 141:123-134.

Savitz, D.A., H. Wachtel, F.A. Barnes, E.M. John, and J.G. Tvrdik. 1988. Case-control study of childhood cancer and exposure to 60-Hz magnetic fields. Am. J. Epidemiol. 128:21-38.

Savitz, D.A., E.M. John, and R.C. Kleckner. 1990. Magnetic field exposure from electric appliances and childhood cancer. Am. J. Epidemiol. 131:763-773.

Savitz, D., C. Boyle, and P. Holmgreen. 1994. Prevalence of depression among electrical workers. Am. J. Ind. Med. 25:165-176.

Scarfi, M.R., F. Bersani, A. Cossarizza, D. Monti, G. Castellani, R. Cadossi, G. Franceschetti, and C. Franceschi. 1991. Spontaneous and mitomycin-C-induced micronuclei in human lymphocytes exposed to extremely low-frequency pulsed magnetic fields. Biochem. Biophys. Res. Commun. 176:194-200.

Scarfi, M.R., F. Bersani, A. Cossarizza, D. Monti, O. Zeni, M.B. Lioi, G. Granceschetti, M. Capri, and C. Franceschi. 1993. 50 Hz AC sinusoidal electric fields do not exert genotoxic effects (micronucleus formation) in human lymphocytes. Radiat. Res. 135:64-68.

Schiffman, J.S., H. Lasch, M.D. Rollag, A.E. Flanders, G.C. Brainard, and D.L. Burk, Jr. 1994. Effect of MR imaging on the normal human pineal body: Measurement of plasma melatonin levels. J. Magn. Reson. Imaging 4:7-11.

Schlesselman, J. 1982. Case-Control Studies. Design, Conduct, Analysis. New York: Oxford University Press.

Schmukler, R., and A.A. Pilla. 1982. A transient impedance approach to nonfaradaic electrochemical kinetics at living cell membranes. J. Electrochem. Soc. 129:526-528.

Schreiber, G.H., G.M.H. Swaen, J.M.M. Mejers, J.J.M. Slangen, and F. Sturmans. 1993. Cancer mortality and residence near electricity transmission equipment: A retrospective cohort study. Int. J. Epidemiol. 22:9-15.

Schwartz, J.L., and G.A.R. Mealing. 1993. Calcium-ion movement and contractility in atrial strips of frog heart are not affected by low-frequency-modulated, 1 GHz electromagnetic radiation. Bioelectromagnetics 14:521-533.

Schwartz, J.L., D.E. House, and G.A.R. Mealing. 1990. Exposure of frog hearts to cw or amplitude-modulated VHF fields selective efflux of calcium ions at 16 Hz. Bioelectromagnetics 11:349-358.

Seegal, R.F., J.R. Wolpaw, and R. Dowman. 1989. Chronic exposure of primates to 60-Hz electric and magnetic fields: II. Neurochemical effects. Bioelectromagnetics 10:289-301.

Seto, Y.J., D. Majeau-Chargois, J.R. Lymangrover, W.P. Dunlap, C.F. Walker, and S.T. Hsieh. 1984. Investigation of fertility and in utero effects in rats chronically exposed to a high-intensity 60-Hz electric field. IEEE Trans. Biomed. Eng. 11:693-702.

Severson, R.K., R.G. Stevens, W.T. Kaune, D.B. Thomas, L. Heuser, S. Davis, and L.E. Sever. 1988. Acute non-lymphocytic and residential exposure to power frequency magnetic fields. Am. J. Epidemiol. 128:10-20.

Shaw, G.M., and L.A. Croen. 1993. Human adverse reproductive outcomes and electromagnetic field exposures: Review of epidemiologic studies. Environ. Health Perspect. 101(Suppl. 4):107-119.

Shiau, U., and A.R. Valentino. 1981. ELF electric field coupling to dielectric spheroidal models of biological objects. IEEE Trans. Biomed. Eng. 28:429-437.

Sibley, D.R., J.L. Benovic, M.G. Caron, and R.J. Lefkowitz. 1988. Phosphorylation of cell surface receptors: A mechanism for regulating signal transduction pathways. Endocr. Rev. 9:38-56.

Sikov, M.R., L.D. Montgomery, L.G. Smith, and R.D. Phillips. 1984. Studies on prenatal and postnatal development in rats exposed to 60-Hz electric fields. Bioelectromagnetics 5:101-112.

Sikov, M.R., D.M. Rommereim, J.L. Beamer, R.L. Buschbom, W.T. Kaune, and R.D. Phillips. 1987. Developmental studies of Hanford miniature swine exposed to 60-Hz electric fields. Bioelectromagnetics 8:229-242.

Slovic, P. 1987. Perception of risk. Science 236:280-285.

Smith, R. 1983. Household magnetic fields and suicide. Health Phys. 44:699-700.

Smith, R.F., and D.R. Justesen. 1977. Effects of a 60-Hz magnetic field on activity levels of mice. Radio Sci. 12:279-286.

Smith, M.T., J.A. D'Andrea, and O.P. Gandhi. 1979. Behavioral effects of strong 60-Hz electric fields in rats. J. Microwave Power 14:223-228.

Smith, S.D., B.R. McLeod, A.R. Liboff, and K. Cooksey. 1987. Calcium cyclotron resonance and diatom mobility. Bioelectromagnetics 8:215-227.

Smith, S.D., B.R. McLeod, and A.R. Liboff. 1993. Effects of CR-tuned 60-Hz magnetic fields on sprouting and early growth of *Raphanus sativus*. Bioelectrochem. Bioenerg. 32:67-76.

Spiegel, R.J. 1976. ELF coupling to spherical models of man and animals. IEEE Trans. Biomed. Eng. 23:387-391.

Spiegel, R.J. 1977. Magnetic coupling to a prolate spheroid model of man. IEEE Trans. Power Appar. Syst. 96:208-212.

Spiegel, R.J. 1981. Numerical determination of induced currents in humans and baboons exposed to 60-Hz electric fields. IEEE Trans. Electromagn. Compat. 23:382-390.

Stern, S., and V.G. Laties. 1985. 60-Hz electric fields: Detection by female rats. Bioelectromagnetics 6:99-103.

Stern, S., and V.G. Laties. 1989. Comparison of 60-Hz electric fields and incandescent light as aversive stimuli controlling the behavior of rats. Bioelectromagnetics 10:99-109.

Stern, S., V.G. Laties, C.V. Stancampiano, C. Cox, and J.O. de Lorge. 1983. Behavioral detection of 60-Hz electric fields by rats. Bioelectromagnetics 4:215-247.

Stevens, R.G. 1987a. Electric power use and breast cancer: A hypothesis. Am. J. Epidemiol. 125:294-300.

Stevens, R.G. 1987b. Electric power use and breast cancer. Am. J. Epidemiol. 125:556-561.

Stollery, B. 1986. Effects of 50 Hz electric currents on mood and verbal reasoning skills. Br. J. Ind. Med. 43:339-349.

Stuchly, M.A., and W. Xi. 1994. Modelling induced currents in biological cells exposed to low-frequency magnetic fields. Phys. Med. Biol. 39:1319-1330.

Stuchly, M.A., and S. Zhao. 1996. Magnetic field-induced currents in the human body in the proximity of power lines. IEEE Trans. Power Delivery 11:102-107.

Stuchly, M.A., J. Ruddick, D. Villeneuve, K. Robinson, B. Reed, D.W. Lecuyer, K. Tan, and J. Wong. 1988. Teratological assessment of exposure to time-varying magnetic field. Teratology 38:461-466.

Stuchly, M.A., D.W. Lecuyer, and J. McLean. 1991. Cancer promotion in a mouse-skin model by a 60-Hz magnetic field: I. Experimental design and exposure system. Bioelectromagnetics 12:261-271.

Stuchly, M.A., J.R.N. McLean, R. Burnett, M. Goddard, D.W. Lecuyer, and R.E.J. Mitchel. 1992. Modification of tumor promotion in the mouse skin by exposure to an alternating magnetic field. Cancer Lett. 65:1-7.

Svedenstal, B.M., and B. Holmberg. 1993. Lymphoma development among mice exposed to x-rays and pulsed magnetic fields. Int. J. Radiat. Biol. 64:119-125.

Takahashi, K., I. Kaneko, M. Date, and E. Fukada. 1987. Influence of pulsing electromagnetic field on the frequency of sister-chromatid exchanges in cultured mammalian cells. Experientia 43:331-332.

Takatsuji, T., M.S. Sasaki, and H. Takekoshi. 1989. Effect of static magnetic field on the induction of chromosome aberrations by 4.5 MeV protons and 23 MeV alpha particles. J. Radiat. Res. 30:238-246.

Tam, C.S., J.M.N. Heersche, T.M. Murray, and J.A. Parsons. 1982. Parathyroid hormone stimulates the bone apposition rate independently of its resorptive action. Endocrinology 110:506-512.

Tan, D.X., L.D. Chen, B. Poeggeler, L.C. Manchester, and R.J. Reiter. 1993a. Melatonin: A potent endogenous hydroxyl radical scavenger. Endocr. J. 1:57-60.

Tan, D.X., B. Poeggeler, R.J. Reiter, L.D. Chen, S. Chen, L.C. Manchester, and L.R. Barlow-Walden. 1993b. The pineal hormone melatonin inhibits DNA-adduct formation induced by the chemical carcinogen safrole in vivo. Cancer Lett. 70:65-71.

Tan, D.X., R.J. Reiter, L.D. Chen, B. Poeggeler, L.C. Manchester, and L.R. Barlow-Walden. 1994. Both physiological and pharmacological levels of melatonin induced by the chemical carcinogen safrole. Carcinogenesis 15:215-218.

Tenforde, T.S. 1991. Biological interactions of extremely-low-frequency electric and magnetic fields. Bioelectrochem. Bioenerg. 25:1-17.

Tenforde, T.S. 1992. Biological interactions and potential health effects of extremely-low-frequency magnetic fields from power lines and other common sources. Annu. Rev. Public Health 13:173-196.

Tenforde, T.S., and W.T. Kaune. 1987. Interaction of extremely low frequency electric and magnetic fields with humans. Health Phys. 53:585-606.

Theriault, G.P. 1990. Cancer risks due to exposure to electromagnetic fields. Recent Results Cancer Res. 120:166-180.

Theriault, G.P., M. Goldberg, A.B. Miller, B. Armstrong, P. Guenel, J. Deadman, E. Imbernon, T. To, A. Chevalier, D. Cyr, and C. Wall. 1994. Cancer risks associated with occupational exposure to magnetic fields among electric utility workers in Ontario and Quebec, Canada, and France: 1970-1989. Am. J. Epidemiol. 139:550-572.

Thomas, A., and P.G. Morris. 1981. The effects of NMR exposure in living organisms: I. A microbial assay. Br. J. Radiol. 54:615-621.

Thomas, J.R., J. Schrot, and A.R. Liboff. 1986. Low intensity magnetic fields alter operant responding in rats. Bioelectromagnetics 7:349-357.

Thomson R.A.E., S.M. Michaelson, and Q.A. Nguyen. 1988. Influence of 60-Hz magnetic fields on leukemia. Bioelectromagnetics 9:149-158.

Tobey, R.A., H.J. Price, L.D. Scott, and K.D. Ley. 1981. Lack of effect of 60-hertz fields on growth of cultured mammalian cells. Rep. No. LA-8831-MS. Los Alamos Scientific Laboratory, Los Alamos, N.M.

Tomenius, L. 1986. 50-Hz electromagnetic environment and the incidence of childhood tumors in Stockholm County. Bioelectromagnetics 7:191-207.

Tucker, R.D., and O.H. Schmitt. 1978. Tests for human perception of 60-Hz moderate strength magnetic fields. IEEE Trans. Biomed. Eng. 25:509-518.

Tynes, T., and A. Andersen. 1990. Electromagnetic fields and male breast cancer [letter]. Lancet 336(8743):1596.

Tynes, T., J.B. Reitan, and A. Andersen. 1994a. Incidence of cancer among workers in Norwegian hydroelectric power companies. Scand. J. Work Environ. Health 20:339-344.

Tynes, T., H. Jynge, and A.I. Vistnes. 1994b. Leukemia and brain tumors in Norwegian railway workers, a nested case-control study. Am. J. Epidemiol. 139:645-653.

Uckun, F.M., T. Kurosaki, J. Jin, X. Jun, J. Bolen, and R.A. Luben. 1995. Exposure of B-lineage lymphoid cells to low energy electromagnetic fields stimulates Lyn kinase. J. Biol. Chem. 270:27666-27670.

U.S. Surgeon-General. 1964. Smoking and Health. Report of the Advisory Committee to the Surgeon General of the U.S. Public Health Service. PHS Publ. No. 1103. U.S. Public Health Service, Washington, D.C.

Vasquez, B.J., L.E. Anderson, C.I. Lowery, and W.R. Adey. 1988. Diurnal pattern in brain amines of rats exposed to 60-Hz electric fields. Bioelectromagnetics 9:229-236.

Vena, J.E., S. Graham, R. Hellman, M. Swanson, and J. Brasure. 1991. Use of electric blankets and risk of postmenopausal breast cancer. Am. J. Epidemiol. 134:180-185.

Verkasalo, P.K., E. Pukkala, M.Y. Hongisto, J.E. Valjus, P.J. Drvinen, K.V. Heikkil, and M. Koskenvuo. 1993. Risk of cancer in Finnish children living close to power lines. Br. Med. J. 307:895-899.

Verreault, R., N.S. Weiss, K.A. Hollenbach, C.L. Strader, and J.R. Daling. 1990. Use of electric blankets and the risk of testicular cancer. Am. J. Epidemiol. 131:759-762.

Villa, M., P. Mustrarelli, and M. Caprotti. 1991. Biological effects of magnetic fields. Life Sci. 49:85-92.

Voigt, L.F., S. Davis, and S. Heuser. 1992. Random digit dialing: The potential effect of sample characteristics on the conversion of nonresidential telephone numbers. Am. J. Epidemiol. 136:1393-1399.

Waksberg, J. 1978. Sampling methods for random digit dialing. J. Am. Stat. Assoc. 73:40-46.

Walleczek, J. 1992. Electromagnetic field effects on cells of the immune system: The role of calcium signaling. FASEB J. 6:3177-3185.

Walleczek, J., and T.F. Budinger. 1992. Pulsed magnetic field effects on calcium signaling in lymphocytes: Dependence on cell status and field intensity. FEBS Lett. 314:351-355.

Walleczek, J., and R.P. Liburdy. 1990. Nonthermal 60 Hz sinusoidal magnetic-field exposure enhances $^{45}Ca^{2+}$ uptake in rat thymocytes: Dependence on mitogen activation. FEBS Lett. 271:157-160.

Wang, W., T.A. Litovitz, L.M. Penafiel, and R. Meister. 1993. Determination of the induced ELF electric field distribution in a two-layer in-vitro system simulating biological cells in nutrient solution. Bioelectromagnetics 14:29-39.

Ward, E.M., S. Kramer, and A.T. Meadows. 1984. The efficacy of random digit dialing in selecting matched controls for a case-control study of pediatric cancer. Am. J. Epidemiol. 120:582-591.

Wartenberg, D., and C. Chess. 1992. Risky business: The inexact art of hazard assessment. Sciences March-April:16-21.

Wartenberg, D., and D.A. Savitz. 1993. Evaluating exposure cutpoint bias in epidemiologic studies of electric and magnetic fields. Bioelectromagnetics 14:237-245.

Washburn, E., M. Orza, J. Berlin, W. Nicholson, A. Todd, H. Frumkin, and T. Chalmers. 1994. Residential proximity to electricity transmission and distribution equipment and risk of childhood leukemia, childhood lymphoma, and childhood nervous system tumors: Systematic review, evaluation, and meta-analysis. Cancer Causes Control 5:299-309.

Watson, J., and E.M. Downes. 1979. Clinical aspects of the stimulation of bone healing using electrical phenomena. Med. Biol. Eng. Comput. 17:161-169.

Weigel, R.J., R.A. Jaffe, D.L. Lundstrom, W.C. Forsythe, and L.E. Anderson. 1987. Stimulation of cutaneous mechanoreceptors by 60-Hz electric fields. Bioelectromagnetics 8:337-350.

Weiss, N.S. 1981. Inferring causal relationships: Elaboration of the criterion of ''dose-response.'' Am. J. Epidemiol. 113:487-490.

Wertheimer, N., and E. Leeper. 1979. Electrical wiring configurations and childhood cancer. Am. J. Epidemiol. 109:273-284.

Wertheimer, N., and E. Leeper. 1980. RE: ''Electrical wiring configurations and childhood leukemia in Rhode Island'' [letter]. Am. J. Epidemiol. 111:461-462.

Wertheimer, N., and E. Leeper. 1982. Adult cancer related to electrical wires near the home. Int. J. Epidemiol. 11:345-355.

Wertheimer, N., and E. Leeper. 1986. Possible effects of electric blankets and heated waterbeds on fetal development. Bioelectromagnetics 7:13-22.

Wertheimer, N., and E. Leeper. 1989. Fetal loss associated with two seasonal sources of electromagnetic field exposure. Am. J. Epidemiol. 129:220-224.

Whitson, G.L., W.L. Carrier, A.A. Francis, C.C. Shih, S. Georghiou, and J.D. Reagan. 1986. Effects of extremely low frequency (ELF) electric fields on cell growth and DNA repair in human skin fibroblasts. Cell Tissue Kinet. 19:39-47.

Wiley, M.J., P. Corey, R. Kavet, J. Charry, S. Harvey, D. Agnew, and M. Walsh.

1992. The effects of continuous exposure to 20 k-Hz sawtooth magnetic fields on the litters of CD-1 mice. Teratology 46:391-398.

Wilson, B.W., W. Snedden, P.E. Mullen, R.E. Silman, I. Smith, and J. Laudon. 1977. A gas chromatography-mass spectrometry method for the quantitative analysis of melatonin in plasma and cerebrospinal fluid. Anal. Biochem. 81:283-291.

Wilson, B.W., L.E. Anderson, D.I. Hilton, and R.D. Phillips. 1981. Chronic exposure to 60-Hz electric fields: Effects on pineal function in the rat. Bioelectromagnetics 2:371-380.

Wilson B.W., L.E. Anderson, D.I. Hilton, and R.D. Phillips. 1983. Erratum. Chronic exposure to 60-Hz electric fields: Effects on pineal function in the rat [1981, 2:371-380]. Bioelectromagnetics 4:293.

Wilson, B.W., E.K. Chess, and L.E. Anderson. 1986. 60-Hz electric-field effects on pineal melatonin rhythms: Time course for onset and recovery. Bioelectromagnetics 7:239-242.

Wilson, B.W., R.G. Stevens, and L.E. Anderson. 1989. Neuroendocrine mediated effects of electromagnetic-field exposure: Possible role of the pineal gland. Life Sci. 45:1319-1332.

Wilson, B.W., C.W. Wright, J.E. Morris, R.L. Buschbom, D.P. Brown, D.L. Miller, R. Sommers-Flannigan, and L.E. Anderson. 1990a. Evidence for an effect of ELF electromagnetic fields on human pineal gland function. J. Pineal Res. 9:259-269.

Wilson, B.W., R.G. Stevens, and L.E. Anderson, eds. 1990b. Extremely Low Frequency Electromagnetic Eields: The Question of Cancer. Columbus, Ohio: Battelle Press.

Wilson, B.W., K. Caputa, M.A. Stuchly, J.D. Saffer, K.C. Davis, C.E. Washam, L.G. Washam, G.R. Washam, and M.A. Wilson. 1994. Design and fabrication of well confined uniform magnetic field exposure systems. Bioelectromagnetics 15:563-577.

Wolf, F. 1986. Meta-Analysis: Quantitative Methods for Research Synthesis. Newbury Park, Calif.: Sage Publications.

Wolff, J. 1892. Gesetz der Transformation der Knochen. Berlin.

Wolff, S., L.E. Crooks, P. Brown, R. Howard, and R.B. Painter. 1980. Tests for DNA and chromosomal damage induced by nuclear magnetic resonance imaging. Radiology 136:707-710.

Wolpaw, J.R., R.F. Seegal, and R. Dowman. 1989. Chronic exposure of primates to 60-Hz electric and magnetic fields: I. Exposure system and measurements of general health and performance. Bioelectromagnetics 10:277-288.

Xi, W., and M.A. Stuchly. 1994. High spatial resolution analysis of electric currents induced in men by ELF magnetic fields. Appl. Comput. Electromagn. Soc. J. 9:127-134.

Xi, W., M.A. Stuchly, and O.P. Gandhi. 1994. Induced electric currents in models of man and rodents from 60 Hz magnetic fields. IEEE Trans. Biomed. Eng. 41:1018-1023.

Yellon, S.M. 1994. Acute 60 Hz magnetic field exposure effects on the melatonin rhythm in the pineal gland and circulation of the adult Djungarian hamster. J. Pineal Res. 16(3):136-144.

Yost, M.G., and R.P. Liburdy. 1992. Time-varying and static magnetic fields act in combination to alter calcium signal transduction in the lymphocyte. FEBS Lett. 296:117-122.

Yost, M.G., J.D. Wu, G. Lee, L. Hristova, and R. Neutra. 1992. Multivariate modeling of residential wire codes and measured magnetic field parameters. Pp. 27-28 in Project Abstracts: The Annual Review of Research on Biological Effects of Electric and Magnetic Fields. Frederick, Md.: W/L Associates.

Youngson, J.H.A.M., A.D. Clayden, A. Myers, and R.A. Cartwright. 1991. A case/control study of adult haematological malignancies in relation to overhead powerlines. Br. J. Cancer 63:977-985.

Zusman, I.P., H.P. Yaffe, and A. Ornoy. 1990. Effects of pulsing electromagnetic fields on the prenatal and postnatal development in mice and rats: In vivo and in vitro studies. Teratology 42:157-170.

Zwingelberg, R., G. Obe, M. Rosenthal, M. Mevissen, S. Buntenkötter, and W. Löscher. 1993. Exposure of rats to a 50-Hz 30-mT magnetic field influences neither the frequencies of sister-chromatid exchanges nor proliferation characteristics of cultured peripheral lymphocytes. Mutat. Res. 302:39-44.

Glossary

ac   Alternating current, normally considered to vary sinusoidal with constant frequency

ampere (A)   Unit of electric current; 1 A = 1 coulomb per second

anisotropic   Having different properties in different directions; dependent on angle

equation here   Magnetic flux density

bifilar   Being composed of two wires

bipartite thoracic centra   Primordial ossification points within the thoracic vertebra

carcinogen   Chemical or physical agent capable of causing cancer

chromodacryorrhea   A stress response consisting of the release from the eye of a porphyrin-based material secreted by a gland behind the eye, also called Ared tears.@

CI   Confidence interval, 95% CI, if not otherwise specified

CIRRPC   Committee on Interagency Radiation Research and Policy Coordination

coulomb (C)   Unit of electric charge, the charge carried by a single electron is 1.6 H 10^{-19} C

cpm   Counts per minute

equation here   Electric flux density, or displacement vector; its units are coulomb per square meter

Errata: Replace the Glossary, pages 338–341, with the following:

Glossary

ac   Alternating current, normally considered to vary sinusoidal with constant frequency

ampere (A)   Unit of electric current; 1 A = 1 coulomb per second

anisotropic   Having different properties in different directions; dependent on angle

$\overline{B}$ Magnetic flux density

bifilar   Being composed of two wires

bipartite thoracic centra   Primordial ossification points within the thoracic vertebra

carcinogen   Chemical or physical agent capable of causing cancer

chromodacryorrhea   A stress response consisting of the release from the eye of a porphyrin-based material secreted by a gland behind the eye, also called "red tears."

CI   Confidence interval, 95% CI, if not otherwise specified

CIRRPC   Committee on Interagency Radiation Research and Policy Coordination

coulomb (C)   Unit of electric charge, the charge carried by a single electron is 1.6×10^{-19} C

cpm   Counts per minute

$\overline{D}$ Electric flux density, or displacement vector; its units are coulomb per square meter *dc* Direct current, or current with "steady" flow

displacement current Quantity related to the time rate of change of the electric field

DMBA 7,12-dimethylbenz[*a*]anthracene, a known carcinogenic agent

DNA Deoxyribonucleic acid, found primarily in the cell nucleus and forms the molecular basis for heredity and cell function

ε Medium permittivity

ε_0 Permittivity of free space

$\overline{E}$ Electric-field strength (V/m)

electromagnetic fields Coupled electric and magnetic fields

ELF Extremely low frequency, usually associated with frequencies of the order of 3 Hz to 3 kHz

EMF Electromagnetic fields

foot (ft) 1 foot = 0.304 meter

fos An early-response gene

gauss (G) The centimeter-gram-second (cgs) unit of magnetic flux density; $1\ G = 10^{-4}$ tesla (T) (tesla is the SI unit of magnetic flux density); $1\ mG = 0.1\ \mu T$; the earth's static magnetic field is about 0.5 G

Gy (gray) Unit of absorbed dose of ionizing radiation equal to 1 joule of energy deposited per kilogram of tissue

$\overline{H}$ Magnetic-field strength (ampere per square meter)

harmonics Signals of nf_0, where n is an integer and f_0 is the fundamental frequency (e.g., the harmonics of a 60-Hz signal will be 120 Hz, 180 Hz, 240 Hz, and so forth)

Hertz (Hz) The SI unit of frequency, $1\ Hz = sec^{-1}$

heterogeneous Composed of components of different kinds

HIOMT Hydroxyindole-*O*-methyltransferase

homogeneous Composed of components of like kind

impedance The equivalent of electric resistance in an ac circuit element that determines the flow of current

in vitro Meaning "outside the living body," as measured in tissue or cell culture

in vivo Meaning "in the living organism," as measured in animals

isotropic Having one or more properties that are independent of direction; independent of angle

jun Cellular early-response gene

kilo (k) Designating the factor 1,000; as in $kHz = 10^3\ sec^{-1}$

dc   Direct current, or current with Asteady@ flow

displacement current   Quantity related to the time rate of change of the electric field

DMBA   7,12-dimethylbenz[*a*]anthracene, a known carcinogenic agent

DNA   Deoxyribonucleic acid, found primarily in the cell nucleus and forms the molecular basis for heredity and cell function

g   Medium permittivity

g_0 Permittivity of free space

equation here   Electric-field strength (V/m)

electromagnetic fields   Coupled electric and magnetic fields

ELF   Extremely low frequency, usually associated with frequencies of the order of 3 Hz to 3 kHz

EMF   Electromagnetic fields

foot (ft)   1 foot = 0.304 meter

fos   An early-response gene

gauss (G)   The centimeter-gram-second (cgs) unit of magnetic flux density; 1 G = 10^{-4} tesla (T) (tesla is the SI unit of magnetic flux density); 1 mG = 0.1 FT; the earth = s static magnetic field is about 0.5 G

Gy (gray)   Unit of absorbed dose of ionizing radiation equal to 1 joule of energy deposited per kilogram of tissue

equation here   Magnetic-field strength (ampere per square meter)

harmonics   Signals of nf_0, where *n* is an integer and f_0 is the fundamental frequency (e.g., the harmonics of a 60-Hz signal will be 120 Hz, 180 Hz, 240 Hz, and so forth)

Hertz (Hz)   The SI unit of frequency, 1 Hz = sec^{-1}

heterogeneous   Composed of components of different kinds

HIOMT   Hydroxyindole-*O*-methyltransferase

homogeneous   Composed of components of like kind

impedance   The equivalent of electric resistance in an ac circuit element that determines the flow of current

in vitro   Meaning Aoutside the living body,@ as measured in tissue or cell culture

in vivo   Meaning Ain the living organism,@ as measured in animals

isotropic   Having one or more properties that are independent of direction; independent of angle

jun   Cellular early-response gene

kilo (k)   Designating the factor 1,000; as in kHz $= 10^3$ sec^{-1}

mega (M)   Prefix meaning mega, designating the factor 10^6; as in 1 MV $=$ 1,000,000 volts; or as in MHZ $= 1,000,000$ sec^{-1}

milli (m)   Designating the factor 10^{-3}, as in 1 msec $= 10^{-3}$ sec; also used as the symbol for meters (m $=$ meter) in application to length; 1 mm $= 10^{-3}$ m; 1 m $= 3.281$ ft

myc   Cellular early-response gene

F   Prefix meaning micro, designating the factor 10^{-6}; as in 1 :m $= 0.000001$ meter; also, medium permeability, when used as equation here $=$ equation here

F_0 Permeability of free space

n   Prefix meaning nano, designating the factor 10^{-9}; as in 1 nsec $= 1$ H 10^{-9} sec

NAT   N-acetyltransferase

NMU   N-nitroso-*N*-methylurea

OR   Odds ratio

ossification   Formation of bone

permittivity   The permittivity is a complex number consisting of a dielectric constant and a loss factor (related to the conductivity); it is an electric property of the material

piezoelectric   The production of an electric potential by an applied pressure or force

PKC   Protein kinase C

prokaryote   Cellular organism with no distinct nucleus, such as a bacterium or blue-green algae

relative risk (RR)   Quotient of the risk in exposed to risk in unexposed individuals

rms   Root mean square; the rms value of x is $\%x^2$

SCE   Sister chromatid exchanges

semipartite thoracic centra   Primordial ossification points within the thoracic vertebra

SI   SystPme International; internationally adopted system of units, such as the meter, kilogram, coulomb, and second

F   Volume conductivity

soma   The body of an organism, or the body of a nerve cell

sternebra   Primordial sternum of the embryo

terata   Abnormalities in the developing or newborn fetus

tesla (T)   Tesla, an SI unit of magnetic flux density; 1 T $= 10^4$ G *mega (M)* Prefix meaning mega, designating the factor 10^6; as in 1 MV = 1,000,000 volts; or as in MHZ = 1,000,000 sec^{-1}

milli (m) Designating the factor 10^{-3}, as in 1 msec = 10^{-3} sec; also used as the symbol for meters (m = meter) in application to length; 1 mm = 10^{-3} m; 1 m = 3.281 ft

myc Cellular early-response gene

μ Prefix meaning micro, designating the factor 10^{-6}; as in 1 μm = 0.000001 meter; also, medium permeability, when used as $\overline{B} = \mu\overline{H}$

μ_0 Permeability of free space

n Prefix meaning nano, designating the factor 10^{-9}; as in 1 nsec = 1 $\times$ 10^{-9} sec

NAT N-acetyltransferase

NMU N-nitroso-N-methylurea

OR Odds ratio

ossification Formation of bone

permittivity The permittivity is a complex number consisting of a dielectric constant and a loss factor (related to the conductivity); it is an electric property of the material

piezoelectric The production of an electric potential by an applied pressure or force

PKC Protein kinase C

prokaryote Cellular organism with no distinct nucleus, such as a bacterium or blue-green algae

relative risk (RR) Quotient of the risk in exposed to risk in unexposed individuals

rms Root mean square; the rms value of x is $\sqrt{x^2}$

SCE Sister chromatid exchanges

semipartite thoracic centra Primordial ossification points within the thoracic vertebra

SI Système International; internationally adopted system of units, such as the meter, kilogram, coulomb, and second

σ Volume conductivity

soma The body of an organism, or the body of a nerve cell

sternebra Primordial sternum of the embryo

streaming potential Potentials that vary with the movement of the ions in the surrounding fluids

terata Abnormalities in the developing or newborn fetus

tesla (T) Tesla, an SI unit of magnetic flux density; 1 T = 10^4 G

TPA 12-*O*-tetradecanoylphorbol 13-acetate, a phorbol ester known as a cancer promoter

transients Short duration signals containing a range of frequencies and appearing at irregular time intervals

trenimon An alkylating agent used for the treatment of ovarian cancer

vector quantities Quantities that are specified by magnitude *and* direction at all points in space

VLF Very low frequency, common designation for the frequency range from 3 Hz to 3 kHz

volt (V) A measure of electric potential

ω $2\pi f$, where f = frequency

wire code A surrogate means of assessing electric- and magnetic-field exposure on the basis of well-defined wiring configurations

TPA   12-*O*-tetradecanoylphorbol 13-acetate, a phorbol ester known as a cancer promoter

transients   Short duration signals containing a range of frequencies and appearing at irregular time intervals

trenimon   An alkylating agent used for the treatment of ovarian cancer.

streaming potential   Potentials that vary with the movement of the ions in the surrounding fluids

vector quantities   Quantities that are specified by magnitude *and* direction at all points in space

VLF   Very low frequency, common designation for the frequency range from 3 Hz to 3 kHz

volt (V)   A measure of electric potential

T   2Bf, where f = frequency

wire code   A surrogate means of assessing electric- and magnetic-field exposure on the basis of well-defined wiring configurations

Biographic Sketches of Committee Members

CHARLES F. STEVENS, M.D., Ph.D. (Chairman), is a professor and investigator at the Howard Hughes Medical Institute, Salk Institute, La Jolla, California. Dr. Stevens is internationally respected for his research work in neurobiology specializing in synaptic transmission and properties of excitable membranes. He is a member of the National Academy of Sciences and of the American Academy of Arts and Sciences. He is a member and past member of several committees of the National Academy of Sciences and of the Institute of Medicine.

DAVID A. SAVITZ, Ph.D. (Vice Chairman), is a professor and chair of the Department of Epidemiology in the School of Public Health at the University of North Carolina, Chapel Hill. He is an epidemiologist specializing in occupational and environmental exposures in relation to reproductive outcomes and cancer. He has directed two major epidemiologic studies of EMF and cancer, one concerning residential exposures and childhood cancer and the other evaluating magnetic fields and cancer in electric utility workers. Other research activities focus on the causes of preterm delivery and pregnancy complications and health effects of pesticides. He is an editor of the *American Journal of Epidemiology*.

LARRY E. ANDERSON, Ph.D., is a staff scientist at the Pacific Northwest Laboratory, Richland, Washington. He is a specialist in neurochemistry with research experience in axoplasmic transport; protease-isolation and mechanisms of action; growth regulation-neural cells and neurotoxicity; and biological rhythms and pineal gland function. Dr. Anderson has had extensive experience in EMF research, including the development and use of exposure systems for in vitro and whole animal studies. He serves on a number of national and international committees on EMF and health, including the National Council on Radiation Protection (NCRP), International Conference on High Voltage Systems (CICVS), International Member of Radio Science, Commission K (USRI), and the World Health Organization working group (WHO).

DANIEL A. DRISCOLL, Ph.D., is a licensed professional engineer (New York), specialist in electrical and biomedical engineering, and an employee of the State of New York Department of Public Service. He serves as head of the department's electric and magnetic fields task force. He has had extensive experience in the evaluation of environmental electric and magnetic fields and in the interaction of engineering principles, regulatory issues, and public policy.

FRED H. GAGE, Ph.D., is a professor in the Laboratory of Genetics at the Salk Institute, La Jolla, California. He has conducted extensive studies aimed at treatment of central nervous system disorders, such as the use of gene therapy using intracerebral grafting of genetically modified fibroblasts and the use of neural transplants. His recent research focus has been on molecular and cellular approaches to understanding neurological factors in aging and Alzheimer's disease. Dr. Gage is particularly interested in alterations to the central nervous system that may lead to behavioral and/or learning disabilities.

RICHARD L. GARWIN, Ph.D., is a fellow emeritus of the Thomas J. Watson Research Center, IBM Research Division, Yorktown Heights, New York. His research has contributed to many scientific advances, including work involving instruments and electronics for research in nuclear and low-temperature physics, the establishment of the nonconservation of parity and the demonstration of some of its striking consequences, computer elements and systems, superconducting devices, communication systems, the behavior of solid helium, the detection of gravitational radiation, the design of nuclear weapons, and military technology. He has written numerous books, published over 200 articles and has been granted 42 U.S. patents. Dr. Garwin has served on numerous advisory committees and panels, among them the President's Science Advisory Committee (1962-65 and 1969-72) and the Defense Science Board (1966-69). He is a member of the National Academy of Sciences, the National Academy of Engineering, and the Institute of Medicine and is a Fellow of the American Physical Society and the American Academy of Arts and Sciences. He received from the U.S. government the 1996 R.V. Jones Intelligence Award for "scientific acumen, applied with art, in the cause of freedom."

LYNN W. JELINSKI, Ph.D., is director of biotechnology and a professor of engineering at Cornell University, Ithaca, New York, having moved there in 1991 after 11 years at the AT&T Bell Laboratories. Dr. Jelinski's research interests include the application of microscopic MRI (magnetic resonance imaging) and solid state NMR (nuclear magnetic resonance) to problems in biophysics. She has investigated images of rapid arterial blood flow, developed double-quantum sodium imaging technology at biorelevant concentrations, and published extensively on the structure on macromolecules as determined by NMR.

BRUCE J. KELMAN, Ph.D., a diplomat of the American Board of Toxicology, is the national director of health and environmental sciences for Golder Associates, Inc., Redmond, Washington and an adjunct professor at New Mexico State University. His research interests have focused on issues in reproduction and development including the toxicology of both chemicals and radiation (ionizing and non-ionizing). He has published extensively on transplacental movements of materials (metals and chemicals) and the effects of toxic agents on that movement, effects of ultrasound on placental blood flow and function, and the toxicology of static and ELF electric and magnetic fields. His research has included teratology and life span studies of whole animals exposed to static magnetic fields.

RICHARD A. LUBEN, Ph.D., is an associate professor of biomedical sciences and biochemistry at the University of California, Riverside. His research activities have focused on cellular and molecular mechanisms of hormone action, with emphasis on electromagnetic effects on membrane signal transduction processes. He has also studied the molecular endocrinology of calcium and phosphate homeostasis and bone healing. Dr. Luben is a member of the National Council on Radiation Protection and Measurements (NCRP), and also serves on NCRP Subcommittees 89-3 and 89-5 for Bioeffects and RF Radiation. He has served as member and chair of NIH Special Study Section on Bioelectromagnetics and he is a member of the Radiation Study Section. Dr. Luben is president-elect of the Bioelectromagnetics Society.

RUSSEL J. REITER, Ph.D., is a professor in the Department of Cellular and Structural Biology at the University of Texas Health Sciences Center, San Antonio. His research focus is neuroendocrinology, especially the pineal gland and reproductive physiology, brain chemistry, and behavior. Dr. Reiter's recent research interests have included the effects of circadian-rhythm on pineal gland function and its interrelation with the immune system. Dr. Reiter was an author of the Oak Ridge Associated Universities report (1992) on Health Effects of Electromagnetic Fields commissioned by the Committee on Interagency Radiation Research and Policy Coordination (CIRRPC).

PAUL SLOVIC, Ph.D., is president of Decision Research and a professor of psychology at the University of Oregon. He studies human judgment, decision making, and risk analysis. During the past 15 years, Dr. Slovic and his colleagues have developed methods to describe risk perceptions and measure their impacts on individuals, industry, and society. They created a taxonomic system that enables one to understand and predict perceived risk, attitudes toward regulation, and impacts resulting from accidents or failures. Dr. Slovic publishes extensively and serves as consultant to companies and government agencies. He is past president of the Society for Risk Analysis and in 1991 received its Distinguished Contribution Award. He also serves on the Board of Directors for the National Council on Radiation Protection and Measurements. In 1993 he received the Distinguished Scientific Contribution Award from the American Psychological Association, and in 1995 received the Outstanding Contribution to Science Award from the Oregon Academy of Science.

JAN A. J. STOLWIJK, Ph.D., is a professor of epidemiology and acting chair of the Department of Epidemiology and Public Health at Yale University, New Haven, Connecticut. He has published extensively on epidemiology and studies of risk assessment for a broad range of topics including indoor air pollution, the proximity to hazardous waste sites, smoking and indoor radon exposure, and exposure to EMF. His specialties include regulatory systems in physiology, thermal receptor structures, and the construction and application of mathematical models for the study of complex physiological systems. Dr. Stolwijk has been active in the study of effects of heat on animal response and has worked with international groups (WHO) in this area. He also has a broad level of experience in studies of the health effects of exposure to EMF having served on an international working group concerned with the health effects of non-ionizing radiation and chaired an ad-hoc committee for the Connecticut Academy of Sciences and Engineering to review potential health effects of EMF.

MARIA A. STUCHLY is a professor and NSERC/BC Hydro/TransAlta Industrial Research Chair in the Department of Electrical and Computer Engineering, University of Victoria, British Columbia, Canada. Her research interests are in numerical and experimental modeling of physical interactions of electromagnetic fields with living systems. She has published extensively in the areas of tissue electrical properties, radio frequency dosimetry and safety standards, and medical applications. Dr. Stuchly is a fellow of the IEEE.

DANIEL WARTENBERG, Ph.D., is an associate professor in the Department of Environmental and Community Medicine at the Robert Wood Johnson Medical School, where he directs the epidemiology and quantiative methods track in the New Jersey graduate program in public health. His research addresses methodological issues in the conduct of epidemiologic research and risk assessment, including issues of analysis, computerization, data base management, modeling, and geographic patterns. He has published extensively on the statistical evaluation of disease clusters. Issues he has studied include effects of exposure to toxic chemicals, pesticides, and EMF. In addition, he serves as a member of the New Jersey Governor's Commission on Radiation Protection.

JOHN S. WAUGH, Ph.D., is a professor at the Massachusetts Institute of Technology. His research has concentrated on the theory of nuclear magnetic resonance (NMR) and its applications to chemistry, with recent emphasis on NMR spectroscopy at temperatures below 0.01 K and the theory of spin dynamics. He is the author of one book and about 200 research papers. He is a member of the National Academy of Sciences and a Fellow at the American Academy of Arts and Sciences, the American Physical Society, and the American Association for the Advancement of Science. He is the recipient of a number of honors including the Langmuir Award, the Pauling and Richards Medals, and the Wolf Prize.

JERRY R. WILLIAMS, Sc.D., is a professor of oncology at The John Hopkins University Oncology Center, Baltimore, Maryland. His research specialty is in cellular and molecular biology with a focus on physiology and radiobiology. He has published extensively on mutagenesis, radiolabeled antibodies, DNA topoisomerases, and models of sister chromatid exchange. His most recent work has involved sensitization processes in human tumor cells and the use of monoclonal antibodies for diagnosis and treatment of cancer. He has extensive experience in studies of ionizing radiation and the mechanisms of cancer induction, progression, and treatment.

Index

Genes, early response, 53. *See also fos; jun; myc*
Genotoxicity, 52-55, 56-58, 210-219. *See also* Chromosomal aberrations; Mutations
GH. *See* Growth hormone (GH) effects
Glandular effects. *See* Neuroendocrine responses
Gray (Gy), 339
Grounding system, residential, 28, 205, 295, 301
Growth hormone (GH) effects, 104, 114
Growth retardation, fetal. *See* Reproductive studies
Gy. *See* Gray (Gy)

H

Hamsters, studies of, 252
Harmonics, 25
defined, 5, 339
Hazard identification, 192-195
Hazardous chemicals. *See* Chemicals, hazardous
Headaches, 187-189
Health effects. *See* Behavioral effects; Cancer; Central nervous system effects; Developmental studies; Disease; Learning disabilities; Reproductive studies
Heating, electric. *See* Electric heating
Helmholtz coils, 41, 45, 111
Herbicides, 174
Heritable changes, 55-58
Hertz (Hz), 11n, 339
Heterogeneity, 22, 136-137, 143
defined, 339
High frequency, 13, 35
High-voltage transmission lines. *See* Transmission and distribution lines
HIOMT. *See* Hydroxyindole-O-methyltransferase (HIOMT)
HIV-LTR expression, 219
Hodgkin's disease, 279
Home appliances. *See* Appliances, electric
Home exposures. *See* Age of home; Residential exposures

Home wiring, 4, 28. *See also* Transmission and distribution lines
Homogeneity, 143
defined, 339
Hormonal changes. *See* Neuroendocrine responses
Humans, studies of, 245, 254
6-Hydroxy melatonin sulfate, urinary, 254
Hydroxyindole-O-methyltransferase (HIOMT), 98, 249, 339
Hz. *See* Hertz (Hz)

I

ICR. *See* Ion cyclotron resonance (ICR) model
Impedance, 23, 339
In vitro studies, 15, 43, 55-72, 210-219. *See also* Mammalian cells; Prokaryotic cells
defined, 339
on exposure, 6, 44-46, 52-72
extrapolations from, 55
limits of, 53-54
possible effects. *See* Calcium concentrations, intracellular, changes in; Carcinogenicity; Gene expression, changes in; Genotoxicity
reliability issues, 57, 61, 63
In vivo studies, 15
defined, 339
on exposure, 6-8, 40-44
Independent fields, 4
Induced fields and currents, 4, 22, 41, 46-51
Helmholtz induction, 111
Influence analysis, 136-137
Information bias, 120-122, 155-158, 170, 187
Inheritable changes, 55-58
Instrumentation, 26-27
research needs, 204
unavailability of, 164
Intermittent exposure, 56
Interspecies scaling, 44, 48-50
Intracellular calcium concentrations. *See* Calcium concentrations, intracellular, changes in